REFERENCE MATERIALS *for* ENVIRONMENTAL ANALYSIS

Edited by

Ray E. Clement, Ph.D.

Ontario Ministry of Environment and Energy
Laboratory Services Branch
Etobicoke, Ontario, Canada

Lawrence H. Keith, Ph.D.

Radian International LLC
Austin, Texas

K.W. Michael Siu, Ph.D

Institute for National Measurement Standards
National Research Council of Canada
Ottawa, Ontario, Canada

CRC LEWIS PUBLISHERS

Boca Raton New York London Tokyo

Contact Editor: Ken McCombs
Project Editor: Carole Sweatman
Marketing Manager: Greg Daurelle
Direct Marketing Manager Arline Massey
Cover Design: Dawn Boyd
Prepress: Kevin Luong
Manufacturing: Sheri Schwartz

Library of Congress Cataloging-in-Publication Data

Reference materials for environmental analysis / edited by Ray E. Clement, Lawrence H. Keith, K.W. Michael Siu.
p. cm.
Includes bibliographical references and index.
ISBN 1-56670-102-3 (alk. paper)
1. Pollutants--Analysis. I. Clement, Ray. E. II. Keith, Lawrence H. III. Siu, K. W. M.
TD193.R444 1997
628.5—dc20

96-22062
CIP

This book contains information obtained from authentic and highly regarded sources. Reprinted material is quoted with permission, and sources are indicated. A wide variety of references are listed. Reasonable efforts have been made to publish reliable data and information, but the authors and the publisher cannot assume responsibility for the validity of all materials or for the consequences of their use.

Direct all inquiries to CRC Press, Inc., 2000 Corporate Blvd., N.W., Boca Raton, Florida 33431.

Lewis Publishers is an imprint of CRC Press

International Standard Book Number 1-56670-102-3
Library of Congress Card Number 96-22062
Printed in the United States of America 1 2 3 4 5 6 7 8 9 0
Printed on acid-free paper

Preface

In the past 15 years, truly amazing progress has been made in the determination of ultra-trace concentrations of environmental pollutants. Leading the way in these developments has been the maturity of a variety of sophisticated, complex, sensitive, and selective instrumentation. Specifically, modern gas chromatography/mass spectrometry (GC/MS) instruments have permitted sub-picogram detection of organics, such as the chlorinated dibenzo-*p*-dioxins, while inductively coupled plasma/mass spectrometry (ICP/MS) instruments have accounted for similar low-level detection of metals in environmental samples. However, like all analytical tests from the simplest to the most complex, such low-level detection capabilities, by themselves, cannot guarantee the accurate quantitative determination of trace analytes. Real-matrix environmental reference materials are also required.

Despite the obvious importance of environmental reference materials, there is surprisingly little coverage of this topic in available reference books. This book is directed toward helping to fill this gap in the published literature. Based on a symposium held as part of the annual conference of the Canadian Society for Chemistry, Sherbrooke, Quebec, Canada, June 1993, the contributions herein are directed to the working environmental analytical practitioner — not the theorist. The various chapters herein describe how to prepare proper reference materials and how to use these materials to ensure that accurate environmental measurements can be obtained.

A clear majority of the chapters deal with reference materials for trace levels of organic analytes such as the polychlorinated biphenyls (PCBs), chlorinated dibenzo-p-dioxins/dibenzofurans (PCDDs/PCDFs), polycyclic aromatic hydrocarbons (PAHs), and related groups of compounds. This simply reflects the current importance of these analytes in environmental analysis applications. We feel that, taken collectively, the principles presented in this book can be applied to the preparation and use of any environmental reference material.

The Editors wish to thank all of the chapter authors for their hard work and patience. Although it has taken two years to complete this project, the emphasis of the authors on techniques and applications, rather than the theoretical basis of the use of reference materials, ensures that the information presented here is just as applicable today as when this project was started. With the increasing requirements for improved documentation and traceability of analytical results, environmental analytical practitioners cannot afford to operate without the knowledge contained herein. We also acknowledge the superb efforts of the CRC/Lewis editorial and production team in bringing this project to completion, including the project editor, Carole Sweatman, Susan Alfieri, Ken McCombs, and Greg Daurelle; they all exhibited the same emphasis on quality that is the ultimate purpose of this book.

Ray E. Clement
Lawrence H. Keith
K.W. Michael Siu

The Editors

Ray E. Clement, Ph.D., is Senior Research Scientist at the Ontario Ministry of Environment and Energy. After graduating in co-op applied chemistry at the University of Waterloo in 1974, he earned his Ph.D. in 1981. Dr. Clement then joined the Ontario Ministry of the Environment, Dioxin Laboratory and was appointed supervisor in 1984. In 1988 he achieved his current position.

Dr. Clement has over 100 publications; most deal with the analysis, sources, and distribution of dioxins/furans in the environment — he has also published or edited 5 books. In 1991 he received the Francis W. Karasek Award for *achievements in environmental science*, and in 1992 the Canadian Society for Chemistry McBryde Medal for a *significant achievement in pure or applied analytical chemistry.* Dr. Clement also holds adjunct appointments at Carleton University and the University of Western Ontario, where he has taught a third-year course in instrumental environmental analysis.

Dr. Clement was co-chair of both the 9th and 15th International Symposia on Chlorinated Dioxins and Related Compounds, and served on the Exposure Assessment Panel for the recent dioxin reassessment conducted by the U.S. EPA. He is on the editorial boards of *Chemosphere*, *CRC Critical Reviews in Analytical Chemistry*, *Environmental Science and Pollution Research International*, and *The Crucible* — a publication of the Science Teachers' Association of Ontario. He recently organized *Enviro*Analysis '96 — a conference dedicated to all aspects of analysis and monitoring of the environment.

Lawrence H. Keith, Ph.D., is a Corporate Fellow at Radian International LLC in Austin, Texas. A pioneer in environmental sampling and analysis, method development, and handling of hazardous compounds, Dr. Keith has published more than thirty books and presented and published more than one hundred technical articles. Recent publications have involved electronic book formats and expert systems. Dr. Keith serves on numerous government, academic, publishing, and environmental committees and is a past chairman of the American Chemical Society (ACS) Division of Environmental Chemistry. He is chairman of the ACS subcommittee on Environmental Monitoring and Analysis and current editor of the Division of Environmental Chemistry newsletter and its Internet home page (http://acs.environmental.duq.edu/envacs/enchem.htm). Prior to joining Radian in 1977, he was a research scientist with the U.S. Environmental Protection Agency.

K.W. Michael Siu, Ph.D., is a Senior Research Officer in the Institute for National Measurement Standards, National Research Council of Canada (NRC). Dr. Siu was born in Hong Kong in 1953. He received his B.Sc. degree from the University of Hong Kong in 1976 and his M.Sc. (analytical chemistry) from the University of Birmingham, England in 1977. In 1981, he graduated with a Ph.D. degree from Dalhousie University, Canada, under the supervision of Walter A. Aue. He joined the Division of Chemistry, NRC as a Research Associate and became a permanent member of the staff in 1983.

Dr. Siu's research interests are in the areas of chromatography and mass spectrometry. In the past few years, his work has concentrated on two fronts: (1) developing efficient and accurate methodologies for the determination of environmentally significant pollutants, including organochlorine compounds and organometallic species, and (2) investigating the fundamentals of, as well as exploring new application areas for, electrospray mass spectrometry.

Dr. Siu also holds adjunct professorship in the Department of Chemistry at Carleton University and Queen's University. He was a Visiting Lecturer in the Department of Chemistry, University of Hong Kong in the winter of 1995. He was a winner of the Maccoll Prize, awarded by the *Journal of Organic Mass Spectrometry* in 1991. He received the 1994 F.W. Karasek Award from the Faculty of Science Foundation, University of Waterloo, and the 1996 W.A.E. McBryde Medal from the Canadian Society for Chemistry. Dr. Siu has been serving on the Executive Board of the Analytical Chemistry Division, Canadian Society for Chemistry since 1988. He is currently the Chair of that division.

Contributors

Harry Alkema
Environmental Standards and Statistics
Aquatic Ecosystem Protection Branch
National Water Research Institute
Burlington, Ontario, Canada

Nabil Arafat
Environmental Standards and Statistics
Aquatic Ecosystem Protection Branch
National Water Research Institute
Burlington, Ontario, Canada

Keijo I. Aspila
Environmental Standards and Statistics
National Water Research Institute
Burlington, Ontario, Canada

Shier S. Berman
Institute for National Measurement Standards
National Research Council of Canada
Ottawa, Ontario, Canada

Daniel L. Bolt
Cambridge Isotope Laboratories
Andover, Massachusetts

William S. Bowman
Natural Resources Canada
Ottawa, Ontario, Canada

Joel C. Bradley
Cambridge Isotope Laboratories
Andover, Massachusetts

Raymond E. Clement
Laboratory Services Branch
Ontario Ministry of Environment and Energy
Etobicoke, Ontario, Canada

Glenys Foster Roberts
PHILIP Analytical Services
ZENON Laboratories
(a division of Philip Analytical Services Corporation)
Burlington, Ontario, Canada

Catharine A. Fraser
Institute for National Measurement Standards
National Research Council of Canada
Ottawa, Ontario, Canada

Graeme J. Gardner
Institute for National Measurement Standards
National Research Council of Canada
Ottawa, Ontario, Canada

Roger Guevremont
Institute for National Measurement Standards
National Research Council of Canada
Ottawa, Ontario, Canada

M. Coreen Hamilton
Axys Analytical Services Ltd.
Sidney, British Columbia, Canada

Lawrence H. Keith
Radian International LLC
Austin, Texas

Donald E. King
Quality Contexts
Brampton, Ontario, Canada

Cariton Kubwabo
Environmental Health Centre
Health Canada
Ottawa, Ontario, Canada

Maureen E. Leaver
Natural Resources Canada
CANMET
Ottawa, Ontario, Canada

Paulette S. Maxwell
Institute for National Measurement Standards
National Research Council of Canada
Ottawa, Ontario, Canada

Marilyn A. Melton
National Environmental Testing, Inc.
Bartlett, Illinois

Alan W. Nichols
Radian International LLC
Austin, Texas

Barry G. Oliver
ZENON Laboratories
(a division of Philip Analytical Services Corporation.)
Burnaby, British Columbia, Canada

Gregory E. Pronger
National Environmental Testing, Inc.
Bartlett, Illinois

Michael A. Re
Radian International LLC
Austin, Texas

Ian H. Rogers
Institute of Ocean Sciences
Fisheries and Oceans Canada
Sidney, British Columbia, Canada

Michele M. Schantz
Analytical Chemistry Division
National Institute of Standards and Technology
Gaithersburg, Maryland

David B. Sergeant
Fisheries and Oceans Canada
Ultratrace Laboratory
Bayfield Institute
Burlington, Ontario, Canada

K.W. Michael Siu
Institute for National Measurement Standards
National Research Council of Canada
Ottawa, Ontario, Canada

Richard Turle
Environmental Technology Centre
Environment Canada
Ottawa, Ontario, Canada

Bryan Wakeford
Laboratory Services
Canadian Wildlife Service
Hull, Quebec
Canada

Scott N. Willie
Institute for National Measurement Standards
National Research Council of Canada
Ottawa, Ontario, Canada

Stephen A. Wise
Analytical Chemistry Division
National Institute of Standards and Technology
Gaithersburg, Maryland

Jean-Pierre Zajac
Chemical Manufacturing Division
SCP Science Ltd.
Lasalle, Quebec, Canada

Contents

Part I: Making Environmental Reference Materials

Part I
Making Environmental Reference Materials

CHAPTER 1

Guidelines for the Preparation and Verification of Neat Organic Environmental Reference Materials

Michael A. Re

The preparation and verification of neat reference materials are critical in preparing analytical standards. Accurate standards cannot be prepared unless neat materials are of correct identity, of high purity, homogeneous, and stable under proper storage conditions. Multiple techniques, including NMR, GC/MS, FTIR, GC, HPLC, TLC, and elemental analysis, are used to confirm identity, purity, and homogeneity. Stability testing is also essential to ensure reliability of reference materials over time.

Analytical methods for environmental testing rely on a variety of reference materials for accurate reporting of analytical data. Environmental Reference Materials (ERMs) can be classified into three categories:

1. Synthetic Reference Materials
2. Natural Matrix Reference Materials
3. Neat Reference Materials

Synthetic Reference Materials refer to standards that contain known analytes at a specified purity and concentration that are dispersed in solution or onto a solid matrix. Some examples of solution standards include those that are used for calibration of instrumentation, calibration check standards, recovery standards, surrogate and internal standards, and matrix spiking standards. Fortified natural matrix materials can also be considered Synthetic Reference Materials. These types of standards include soil, waste water, fish tissue, or other natural matrices that have been spiked with known amounts of specific analytes. Fortified matrix materials are useful as performance evaluation (PE) standards, for method validation studies (for determination of method detection levels), and in some instances for calibration of analytical instrumentation.

Natural Matrix Reference Materials can be defined as natural soil, water, or biota that have been homogenized and well characterized for a specific set of analytes by multiple analyses. These types of standards are useful for site-specific PE samples, method detection level determination, and as natural blank matrices for use as method blanks or as a basis for preparing fortified matrix standards.

Neat Reference Materials can be defined as solid or liquid materials that have established identity and purity data associated with them. These materials can be highly purified compounds or technical mixtures containing multiple components. Neat Reference Materials provide the basis for the preparation and verification of all Synthetic Reference Materials. The quality of any type of synthetic standard will only be as good as the quality of the neat material that is used to prepare the standard. For example, if the purity data for an analyte is stated to be 99% and the actual purity is only 85%, a gravimetric preparation of a solution standard of the analyte would result in a significant error in concentration of the solution standard even if the analyte has been accurately weighed and diluted. If a neat material is misidentified when it is prepared, the preparation of a standard solution may result in a very accurate solution standard of the wrong analyte. These are simplified illustrations of errors that occur often in the preparation of analytical standards, with the root cause of the error being the quality of the neat material.

Government agencies that set regulations for environmental testing and private industries that require environmental testing continue to establish guidelines for analytical reference materials. At the very basis of these requirements and guidelines is the establishment of traceability of reference materials. One definition of traceability as it applies to reference materials has been given as the "property of result of a measurement of the value of a standard whereby it can be related, with a stated uncertainty, to stated references, usually national or international standards, through an unbroken chain of comparisons."[1] To provide an unbroken chain of comparisons for any synthetic standard implies that traceability begins with establishing measurements of the neat material used in preparing the standard.

In establishing traceability, certain objectives should be considered during the preparation and verification process of neat ERMs:

- Develop a specific plan for preparation of the neat compound.
- Set specifications for identification.
- Set specifications for purity.
- Set specifications for homogeneity.
- Provide complete documentation.

In addition to establishing traceability, it is also very important to establish stability of ERMs, including neat reference materials. Analytical laboratories, because of specifications set by their clients, are requiring expiration dates to be established for analytical standards. Expiration dates for manufactured reference materials are best determined by an ongoing stability testing program. Using historical data from stability testing will provide valid expiration dates.

LABORATORY SYNTHESIS AND PURIFICATION OF NEAT ORGANIC ERMs

To correctly establish traceability of neat materials, it is very important to know the source of the material. In virtually all instances, neat organic materials will have been manufactured either in a laboratory setting or in a bulk process. The sources for banned environmental contaminants (such as pesticides that are no longer in use) are becoming very limited, and small amounts of these materials are now being prepared in laboratory settings for use as reference materials. In other cases, when the source of contaminants is the result of by-products from industrial processes, there are no sources for authentic materials. These materials are typically prepared in a laboratory setting. Laboratory syntheses must be well planned and executed properly to ensure that the goals for establishing correct identity and purity are successfully met.

Preparation of neat ERMs in the synthesis laboratory should begin with an established design basis. In designing the synthesis, it is extremely important to perform an extensive literature review of existing methods of preparation. In many instances expired patents exist for environmental contaminants such as banned pesticides. Patents often provide experimental detail for preparation of specific compounds. In other cases, patents may provide information on how to prepare a technical mixture but not specific compounds. A better synthetic method for a specific compound may be found in other chemical literature. It is helpful to use multiple sources, such as technical papers in chemical journals, and review articles as well as expired patent information to develop the best design for a specific compound.

Many environmental contaminants have multiple isomeric forms. One or more specific isomers may be present as a contaminant; it is, therefore, very important to have reference materials of the isomers of interest. When preparing ERMs it is useful to develop isomer-specific reactions whenever possible. Using an isomer-specific reaction helps to provide correct unambiguous verification of structure assignments. A good example of the successful use of isomer-specific synthesis is in the preparation of polychlorinated dibenzo-p-dioxins. 2,3,7,8-Tetrachlorodibenzo-p-dioxin (2,3,7,8-TCDD) has been prepared in the laboratory by multiple methods, ranging from direct chlorination of unsubstituted dibenzo-p-dioxin to condensation of a substituted catechol with a substituted nitrobenzene[2,3] (see Figure 1). Direct chlorination (A) results in a mixture of isomers that are very difficult to separate and identify. Using the catechol route (B) yields the desired isomer and provides a good basis for establishing identity of the synthetic product.

Other factors should be considered in the design of ERM preparations; proposed preparative methods should have specified control points and a predetermined set of technical specifications. Specified control points may include setting identity and purity criteria for key intermediates of a synthetic scheme. Acceptable data and documentation of intermediate control points will provide basic traceability of the preparative method and help establish the unbroken chain of traceability of the ERM. Technical specifications of the final product should also be proposed in the design process. This may include setting goals for final product purity and amount of

A. Direct Chlorination (non-specific reaction)

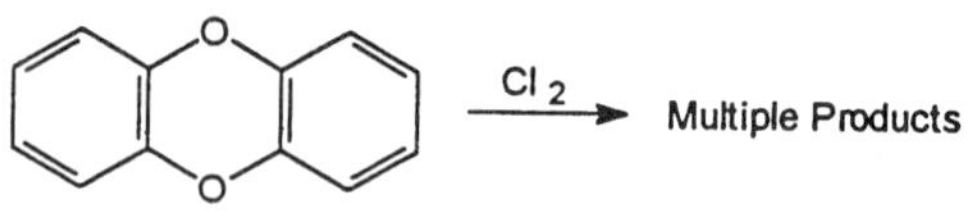

B. Condensation (isomer-specific reaction)

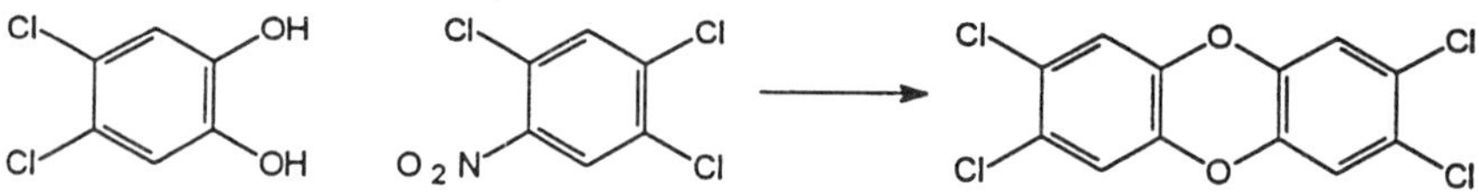

Figure 1 Two methods for preparation of 2,3,7,8-TCDD.

material to be prepared. It should also include defining methods of analysis of the final product both for purity and identity confirmation.

Due to the hazardous nature of many ERMs, it is essential to consider needs for special facilities or equipment and safety precautions during the design process. Special facilities may include a laboratory specifically designed for the safe handling of hazardous material. Specially designed ventilation hoods or protective clothing may be necessary. Extensive research of the chemical, physical, and toxicological properties of all materials to be used should be part of the design basis.

VERIFICATION OF NEAT ENVIRONMENTAL REFERENCE MATERIALS

A critical part of the preparation process is the verification of identity, purity, homogeneity, and stability of the final product. There are many methods and analytical techniques available to perform these verifications. The use of multiple methods to establish each parameter is necessary to provide complete assurance that the design basis criteria have been met.

Identity

The determination and verification of compound identity can be accomplished using a variety of analytical methods each of which can be classified as either a primary or secondary method. Primary methods include spectral analysis using mass spectrometry (MS), nuclear magnetic resonance (NMR), and infrared spectroscopy (IR). These methods provide direct evidence for structure elucidation, such as the determination of molecular weight and the presence of functional groups. Secondary methods use physical or chemical properties as a source of verification and include melting point (MP) or boiling point (BP) determination, elemental analysis, refractive index, and gas chromatography (GC). Melting point, boiling point, and refractive index depend on the availability of literature values for comparison. Comparing to known values provides evidence that the prepared compound is consistent with literature references. Gas chromatography can be used if an authentic reference material is available for comparison of retention times. None of the methods used for identity determination should be used independently. Each method provides specific information regarding the structure of the compound, and the use of a combination of methods is essential to provide complete traceability of the compound.

The most common occurrence of misidentification of neat ERMs is incorrect isomer assignments. As discussed above, isomer-specific synthesis is a good approach to avoid this problem. In some cases isomer-specific reactions may not be an option, but in either situation specific isomers must be verified using the various analytical methods. *Cis*- and *trans*-Chlordane are two environmental contaminants that are common target analytes in a number of analytical methods. Preparation of neat ERMs of the two analytes can be accomplished using isomer-specific reactions or by separation of a technical mixture of Chlordane. Primary methods of identification include both MS and NMR. Mass spectra of the two compounds are shown in Figure 2. Examination of the fragmentation patterns indicate only minor differences in the two spectra, and determination of isomers by MS is virtually impossible. ^{1}H NMR spectra for each compound are shown in Figure 3. A detailed analysis of each spectrum indicates differences in coupling constants of *cis*- and *trans*-hydrogens, which provide positive confirmation of the two isomers. Literature melting point values are also available, but the difference in melting point range of the two compounds (*cis*-Chlordane, mp = 106–107°C and *trans*-Chlordane, mp = 103–105°C) is not enough to provide conclusive evidence. GC retention times of *cis*- and *trans*-Chlordane with relation to each other have been documented for various GC columns; these data can also be used to help distinguish the two isomers. In this example multiple methods are used to provide supporting data, but ^{1}H NMR is necessary to provide conclusive structure confirmation.

Purity

Determination of the purity of neat ERMs can also be performed using a variety of methods that can be classified as primary and secondary methods. The primary methods provide quantitative values that can be assigned to the compound. The most common method is gas chromatography with a flame ionization detector (GC-FID). A large number of environmental pollutants can be analyzed by GC. FID is a nonselective detector that will detect any combustible material. Another primary method is High Performance Liquid Chromatography (HPLC). Ultraviolet (UV) and refractive index (RI) are the two most common HPLC detectors. UV detection is somewhat limited because it is dependent on the UV absorbtivity at a given wavelength of both the compound of interest and any impurities that may be present. HPLC is often used for analysis of nonvolatile or highly polar compounds that are not easily analyzed by GC. A third primary method is Differential Scanning Calorimetry (DSC). DSC provides quantitative values for purity by measuring temperature differences of a sample and an inert reference compound as the temperature is

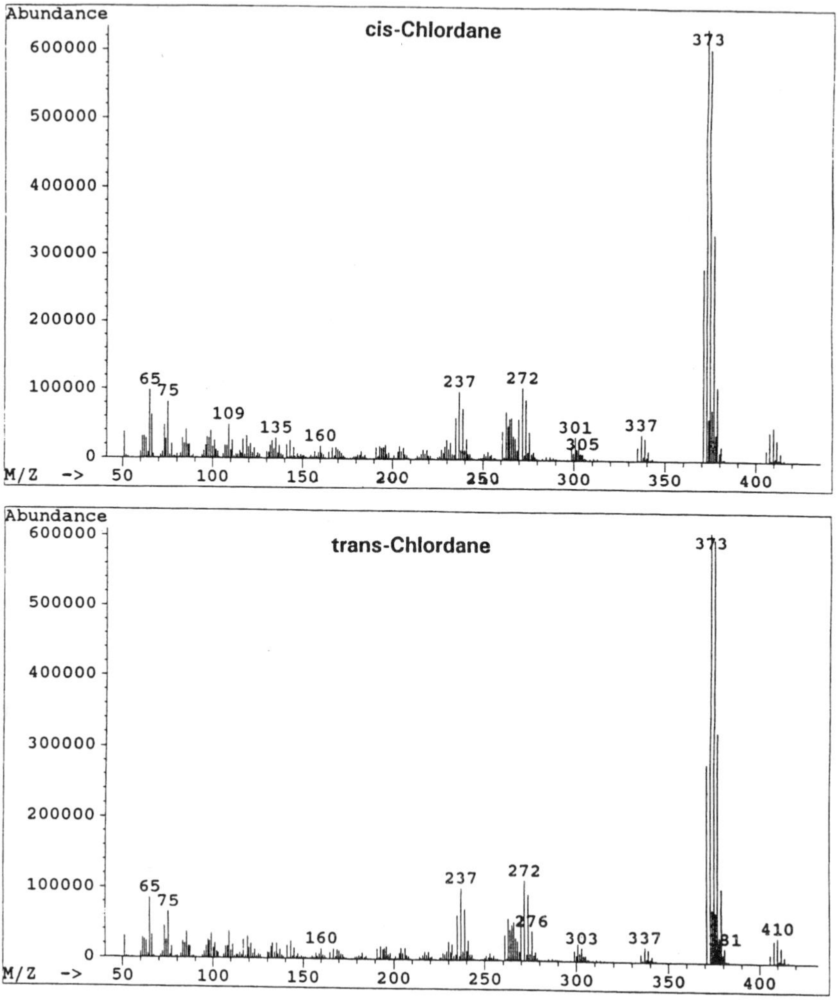

Figure 2 Mass spectra of *cis*- and *trans*-Chlordane.

gradually increased. Figure 4 illustrates a comparison of GC-FID and DSC analysis of a common polycyclic aromatic hydrocarbon (PAH) pollutant. Secondary methods, which provide qualitative purity information, include thin layer chromatography (TLC), elemental analysis, and melting point and boiling point determinations.

As was discussed for identity analysis, purity determinations rely on multiple methods of analysis. Each method measures a different property of the compound relating to purity. GC-FID will only detect materials that are volatile enough to pass through a column and that can be detected by FID. For example, if a compound were contaminated with silica gel (commonly used for purification of ERMs), that impurity would not be detected because silica gel will not pass through a GC. Secondary methods such as elemental analysis or melting point determination would qualitatively indicate that an impurity is present.

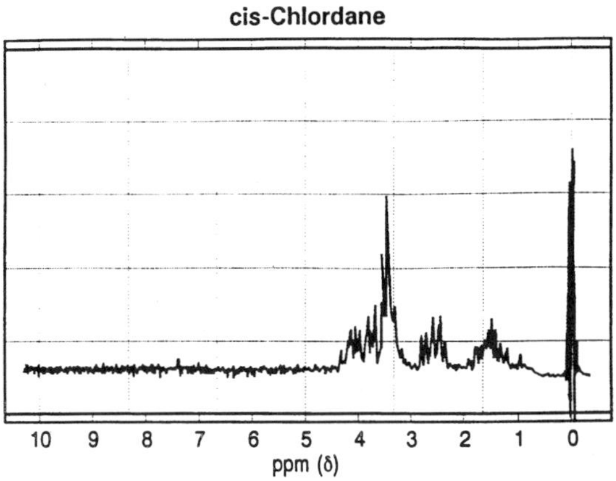

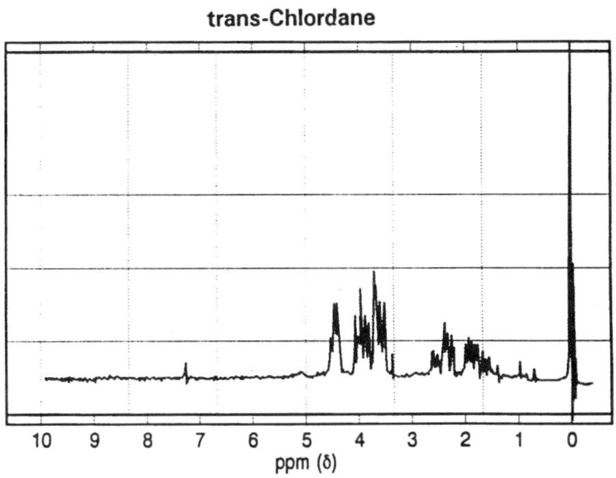

Figure 3 60 mHz ^{1}H NMR of *cis*- and *trans*-Chlordane.

Homogeneity

When neat ERMs are prepared and subunits are to be distributed from the bulk lot, it is very important that each of the subsamples be the same in composition and purity. Homogeneity data provide another link in the traceability chain because they indicate and document that the purity and composition of the material are the same throughout the batch and that initial purity and identity data are valid for each subsample taken from the lot. Some common sources of batch nonhomogeneity include concentration of impurities during purification procedures, contamination of a portion of the batch during handling, or localized compound decomposition (for example, near the surface where the compound is exposed to light or air). A common method for analyzing for batch homogeneity is to randomly remove representative samples from the top, middle, and bottom portions of the container and perform a purity assay such as analysis by GC-FID. Table 1 illustrates homogeneity

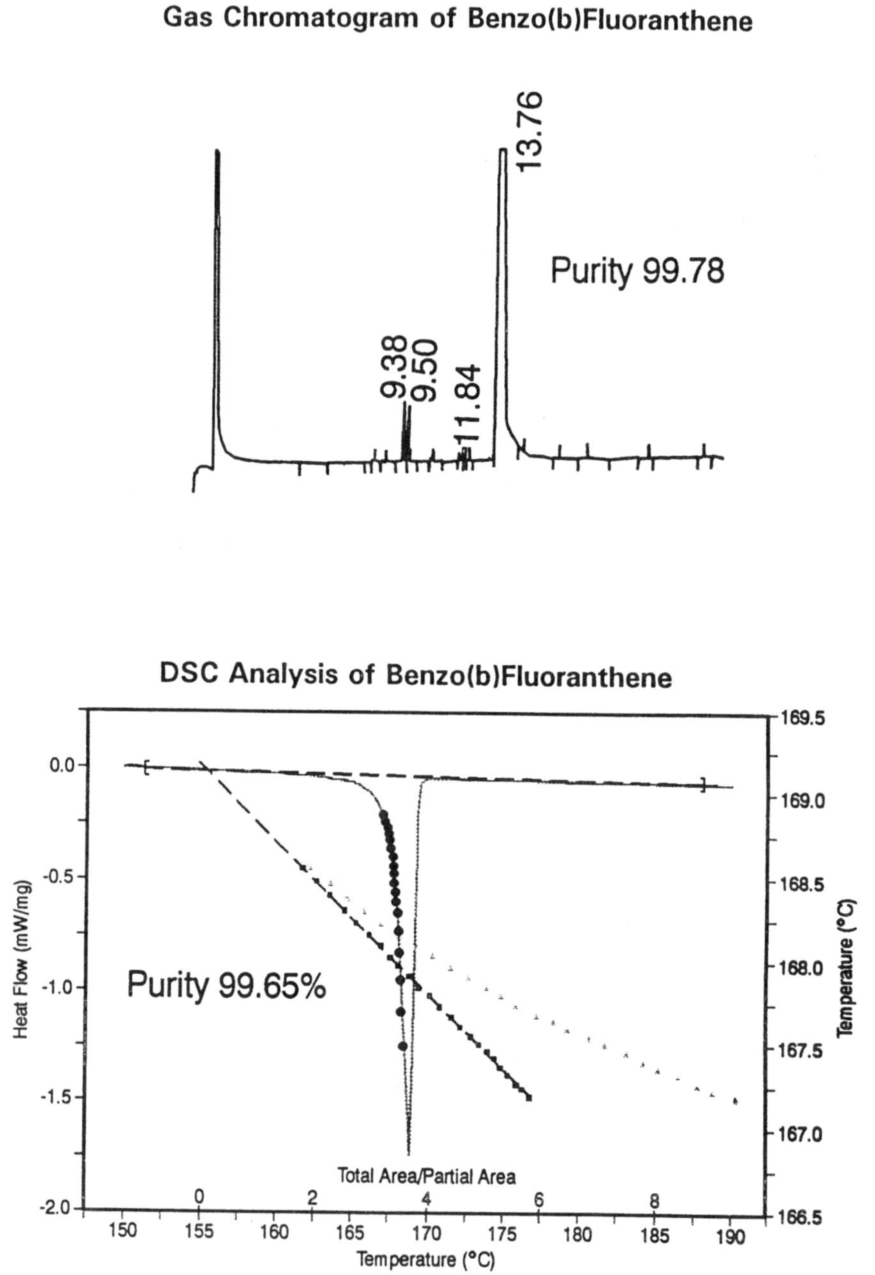

Figure 4 GC/FID and DSC analysis of benzo[*b*]fluoranthene.

analyses for some representative neat ERMs using GC-FID. Nonhomogeneity is a possibility if any of the random analyses for a compound varies from the mean of all analyses of the compound by more than 0.5%.

Table 1 Homogeneity Analyses by GC/FID for Some Typical Neat ERMs

	% Purity				
Compound	Analysis 1	Analysis 2	Analysis 3	Mean	% RSD
Dibenz[*a,h*]anthracene	99.31	99.15	99.22	99.23	0.08
Mirex	99.80	99.76	99.77	99.78	0.02
Dibenz[*a,j*]acridine	99.96	100.00	99.97	99.98	0.02
Heptachlor[*2,3-exo*] epoxide	99.87	99.87	99.86	99.87	0.006
Endosulfan I	99.69	99.71	99.67	99.69	0.02
Endosulfan II	99.85	99.77	99.65	99.76	0.10

Stability

Stability testing for neat ERMs is essential to assure that chemical degradation over time is not occurring. Testing is also necessary to accurately determine expiration dates. Expiration dates are often assigned on a "best guess" basis when stability data are not available. These assignments should be followed up with an extensive stability testing program. It is also important to determine proper storage conditions for specific reference materials. Storage can often play a major role in the stability of a compound. Some classes of compounds, such as polycyclic aromatic hydrocarbons (PAHs), dioxins, and dibenzofurans, are quite stable when stored in Teflon®-lined screw-cap containers at ambient temperature. Other compounds may be sensitive to temperature, light, or air, and special storage conditions are required. A specific compound may be very stable over a long period of time when stored at 4°C under an argon atmosphere, but when it is stored at ambient temperature, the compound readily decomposes. This type of stability data is often available in the chemical literature, and these factors must be considered as part of the design basis planning.

The frequency of stability testing will often vary depending on the compound or class of compounds. Typical time intervals for testing are 6 months and 1 year after preparation. Continued testing is then typically performed on an annual basis thereafter. Early initial testing is extremely important to determine short-term stability of the ERM; therefore, some producers of neat reference materials will perform stability testing 1 month after preparation. A 3-month analysis will also be performed if there are any indications of chemical degradation problems in the 1-month analysis.

Table 2 illustrates some examples of 1-year stability results for some typical neat reference materials. The purity data were acquired using GC-FID, and both initial purity and 1-year results are shown. In all of the examples, the slight differences between initial and 1-year results can be attributed to analytical variability, and no changes in chemical purity have occurred. A decrease in the purity value for a compound of more than 1% of the initial purity assay may indicate that a stability problem exists.

Because of the reactivity of some classes of reference materials, stability of the neat ERM may be a problem. These materials may be relatively stable in solution but instable in the neat form. A good example of this can be seen with stability studies for the pesticide Aramite® in the neat form and in solution. In this study,

Table 2 1-Year Stability Results for Some Typical Neat ERMs

Compound	Initial purity (%)	Analyzed purity after 1 year (%)
Benzo(*b*)fluoranthene	99.94	100.00
trans-Chlordane	99.93	99.96
Chlordecone	99.13	99.38
cis-Nonachlor	99.06	99.29
trans-Nonachlor	99.42	99.64

Table 3 Stability Testing of Aramite in Solution and in Neat Form

Time since initial analysis	% purity of neat material (stored at –20°C)	% purity 2000 μg/mL solution in hexane (stored at –20°C)
Initial	99	—
2 Days	97	97
2 Months	88	97
3 Months	86	97
4 Months	83	97
7 Months	74	97
15 Months	62	97

Aramite was synthesized and immediately analyzed after purification. After only 2 days, the material had degraded from a purity of 99% to 97%. At this point, a standard solution in n-Hexane at 2000 μg/mL was prepared with a portion of the neat material, distributed into amber ampules, and flame-sealed. The remainder of the neat material was placed into amber ampules under an argon atmosphere and flamed-sealed. The standard solution and the neat material were then stored at –20°C. Table 3 summarizes results of the stability analysis for each reference material. The data indicates a very rapid decrease in purity of the neat material while the purity of the standard solution remained unchanged. It was determined from this study that a neat ERM of high purity Aramite is impractical and that the best alternative is to immediately prepare a standard solution of the purified neat material.

A second but very different example of compound instability is shown with Endrin, another chlorinated pesticide. In this case, instability of the compound was due to trace impurities present in the final product and not due to instability of the compound itself. Laboratory preparation of Endrin by one method includes final purification using recrystallization. A solvent mixture of methanol and methylene chloride was used to achieve crystals that were analyzed to be 99% pure. Stability analysis indicated that after 1 year the material had degraded to a purity of less than 90%. This was unusual since previous batches of Endrin had no history of stability problems. It was concluded that the chemical degradation was likely due to residual amounts of hydrochloric acid (HCl) present in the final product. Endrin is an acid-sensitive compound, and the common breakdown products of Endrin were observed in the stability assays. Methylene chloride often contains trace amounts of HCl, which would be the most likely source of the impurity. Trace amounts of acid were removed by rinsing a solution of purified Endrin with dilute aqueous base. Stability testing of the new material indicated no stability problems.

The Endrin stability problem is a good example of why historical data should not be the only information used in determining stability. It is necessary to perform stability testing on all new lots of reference materials and not just on the initial batch. Suppliers of ERMs should have an ongoing stability program to test all current batches of reference materials.

DOCUMENTATION

Users of reference materials often require and should expect to receive documentation that details the verification and traceability of the reference material from suppliers of ERMs. Documentation is the only way that traceability can be demonstrated. Producers of reference materials should have a system in place that provides internal records as well as external documentation to users. Internal documentation offers the producer permanent records for all past and existing products. Archive records of past products are invaluable in providing historical information about a product, including original verification data as well as stability data. Internal records may include laboratory notebooks and other documentation, such as quality control and compound preparation worksheets. Original data including all spectra should also be part of the permanent records. Each batch of a product should have a unique identifier associated with it. This is typically accomplished by assigning a specific lot number for each batch. The identifier should provide a way to locate all pertinent information about the product.

External documentation is the information that is received by a user of an ERM accompanying the reference material. The documentation is typically in the form of a Certificate of Analysis (COA) or a technical data sheet. A COA is a document that states that the product meets a set of specifications. COAs will vary from supplier to supplier, but they should include information regarding the traceability of the product, such as the unique product identifier. Often technical data will accompany a COA. This may include copies of spectral and other analytical data. If these data do not come with the product, a user should be able to request this information from the supplier. Since analytical results provide users with necessary traceability data, suppliers should always make this information available. A COA is as important as the actual reference material to a user. The COA will be the permanent record for the user after the ERM has been consumed. This should provide the user with defensible records that are often needed during internal and external quality assurance audits as well as in court cases. Users of ERMs should expect this type of documentation any time they purchase reference materials.

There are an increasing awareness of and a growing concern about contaminants in our environment. With this concern more and more demands will be made for environmental testing and remediation. Environmental reference materials play a significant role in analytical measurements, and high quality reference standards are necessary for accurate analysis. The quality of the reference standard always begins with the neat materials that are used to prepare synthetic standards. The traceability of these neat materials is an essential part of producing accurate standards.

REFERENCES

1. Terms and Definitions Used in Connection with Reference Materials, ISO Guide 30 (1992).
2. **A. S. Kenda, J. J. Wade, D. Ridge, and A. Poland,** Synthesis and Fourier transform Carbon–13 nuclear magnetic resonance spectroscopy of new toxic polyhalodibenzo-p-dioxins, *J. Org. Chem.*, 1974, 39, 931.
3. **M. L. Taylor, T. O. Tiernan, B. Ramalingam, D. J. Wagel, J. H. Garrett, J. H. Solch, and G. L. Ferguson,** Synthesis, isolation, and characterization of the tetra-chlorinated dibenzo-*p*-dioxins and other related compounds, in: *Chlorinated Dioxins and Dibenzofurans in the Total Environment II*, L. H. Keith, C. Rappe, and G. Choudhary, Eds., Butterworth, Ann Arbor Science, Boston, 1985, 17–35.

CHAPTER 2

The Preparation and Validation of Isotope Dilution Standards: A Case Study in PCBs

Daniel L. Bolt and Joel C. Bradley

INTRODUCTION

Isotope dilution is a well established technique for the measurement of pollutants in the environment. This method calls for the addition of a known quantity of a stable isotope-labeled analog of the analyte in question. The labeled and unlabeled substrates can be measured precisely in the mass spectrometer. Since the amount of the labeled analog is known, the naturally occurring pollutant can be quantified precisely. This method has been used extensively for low level quantitation and has been reported on in various articles.[1,2]

The isotope dilution method of quantitation has been used widely in analysis of chlorinated and brominated dibenzo-p-dioxins and dibenzofurans. As in the cases of these "dioxins" and "dibenzofurans", PCBs are a series of many related congeners and isomers. Each formulation of PCBs can be different and may contain varying amounts of these individual components. Furthermore, because of their widespread use around the world, it has become necessary to measure levels of PCB isomers in a variety of diverse and complex matrices, including soil, sludge, waste oil, fish tissues, and even human serum.

Recently, interest has been focused on certain specific isomers, which have been identified as more critical because of their higher toxicities. Concerns over certain "toxic congeners", such as coplanar and mono-ortho substituted PCBs, have led to methods which call for isomer-specific quantitation.

This new isomer-specific approach is in contrast to traditional measurements of total aroclors or decachlorobiphenyl following perchlorination of all biphenyl isomers. Isomer-specific analysis has been one of the driving forces that has led to increased use of ^{13}C-labeled standards. Other benefits of isotope dilution methods include increased specificity, precision at low levels, and the ability to produce lower detection limits in a wide variety of complex matrices.

In response to the increasing need for isotope dilution analysis in the PCB area, Cambridge Isotope Laboratories has synthesized a wide range of the most interesting carbon-13-labeled PCB analogs. In order to ensure that these would be maintained at the highest levels of quality, a complete program of quality control and quality assurance was developed for both labeled and unlabeled PCB analogs.

The original protocol specified gravimetric preparation of the isotope-labeled solutions at 40 ppm to be compared to the corresponding commercially available unlabeled PCB solutions. It further required that the finished product be compared to a second source standard as a final check. In practice, however, there was great disparity among the concentrations of the commercial solutions. A closer check of these unlabeled solutions revealed that there was very little consistency from any existing manufacturer or within the group as a whole. (See Table 1 for examples.) Therefore, a search for unequivocal standards was undertaken.

At that time, there was only one SRM from NIST available that contained individual PCB congeners (SRM-1585; see Table 2 for details). This was a good standard and a place to start, but it contained only 8 of a possible 209 congeners. A protocol was needed to produce reliable unlabeled standards for as many isomers as possible. Therefore, although the NIST SRM was incorporated into the final protocol, it was not the complete answer to the suddenly complex question of how to prepare very reliable isotope-labeled standards of individual PCB congeners.

In fact, no commercial source of unlabeled standards of sufficient reliability and number of congeners could be identified. Consequently, CIL needed to create a program to produce high quality unlabeled reference standards of individual PCB congeners. Once these were in hand, the ultimate goal of producing isotope-labeled standards of the highest quality was achievable.

Table 1 Variance Among Commercially Available Unlabeled Standards

Values agreed within:	5%	3 of 22 (14%)
	10%	5 of 22 (23%)
	20%	8 of 22 (36%)
	50%	6 of 22 (27%)

Table 2 NIST SRM 1585: Chlorinated Biphenyls in 2,2,4-Trimethylpentane

Chlorinated biphenyl	Concentration µg/g	Concentration µg/mL at 23.0°C
4-Chlorobiphenyl	43.3	29.9
4,4′-Dichlorobiphenyl	9.53	6.57
2,2,4′-Trichlorobiphenyl	3.70	2.55
2,2′,5,5′-Tetrachlorobiphenyl	7.72	5.32
3,3′,4,4′-Tetrachlorobiphenyl	6.62	4.56
2,2′,4,5,5′-Pentachlorobiphenyl	5.24	3.61
2,2′,3,4,4′,5-Hexachlorobiphenyl	2.37	1.63
2,2′,4,4′,5,5′-Hexachlorobiphenyl	3.06	2.11

GENERAL PROCEDURES

At CIL, whenever any standard solution is prepared, certain procedures are adhered to:

1. Isomers are synthesized according to the most unambiguous and well-documented literature routes. All end-products are characterized by GC retention times and coinjection with known isomers. Materials are purified by column chromatography, HPLC, and recrystallizations until the purity is greater than or equal to 98% by GC/FID and GC/MS. Other analytical tests that are routinely performed to confirm chemical and isotopic purity include HPLC, ^{1}H or ^{13}C NMR, and/or melting point determinations.
2. All labeled starting materials and finished products are analyzed by GC/MS vs. standards of known isotopic enrichment. The minimum acceptable average carbon-13 enrichment for any such materials is 99%. Selected ion monitoring is done on the mass that corresponds to unlabeled material in the final product, to ensure that no native material is present.
3. Solvents are redistilled and checked for purity by GC/FID and GC/MS.
4. Balances are maintained by a qualified balance technician, are standardized by comparison to NIST class "s" weights, and made ready for gravimetry.
   Once all of the above is done and approved, the standard can be made. Here also, a standard operating procedure (SOP) is applied.
5. Triplicate, independent weighings are performed by three staff scientists.
6. Solutions are made of these three weighings in Class A volumetric flasks using the redistilled solvent prepared previously.
7. Triplicate back-to-back injections are made on a Hewlett-Packard 5971 GC/MS.
8. Appropriate statistical controls are applied (as defined in the individual SOPs).
9. Validation is performed vs. Standard Reference Materials (SRMs) wherever such materials are available.
10. Ampouling and sealing are done in a manner that maintains the integrity and stability of the standard solutions.
11. Useful Technical Data Sheets (TDS) and Material Safety Data Sheets (MSDS) are prepared.

In addition to the preparative steps mentioned above, a system has been established to track lot-to-lot variability for future preparations as well as shelf life and stability of each solution.

CARBON-13 LABELED PCBs

The principles and procedures outlined above have been applied to preparation of ^{13}C-labeled PCB standards routinely available from CIL. In this case, it was well known that the users of these standards fully intended to do isotope dilution quantitation and that the desired detection limits would be extremely low. Although these compounds are stable and persistent in the environment, they are flame-sealed in amber glass ampoules in a nonhalogenated solvent to minimize reactivity. These PCB congeners are not very soluble in most solvents. Therefore, a low concentration of 100 ppm was chosen for the unlabeled standards (to be prepared first) and 40 ppm was chosen for the labeled analogs. The solvents chosen were isooctane for the unlabeled and *n*-nonane for the ^{13}C-labeled standards.

SPECIFIC PROCEDURES PRIOR TO WEIGHING

1. Pure compound is synthesized by the most unambiguous route identified.
2. Complete compound identification, including isomer assignment, is performed. The coupling mechanisms typically used are well understood, and their products are predictable. A GC/FID analysis is performed at the time of synthesis, however, to confirm purity and establish absolute retention times by means of coinjection with a commercially available standard. Also, if needed, FT-IR, ^{13}C, and ^{1}H NMR can be used for structural confirmation.
3. Reagent grade solvents (isooctane and *n*-nonane) are redistilled in-house to ensure the highest possible purity.
4. Balance is calibrated vs. NIST class "s" weights.

SPECIFIC PROCEDURES FROM WEIGHING TO TECHNICAL DATA SHEETS

Triplicate, independent weighings are carried out by three senior staff scientists as follows:

1. Balance autocalibration is performed.
2. A 10 mg class "s" weight is weighed to confirm the balance calibration.
3. The weighing paper is weighed.
4. The balance is tared.
5. The individual pure congener is weighed at 10 ± 0.1 mg
6. The material is carefully transferred to a 100 mL Class A volumetric flask, and the paper is washed with redistilled solvent.
7. The weighing paper is air dried.
8. The paper is reweighed (must be within 0.1 mg of its original weight).
9. The 10 mg class "s" weight is reweighed (must be 10 ± 0.1 mg).
10. The volumetric flask is filled with approx. 80 mL of redistilled solvent.
11. The flask is sonicated in a warm bath for at least 15 min and then allowed to return to room temperature.
12. The flask is brought to volume with redistilled solvent.

This procedure, when done in triplicate, leaves three 100 mL solutions (A,B,C) independently prepared at the nominal concentration of 100 μg/mL. These three solutions are then brought to the quality control department for analysis.

QUALITY CONTROL ANALYSIS

Triplicate, back-to-back injections are made on a Hewlett-Packard 5971 GC/MS. The sequence is run alternating the flasks as follows, to minimize any effect of any change in the instrument over the run time of the analysis:

Flasks	Weighing "A"	Weighing "B"	Weighing "C"
Injection #	1	2	3
	4	5	6
	7	8	9

One microliter is injected under identical acquisition and instrument parameters. Total areas are calculated using selected ions, and those areas are assigned injection numbers and flask designations.

STATISTICAL CONTROLS

The area counts from the GC/MS are subjected to the following statistical requirements:

1. %RSD within solution "A" (injection #s 1,4,7) must be less than 5%.
2. %RSD within solution "B" (injection #s 2,5,8) must be less than 5%.
3. %RSD within solution "C" (injection #s 3,6,9) must be less than 5%.
4. Overall %RSD for injections 1–9 must be less than 5%.

Very few problems were encountered meeting these statistical requirements (see Figure 1). When these requirements are met, the three solutions are considered not statistically different and are combined to make one reliable 100 ppm standard solution.

As stated earlier, once the reliable unlabeled standard of known concentration is in hand, the preparation of labeled standards becomes a simple task. A single weighing is performed of the crystalline labeled PCB congener to a nominal concentration (typically 40 ppm). Aliquots of the two solutions (unlabeled at 100 ppm and isotope-labeled at 40 ppm) are mixed together and injected simultaneously on the GC/MS. Again, this analysis is done in triplicate, and the areas of the appropriately monitored ions are compared. The concentration of the labeled standard is determined by direct comparison to the 100 ppm unlabeled standard, and an assay for the labeled standard is assigned. Once the triplicate analysis is complete, the three assays must agree within 5%. If this statistical requirement is met, the mean of the three assays becomes the actual lot analysis for that particular carbon-13 PCB congener. The guidelines listed in the catalog call for the standards to be 40 ± 4 ppm, but this lot assay assignment gives an actual value (e.g., 39.2 ppm). Current lot assays are shown in Figure 2.

CONCENTRATION VERIFICATION

The next step in the process calls for verification vs. known standards. The NIST standard that was available at the time (SRM 1585) was used as verification for the eight congeners that it contained. It was used in much the same manner as our own in-house unlabeled standards. That is, an addition of the labeled and SRM was done,

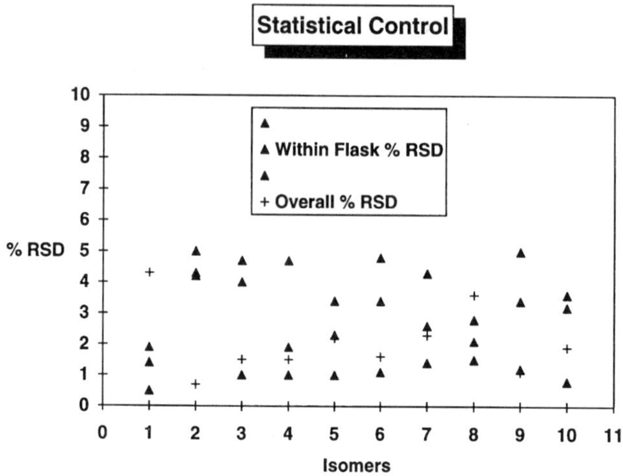

Figure 1 All steps of the quality control are carefully monitored using statistical controls. The within-flask precision as well as the overall agreement must be ≤5% RSD for the triplicate unlabeled standards preparation.

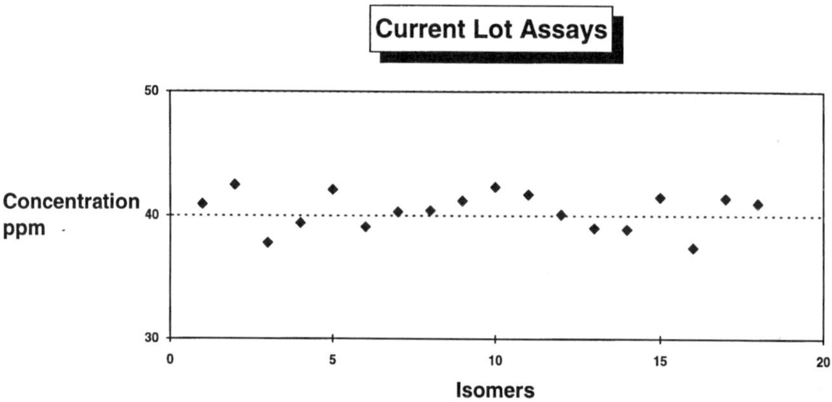

Figure 2 The specifications for the isotope-labeled products are 40 ± 4 ppm. The results shown are actual lots of current CIL 13C-PCB standards as compared to the 100 ppm unlabeled reference standard.

and the appropriate ions were monitored to establish a value for the labeled standard vs. the SRM. Again, the statistical control employed was 5% variance from the assay value. All of the labeled standards prepared according to the protocols described above easily met this criterion for the available PCB congeners. It should be noted that NIST has announced that another SRM with a larger number of PCB congeners will be available shortly. CIL will incorporate these SRMs into the protocols whenever they become available.

Finally, to ensure against in-house instrument or technician bias, the finished solutions of both labeled and unlabeled PCB congeners were sent out to an independent laboratory for formulation and analysis. This laboratory was asked to assign

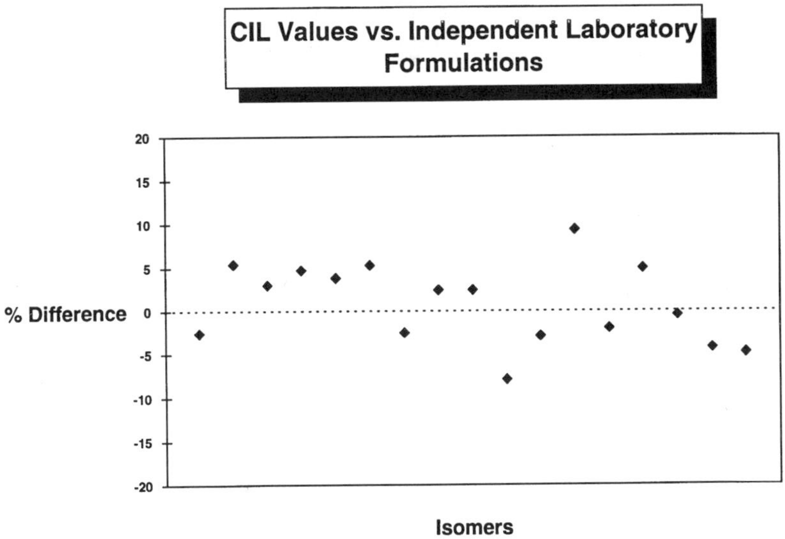

Figure 3 Independent formulation and quantitation by an outside analytical laboratory allows assessment of any technician or instrumental bias.

a value to the labeled standard, assuming a concentration of 100 ppm for the unlabeled standard solution. This time, owing to the larger variable of independent solution preparation and measurement, a statistical control of 10% variance was applied. Again, the agreement was excellent. (See Figure 3.)

TECHNICAL DATA/MATERIAL SAFETY DATA SHEETS

For each available ^{13}C-labeled PCB standard produced under this program, technical data sheets (TDS) and material safety data sheets (MSDS) are prepared. The spectral identifications as well as the chromatographic and any other purity assays are supplied. Further, a cover sheet is provided, which describes the lot number, date prepared, and structural formula of the material, as well as the actual lot analysis in ppm. Also, based on the experiences related above, CIL has made available the unlabeled 100 ppm solutions as commercial products. An example of a typical TDS coversheet can be seen in Figure 4.

CONCLUSIONS

Isotope dilution measurements of individual PCB congeners in the environment have led to a need for high quality quantitative ^{13}C-labeled standards of various PCB isomers. In the course of developing such standards, it became obvious that there were problems with the quality of unlabeled PCB standards commercially available. Although they were abundant and easy to obtain, they did not seem to correlate with

Technical Data Sheet

Compound:	2,3,3',4,4'-Pentachlorobiphenyl (IUPAC #105)
Isotope:	$^{13}C_{12}$,99% ^{13}C
Catalog Number:	EC-1420
Lot Number:	CJ-007

Structure:

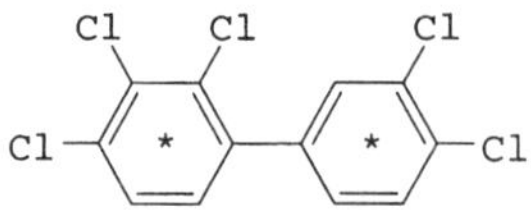

$^{13}C_{12}H_5Cl_5$ (MW 336)

Chemical Purity:	98%
Concentration:	40±4 micrograms/milliliter
Actual Lot Analysis:	**40.5 micrograms/milliliter**
Amount:	3 milliliters
Solvent:	n-Nonane

Figure 4 Typical technical data cover sheet.

one another or with the established SRM (NIST SRM 1585). Unfortunately, this SRM, although useful, contained only 8 congeners and did not address the other 201 congeners of PCB.

In response to this problem, Cambridge Isotope Laboratories developed procedures to produce, under statistical control, high quality, reliable unlabeled standards. With these in hand, isotope-labeled solutions have been prepared and a lot assay assigned vs. the unlabeled reference standards. These standards have compared favorably with the available NIST standards. Further, the solutions of these standards are subject to independent formulation and assay by an independent analytical laboratory to assess the product for in-house instrumental or technician bias. These quantitative unlabeled standards are now used to assay on a lot-by-lot basis the individual carbon-13–labeled PCB standards produced routinely by CIL. These lots are accompanied by technical data sheets showing the actual lot analysis of the individual isomer and a material safety data sheet.

Most importantly, by following this protocol, high quality reference standards can be produced, which can then be used to manufacture reliable isotope-labeled standards for quantitative measurements.

ACKNOWLEDGMENTS

The authors wish to thank the following people for helpful discussions and other contributions to this work.

Dr. Doug Brown, Mylan Laboratories
Dr. Mike Re, Radian Corporation
Dr. David Wagner, Metabolic Solutions, Inc.

REFERENCES

1. **J. F. Pickup and K. McPherson,** *Anal. Chem.*, 48 (1976) 1885.
2. **B. N. Colby and M. W. McCaman,** *Biomed. Mass Spectrom.*, 6 (1979) 225.

CHAPTER 3

Preparation and Verification of ICP and AA Reference Standards

Jean-Pierre Zajac

INTRODUCTION

With the increasing demand for environmental analysis, reference standards for metals are becoming significantly more important. The need for laboratories to have certified standards to do their analyses and the fact that some metals are difficult to put into solution contribute to the popularity of prepared standards.

The intent of this chapter is to present the reader with the basic steps involved in the preparation of standards, which is of benefit to people buying prepared standards and to those who make their own.

DIFFERENCE BETWEEN AA AND ICP REFERENCE STANDARDS

Before proceeding with details concerning the actual preparation and verification of standards, it is important to understand the difference between an AA Standard and an ICP Standard.

Essentially, the distinction between them is in the purity of starting materials. Atomic Absorption (AA) Spectroscopy is an absorption technique; metals present in a solution absorb energy at a specific wavelength. In contrast, ICP Spectroscopy is an emission technique in which metals present in solution emit light when excited in the plasma.

As can be seen from Figures 1 and 2, more lines are produced in the emission technique than in the absorption technique, generating a spectrum that is much more complex.

To illustrate a point, imagine a calibration standard with a high concentration of metallic impurities present in solution. If it is a standard used for AA, it is reasonable to say that in most cases, only the main element will absorb light at a specific wave-

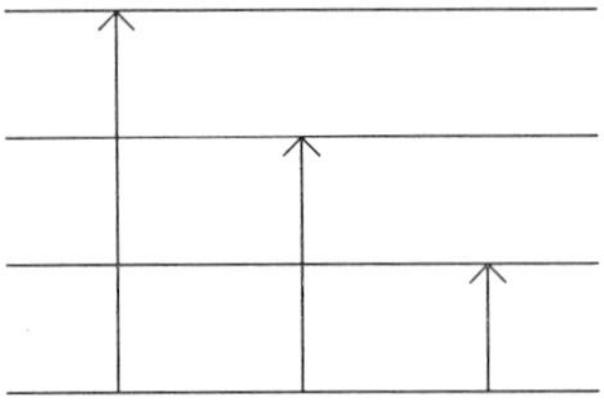

Figure 1 Absorption (3 lines).

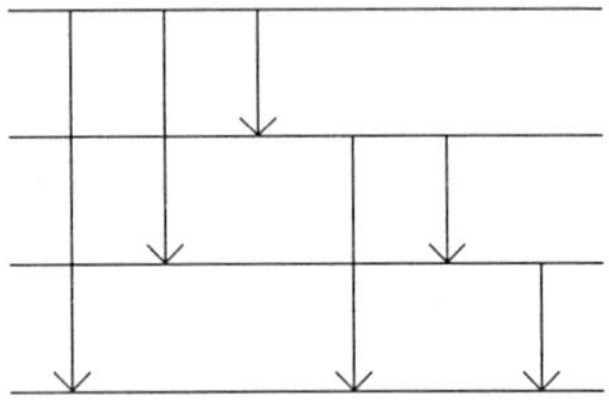

Figure 2 Emission (6 lines).

length; therefore, impurities would not present a significant problem. On the other hand, if it is a standard used for ICP, all the metals present in solution will be excited and will emit light. There is a strong possibility that these impurities would cause interferences with the main element, creating incorrect results during analysis. In addition, should the standard solution be used for matrix matching at high concentrations, the level of impurities becomes even more critical, as it could cause matrix effects.

STATEMENT 1: You can use an ICP Standard in AA Spectroscopy, but you cannot use an AA Standard in ICP Spectroscopy.

Also, with the increasing popularity of ICP/MS for environmental analyses, purity becomes increasingly critical as detection limits are becoming better and better.

There are ICP users using AA Standards for the calibration of their instruments who indicate they never have had a problem. However, the accuracy of their analyses, unless each standard is verified for purity, cannot be assured.

CHOICE OF RAW MATERIALS

The most important component of a standard is the raw material used to prepare the standard. As mentioned before, purity is the difference between an AA and an ICP Standard.

Typically, material with 99.9 or 99.99% purity can be used for AA Spectroscopy, and material with 99.999% plus purity is used for ICP Spectroscopy. Purity is usually expressed as Total Metallic Impurities (TMI). In a 99.999% pure material, there will be a maximum of 10 ppm of metallic impurities present in the raw material, for a TMI of 10.

STATEMENT 2: The purity of the raw material should be the most important factor when choosing a raw material for the preparation of an ICP Standard.

Once the purity of the material is selected, the compound to be used must be determined.

The easiest starting material to use is the pure metal, as one avoids stoechiometry problems. If your material is 99.999% pure, you can assume that if you weigh out 1 g of material, you will obtain 0.99999 g of the main element. The concern with pure metal is the need to be sure there is no oxidation on the surface. If oxidation is present, a problem arises due to the mixture of pure metal and the oxide. Using chunks of metal instead of powder reduces surface area, and consequently, the effect of oxides in calculating the weight of raw material. If oxidation is present on the surface of the metal, chunks or pieces can be cleaned using a dilute acid solution. Naturally, powders are virtually impossible to thoroughly clean.

If powder is the only form available, it is important to store it in such a way that it will not be attacked by air (e.g., storage under nitrogen).

STATEMENT 3: Whenever possible, the pure metal in pieces and not the powder form should be used in the preparation of an ICP or AA Standard.

Unfortunately and for different reasons, it is not always possible to use the pure metal as a starting material. You then must use other forms of materials. The most popular are the salts or the oxides of the metal of interest.

The main problem associated with these compounds is a stoechiometric one; that is, the actual amount of the main element in the compound does not necessarily correspond to the theoretical amount.

The following examples demonstrate this phenomenon:

Material	Theoretical value	Actual value	Diff. (%)
$(NH_4)_2WO_4$	64.75% W	67.77% W	3.02%
Sc_2O_3	65.20% Sc	64.44% Sc	0.76%
$CoCO_3$	49.55% Co	44.78% Co	4.77%

As demonstrated, it is critical to verify the actual concentration of the main element in the compound before preparing a standard; otherwise a standard with an incorrect concentration could be obtained. A way to eliminate this verification is to obtain primary standard grade material, which usually has a defined stoichiometry.

STATEMENT 4: It is absolutely necessary to verify the actual concentration of the main element in a compound if this starting material is not the pure element itself or if the compound is not a primary standard.

DIGESTION OF THE RAW MATERIALS

Once the appropriate raw materials have been selected, they must be put into solution. Most compounds will be soluble in nitric acid, hydrochloric acid, or a mixture of the two. Some salts will be completely soluble in water.

Again, as purity is very important for ICP Standards, the grade of acid used in the preparation of AA and ICP Standards may be different. Typically, electronic grade or ACS grade acids will be used in the preparation of AA Standards, while high purity acids are used in the preparation of ICP Standards.

"High purity" acids should not contain impurities greater than the ppb levels (1 to 2 ppb) (ppt levels for ultrapure grade). These acids are available commercially from different sources.

It should be noted that there is no real advantage to using ultrapure acids if these acids are not used in a clean room environment. For this reason, most laboratories use "high purity" acids and not "ultrapure" acids.

Water used in the preparation of AA and ICP Standards should have a resistivity in the vicinity of 18 MΩ/cm (ASTM type 1). This water is obtained through a purification system that is commercially available.

The water system in use at SCP SCIENCE employs multiple stages, including reverse osmosis, chlorine removal, charcoal filtering to remove organics and organometallics, dual ion-exchange cartridges to remove ionic compounds, a UV light to kill any bacteria present in the water, and finally a 0.06 μm final polishing filter to remove colloids and other traces. The final water is in the range of 18 MΩ/cm.

STATEMENT 5: Only high purity acid and 18 MΩ/cm water should be used in the preparation of ICP Standards.

Once the requirements for the raw materials, acids, and water have been met, it is possible to proceed to the acid digestion of the materials.

The first step is to accurately weigh the amount of material necessary to produce the standard at a given concentration and volume. This amount of material is then transferred to a beaker, and the acid, water, or mixture of the two necessary for digestion is added.

The important step to consider at this stage is the choice of beaker material. Glass beakers should not be used for digestion, as glass by its nature contains many elements, providing a real possibility of contamination even when the beaker is cleaned between digestions.

The best choice for beaker material is Teflon®. Teflon is resistant to almost anything and is made from pure material containing very low levels of impurities. This reduces the possibility of contamination. Nevertheless, for additional safety in preventing contamination, we suggest using the following cleaning process for Teflon® beakers, which has been used successfully at SCP SCIENCE.

Add high purity nitric acid to the Teflon beaker, and bring the acid to a boil. Allow the acid to reflux in the beaker for approximately 30 min. After emptying the beaker, place it in an acid bath containing a 50% V/V mixture of high purity nitric acid and water. Soak the beaker overnight, and then rinse it thoroughly with 18 MΩ/cm double deionized water. Only after this process can the beaker be considered ready for digestion use.

After completion of the digestion, the solution is filtered by gravity using a high purity filter paper held in a Teflon funnel. The filter paper used at SCP SCIENCE has been washed with 18 MΩ/cm H_2O in the manufacturing process and has not shown any traces of impurities after leaching with acid. The filtered solution is then completed to actual weight and is ready for testing.

It is important to specify that all containers used for storing ICP Standards should be precleaned. The precleaning process in use at SCP SCIENCE consists of an acid leach using a solution of 50% V/V nitric acid to water for 1 week. Before use, the containers are rinsed thoroughly with double deionized 18 MΩ/cm water. The leaching process is necessary to prevent contamination from the container or the interaction between elements contained in the container and the actual product.

STATEMENT 6: Only containers that have been precleaned should be used to store ICP Standards to prevent contamination or precipitation problems.

VERIFICATION OF IMPURITIES

As previously mentioned, the level of impurities is very important in ICP Standards; consequently, it is important to verify the final solution for impurities. In theory, there should be a very low level of impurities present in the final solution if all previous steps have been followed. However, it is recommended to verify to be absolutely certain.

An impurity check is performed on the ICP spectrometer for each standard solution. Should for any reason an element be detected, it is analyzed to determine the actual concentration and then reported on the Certificate of Analysis.

A straightforward way of checking for impurities is to run a blank for the list of elements to be verified, and then run the actual standard solution. Any significant differences between the blank and the standard will indicate a contamination. Of course, you need to choose lines that are sufficiently sensitive and without interferences. This represents an interesting challenge!

It is important to report impurities present in the solution, especially when multi-element standards are to be prepared. To demonstrate a potential problem, use the example of the preparation of a ten-element solution starting with pure single element standards. If each pure single element standard contains detectable levels of impurities, it is possible that one or many of the impurities could be the same as some of the elements in your multi-element standard. If so, an element or two, or even more, may have higher concentrations than expected.

PREPARATION OF A STANDARD SOLUTION FOR TESTING

When the standard has been completed to volume, we have a standard with a theoretical or calculated concentration and a need to verify that concentration. There are essentially two ways of verifying the concentration of a standard — by ICP or by classical wet assay. Although the classical wet assay is well known and well accepted, it is much faster to certify standards by ICP Spectroscopy, providing that the precision requirements are well defined and that a sufficient number of measurements are made to meet that precision. The following describes the ICP Spectroscopy method.

The first steps involved in testing by ICP Spectroscopy are (1) the preparation of the calibration standards and (2) the preparation of the sample to be tested. In order to accomplish these steps, volumetric glassware, such as pipets and flasks, are used.

It is important to clean all glassware before use to prevent possible contamination. Cleaning is carried out using high purity nitric acid followed by at least seven rinses with 18 MΩ/cm water. Pipets must be dried before proceeding.

It is also critical to calibrate pipets and volumetric flasks on a regular basis to ensure that they meet Class A specifications. As can be seen from the example below, one cannot assume that the nominal value will correspond to the actual value.

Pipets Class A: Specification: 10 mL ± 0.02 mL

Volume (nominal)	Volume (calibrated)
10.00 mL	10.013 mL
10.00 mL	10.017 mL
10.00 mL	10.013 mL
10.00 mL	10.016 mL
10.00 mL	10.014 mL

Volumetric Flasks 100 mL Class A: Specification: 100 mL ± 0.08 mL

Volume (nominal)	Volume (calibrated)
100.00 mL	99.94 mL
100.00 mL	99.93 mL
100.00 mL	99.98 mL
100.00 mL	99.99 mL
100.00 mL	99.94 mL

STATEMENT 7: It is absolutely necessary to calibrate all volumetric glassware used in the preparation of standards to avoid erroneous results.

Another important point in preparing ICP calibration standards is that the matrix of the calibration standards must match the matrix of the test solution. As can be seen in the following example, a small difference in acid concentration can cause a significant change in the actual concentration.

Example 1: Nitric acid

%HNO_3	Cr	Se	Sr
5	100.8	100.4	99.3
4	101.8	101.6	100.9
3	102.1	102.2	101.8
2	103.2	102.5	102.4
1	103.9	103.8	103.5
0	106.9	106.8	106.6

Note: All concentrations are in μg/mL.

In this example, the original calibration was performed with a standard containing 5% nitric acid. This was followed by analyzing the same element concentration with different concentrations of acid. It is clear that the acid concentration is critical.

The acid concentration is particularly important when preparing multi-element standards. For example, in preparing a 100 μg/mL solution consisting of 5 elements with a final acid concentration of 5% V/V, it is common to use five pure single element standards of 1000 μg/mL with, for example, 4% V/V nitric acid in each solution. To prepare a 100 mL volume, one will pipet 10 mL of each solution and complete to 100 mL.

A common error made is to add 5 mL of nitric acid and then to add the 5 × 10 mL aliquots of pure, single element standards. After completing to 100 mL, an incorrect conclusion would be that a multi-element standard with a final acid concentration of 5% V/V has been achieved.

In fact, if in each aliquot there is 4% V/V of nitric acid present, this represents a total of 2 mL of acid plus the 5 mL added, for a total of 7 mL, or 7% V/V.

This difference of 2% could lead to erroneous results, as is seen in the previous example.

STATEMENT 8: To prevent erroneous results, it is critical that the matrix of the standard match, as closely as possible, the matrix of the solution to be analyzed.

TESTING BY ICP SPECTROSCOPY

Once all calibration standards and the solution to be analyzed have been prepared, testing on the ICP Spectrometer can begin.

Important factors to consider when testing by ICP Spectroscopy is the need for calibration standards to generate calibration curves as well as the need for a second standard to verify your calibration. This second standard is usually called a Quality Control Standard. When the Quality Control Standard meets your specification limits for concentration, testing can begin.

Most analysts will calibrate their instruments with commercially available standards and verify the calibration curves with reference standards, such as NIST standards. Because, at SCP SCIENCE our standards are directly traceable to NIST, we use pure single element ICP Standards, Series 3100 from NIST (National Institute of Standards and Technology), to generate calibration curves and a second commercially available standard for each element to verify calibration curves. All solutions are run in duplicate, using a bracketing technique to ensure the highest accuracy possible. The precision required has to be within 0.5% or better of the nominal value of the standards.

In order to achieve accurate results, it is obvious that the ICP instrument must be maintained in excellent operating condition. Each manufacturer's instrument is different in concept and requires different maintenance procedures. It is not the intent of this chapter to discuss maintenance or operating procedures of ICP instruments, but only to stress that ICP instruments must be maintained on a regular basis to achieve accurate analysis. A common way to monitor the performance of an instrument is to use the Quality Control Standard mentioned previously. By testing the QC Standard over a period of time (months), the analyst is able to determine trends.

CERTIFICATE OF ANALYSIS

Finally, after completion of the testing, results need to be reported. A few points regarding the information that should appear on a Certificate of Analysis are worthy of mention.

The first and very important point is the lot number, which will identify the product. Lot numbers are used to trace all information pertinent to the actual standard. The expiration date for the specific lot must also appear on the certificate.

The actual concentration of the element, as well as the matrix and trace impurities, must appear on the certificate to provide the user the information required to prepare multi-element standards. A note indicating the precision of the measurement must also be included.

Finally, the method of testing should be indicated, together with actual traceability information (NIST if applicable).

CONCLUSION

The information in this chapter should be useful to those people preparing or using calibration standards with AA or ICP instruments; by following the guidelines described in this chapter, successful analysis should be achieved.

ACKNOWLEDGMENT

I would like to thank Dr. Ray Clement for giving me the opportunity to write this chapter, Dr. Otto Hermann for his technical help, and Mr. George Feilders for editing the text.

CHAPTER 4

The Preparation and Stability of PCDD/PCDF Soil Standards

Michael A. Re, Lawrence H. Keith, and Alan W. Nichols

Fortified and natural matrix soil standards were prepared. Consensus values of analyte concentrations were determined by an international interlaboratory study. The fortified standard was prepared using the 17 toxic PCDD/PCDF isomers. Analyte identity, purity, and homogeneity were confirmed by following rigorous QC protocols. A study is currently underway to determine long-term stability of the standards.

Matrix-based environmental reference materials (ERMs) encompass a variety of both natural matrix and synthetic reference materials. These types of standards can be broadly defined as well-characterized environmental materials that contain or have been screened for target analytes. Some common environmental matrices include soil, water, and biota such as fish or vegetation. Matrix-based synthetic reference materials are prepared by fortifying a natural environmental matrix with known amounts of target analytes. Natural matrix reference materials are typically homogenized and characterized for target analytes. These standards are commonly used as quality control samples for laboratory quality assurance programs. They are used as performance evaluation (PE) samples for both internal and external laboratory audits. Matrix standards will also play a major role in defining method detection levels (MDLs) if newly proposed definitions are accepted.

Polychlorinated dibenzo-*p*-dioxins (PCDDs) and dibenzofurans (PCDFs) are environmental contaminants that have generated great interest for many years due to the potential health risks associated with them. Compounds that contain chlorine in the 2, 3, 7, and 8 positions of the dibenzo-*p*-dioxin and dibenzofuran ring systems are considered to be the most highly toxic. There are 17 toxic congeners, including 7 dioxins and 10 dibenzofurans. There are a number of potential sources of these pollutants, including by-products from chemical processes and formation during incineration processes. Because of the potential health risks and number of possible sources for contamination, analytical testing is done on a variety of natural matrices, including soil, water, and biological tissue. While calibration standards have been available commercially, the availability of matrix standards has been limited. Since soil is commonly tested for dioxins and furans, unfortified (blank) and fortified standards of a representative soil would be a useful addition to quality assurance programs.

There are a number of important objectives that should be achieved when preparing matrix standards. When preparing a synthetic reference material such as a fortified soil, the traceability of each of the analytes must be well established. The origins of materials used for fortification should be known and documented. Demonstrating homogeneity is also very important for matrix standards. Because most environmental matrices are complex, it is critical that methods to homogenize and analyze them be developed and verified prior to their production.

Another factor that establishes the overall usefulness of the standard is to demonstrate that analytical data is reproducible. One way to demonstrate reproducibility is to use an interlaboratory study. This can provide data on overall variability between laboratories and can also be used to establish consensus values and tolerance limits for the analytes of interest. Another requirement is the demonstration of short-term and long-term stability of any matrix-based standard.

TWO TYPES OF SOIL STANDARDS

The preparation of soil standards can lead to two kinds of reference materials: natural matrix and fortified matrix ERMs. The first step in preparation of fortified standards is to collect and homogenize a representative soil. The soil is then characterized for target analytes. At this point the soil can be used as a natural matrix ERM for QC blanks or as unknowns if the soil contains target analytes "naturally". The blank soil can also be used as the matrix to prepare fortified (synthetic) standards by adding known amounts of target analytes.

TRACEABILITY OF ANALYTES

To successfully establish traceability of an ERM, it is essential to establish traceability of all materials that are used to produce the ERM. This includes establishing identity and purity of neat materials. When solution standards are used to fortify the matrix, traceability of analyte identity and concentration of the solution must be known and documented.

In preparing a soil matrix standard containing the 17 toxic dioxin and furan congeners, traceability was established for each congener. This began with the synthesis of each of the compounds. In many instances, isomer-specific reactions were used. This was the first step in establishing identity of the compounds. Once the compounds were prepared, multiple methods, including GC/MS, melting point (MP), and GC retention times were used to establish identity. Purity was established using gas chromatography with a flame ionization detector (GC-FID). Other secondary methods, including thin layer chromatography (TLC) and melting point (MP), were also used to confirm purity.

Standard solutions of each of the 17 congeners were used to fortify the soil; therefore, it was also important to have traceability of each of the solutions. As stated above, showing analyte identity and showing concentration for each solution are important factors to consider. Analyte identity of each solution was confirmed using GC/MS. Methods for accurately preparing solution standards of dioxins and furans have been described.[1] The methods begin with gravimetric preparations, followed by extensive quality control testing for solution concentration and homogeneity. An interlaboratory study was also conducted to establish consensus values for each of the 17 standard solutions.[2] Obtaining these data for each analyte provides the unbroken chain of comparisons necessary to establish basic traceability.

SOIL SELECTION AND PREPARATION

The selection of soil to be used for a matrix standard is very important. There are a wide variety of soils, ranging from sand to clay, and their properties are vastly different. A representative soil that is somewhere between sand and clay is probably the best choice for a general use soil standard. An East coast sandy loam was used for the dioxin/furan standard because it exhibits those properties.

The initial preparation of soil included collection, drying, sterilizing, and homogenization. During the collection procedure, rocks and sticks were removed. Drying and sterilization were achieved by heating the soil to 170°C. It is desirable to have particles of similar size to facilitate more efficient spiking of analytes and to obtain a more homogeneous sample. The soil was sifted through a fine mesh screen, which gave a more homogeneous particle size.

HOMOGENEITY

Establishing homogeneity is an important part of preparing matrix standards. All samples taken from a batch of standard must be identical. One method for demonstrating homogeneity is to add a known analyte to the matrix and then analyze random samples after mixing. A simple but very effective method was developed to demonstrate homogeneity of the soil standards. Disodium fluorescein was added to a small portion of soil to form a premix. This premix was then mixed with the total batch of soil. Random samples were then removed from different areas of the mixing container and analyzed. Samples could be quickly analyzed by agitating the soil with water, filtering, and measuring the absorbance of disodium fluorescein. Disodium fluorescein was chosen because of the ease and sensitivity of the method and because it will not interfere with GC/MS analysis of the soil. Typical preparative methods of soil samples include an extraction process that involves an aqueous rinse. Disodium fluorescein would be removed in this step because it is soluble in water. Typical results for adding disodium fluorescein to as much as 100 kg of soil and analyzing random samples from the top, middle, and bottom of the mixing chamber yielded relative standard deviations in the range of 2 to 4%.

Table 1 Sample Homogeneity of Fortified Soil (Top, Middle, and Bottom Aliquots)

Analyte	Mean measured concentration (ng/g)	Standard deviation (ng/g)	Relative standard deviation (%)
	Dioxins		
2378-TCDD	0.478	0.045	9.4
12378-PCDD	1.073	0	0
123478-HxCDD	0.975	0	0
123678-HxCDD	0.863	0.018	2.1
123789-HxCDD	0.925	0.025	2.6
1234678-HpCDD	1.350	0.092	6.8
OCDD	4.020	0.245	6.1
	Furans		
2378 TCDF	0.460	0.033	7.3
12378-PCDF	0.810	0.043	5.3
23478-PCDF	0.817	0.043	5.2
123478-HxCDF	0.890	0.082	9.2
123678-HxCDF	0.878	0.063	7.1
123789-HxCDF	0.803	0	0
234678-HxCDF	0.907	0.054	6.0
1234678-HpCDF	1.332	0.095	7.1
1234789-HpCDF	1.227	0	0
OCDF	1.883	0.173	9.2

Mixing and homogenizing large amounts of soil is best accomplished using a large mixing barrel. The dioxin/furan soil standards were mixed in a specially designed barrel, which was placed on a barrel roller and rotated until homogeneity was achieved. Using the disodium fluorescein method offered a quick procedure for periodically checking the soil for consistency. This method was used to prepare both the blank and the fortified soil.

A second method was used to independently demonstrate homogeneity of the fortified soil. This procedure involved the GC/MS analysis of random samples of the fortified soil for the 17 isomers. Showing consistency of results from random analysis for all of the isomers is also a very good indication of homogeneity. Results from analysis of samples from the top, middle, and bottom portions of the mixing barrel are shown in Table 1. The overall average relative standard deviation for all analytes was 5.9%.

FORTIFICATION OF BLANK SOIL

Once a natural soil has been homogenized and screened for target analytes, it can be used as a natural blank ERM or used to prepare a fortified soil standard. The most important aspect of adding analytes to any natural matrix is to ensure that the analytes are distributed evenly throughout the standard. One method for adding chemicals to soil is a two-step process. The first step involves preparing a solution containing the analytes and adsorbing the mix onto a small portion of the soil to create a premix. In the second step the premix is then blended into the remainder of the soil. This method provides more consistent mixing than a single step process.

The procedure used to prepared the fortified dioxin/furan standard was slightly more complex. To obtain the best consistency of mixing, three premixes were prepared. A solution containing all 17 congeners was divided into three equal portions. The three portions of solution were then added to three portions of soil, and the solvent was removed from each. Additionally, three separate aliquots of soil were spiked with disodium fluorescein. The six premixes were then blended with a large portion of blank soil. Mixing was initially checked for completion by measuring for disodium fluorescein. After acceptable results were obtained, homogeneity was then independently confirmed by the GC/MS method.

REPRODUCIBILITY OF ANALYTICAL DATA

An ERM will only be useful if reproducible data can be generated from multiple analyses. As discussed above, one way to demonstrate reproducibility is to use interlaboratory comparisons. Typically, a group of highly qualified laboratories are chosen to participate, and standards are provided as blind samples. Each laboratory will then analyze the standards as unknown samples and report their results to the organizer of the study. Results from all laboratories are compiled and extensive statistical data analysis is performed.

Ten laboratories participated in the study on the dioxin/furan soil standards. Standard dioxin analytical methods were used by all laboratories which included both high resolution and low resolution mass spectrometry. Each laboratory was supplied with samples of blank and fortified soil. Raw data from each of the laboratories was used to calculate consensus values for each of the 17 target analytes in both standards. In addition upper and lower confidence limits were calculated. Results for the unfortified (blank) and fortified soil are shown in Tables 2 and 3, respectively. Table 3 also includes target values for each of the analytes. This is the theoretical value for each of the analytes that was initially added to the blank soil. These data are very valuable because they give laboratories some guidelines as to what results they should expect when they analyze the soil standards.

STABILITY TESTING

Another important consideration in preparing analytical standards is the determination of both short-term and long-term stability. Fortified natural matrix standards are typically prepared in large batches so that remakes of the standards are less frequent. This is done because of the complexity of preparation and analysis of natural matrix ERMs. Shelf life of the standard becomes very important. Stability testing of ERMs is usually done in two phases. First, an accelerated short-term stability analysis is performed to identify any potential stability problems. The second phase is a long-term testing program to measure overall stability of the

Table 2 Interlaboratory Consensus Values and Tolerance Limits for Unfortified (Blank) Soil

Analyte	Consensus value[a] (ng/g)	Range of expected values[b]	
		Lower	Upper
	Dioxins		
2378-TCDD	0.007 ± 0.003	ND	0.025
12378-PCDD	0.009 ± 0.005	ND	0.042
123478-HxCDD	0.010 ± 0.005	ND	0.041
123678-HxCDD	0.010 ± 0.004	ND	0.036
123789-HxCDD	0.010 ± 0.005	ND	0.043
1234678-HpCDD	0.023 ± 0.005	ND	0.060
OCDD	1.017 ± 0.07	0.507	1.53
	Furans		
2378-TCDF	0.006 ± 0.002	ND	0.019
12378-PCDF	0.007 ± 0.003	ND	0.030
23478-PCDF	0.008 ± 0.003	ND	0.031
123478-HxCDF	0.009 ± 0.003	ND	0.032
123678-HxCDF	0.008 ± 0.003	ND	0.032
123789-HxCDF	0.008 ± 0.003	ND	0.028
234678-HxCDF	0.012 ± 0.005	ND	0.048
1234678-HpCDF	0.012 ± 0.005	ND	0.045
1234789-HpCDF	0.010 ± 0.005	ND	0.048
OCDF	0.021 ± 0.004	ND	0.049

[a] Refers to mean value and confidence limit (at the 95% confidence level) of interlaboratory study based on triplicate analyses by ten laboratories.

[b] These values represent the upper and lower 95% tolerance limits which indicate the range of results that 95% of the laboratories should obtain 95% of the time for a single sample analysis. ND — Not Detected.

standard over time. Typical protocols for testing include analyses at 6 months and 1 year after preparation and on an annual basis thereafter.

This two-phased stability testing program was used for the dioxin/furan soil standards. In this study the soil standards were stored at ambient temperature for the first 3 months after preparation. The standards were then stored long term at 4°C. The initial storage at ambient temperature was performed to help accelerate any analyte degradation to identify any potential stability problems. The fortified soil standard was analyzed at 3 months and then again at 6 months and 1 year after preparation. The results of these analyses are shown in Table 4. These data indicate that no degradation is occurring with any of the analytes over time.

CUSTOMIZED STANDARDS

One of the problems that can exist with performance evaluation (PE) samples is that laboratories recognize them because identity and concentration are constant. For example, a laboratory may recognize a "blind" PE sample after one or two analyses results in the same answer. The laboratory will know what values to expect

Table 3 Interlaboratory Consensus Values and Tolerance Limits for Fortified Soil

Analyte	Target value[a] (ng/g)	Consensus value[b] (ng/g)	Range of expected values[c] Lower	Upper
		Dioxins		
2378-TCDD	0.5	0.46 ± 0.03	0.26	0.67
12378-PCDD	1.0	0.96 ± 0.06	0.56	1.37
123478-HxCDD	1.0	0.90 ± 0.06	0.50	1.29
123678-HxCDD	1.0	0.87 ± 0.05	0.52	1.21
123789-HxCDD	1.0	0.90 ± 0.06	0.46	1.33
1234678-HpCDD	1.5	1.39 ± 0.10	0.71	2.07
OCDD	3.5[d]	3.51 ± 0.22	1.98	5.03
		Furans		
2378-TCDF	0.5	0.45 ± 0.03	0.26	0.64
12378-PCDF	1.0	0.87 ± 0.04	0.59	1.15
23478-PCDF	1.0	0.86 ± 0.06	0.41	1.31
123478-HxCDF	1.0	0.88 ± 0.05	0.53	1.23
123678-HxCDF	1.0	0.95 ± 0.09	0.34	1.56
123789-HxCDF	1.0	0.82 ± 0.06	0.39	1.26
234678-HxCDF	1.0	0.91 ± 0.06	0.48	1.35
1234678-HpCDF	1.5	1.27 ± 0.11	0.52	2.01
1234789-HpCDF	1.5	1.12 ± 0.12	0.25	1.98
OCDF	2.5	2.25 ± 0.15	1.17	3.33

[a] The target value represents the amount of each analyte that was initially added to the soil. The difference between these values and the consensus values should be a good indication of the expected percent recovery of the analytical procedure.
[b] Refers to mean value and confidence limit (at the 95% confidence level) of interlaboratory study based on triplicate analyses by ten laboratories.
[c] These values represent the upper and lower 95% tolerance limits which indicate the range of results that 95% of the laboratories should obtain 95% of the time for a single sample analysis.
[d] Includes average value of 1.0 ng/g present in native soil before fortification.

from future analyses. One of the advantages of preparing blank and fortified standards from the same soil (or any other matrix) is that the two standards are identical with the exception of the analytes added to the fortified soil. This makes it possible to use a combination of blank and fortified soil to custom blend PE samples. By using varying amounts of the two standards, the analyte concentration can easily be adjusted. This makes a PE sample more difficult to recognize. Some examples of custom mixing of blank and fortified dioxin/furan soil standards are shown in Table 5. The physical appearance of each of the custom standards would be the same since all were prepared from the same original soil.

NATURAL MATRIX REFERENCE MATERIALS FOR THE FUTURE

There is an increasing demand for the analytical testing industry to provide highly accurate and precise data when testing our environment. Improved methodologies for sampling and analysis are resulting in more meaningful and accurate

Table 4 Stability Results for Fortified Soil

Analyte	Consensus value (ng/g)	Analyzed value (ng/g)[a] 3 month results	Analyzed value (ng/g)[a] 6 month results	Analyzed value (ng/g)[a] 12 month results
		Dioxins		
2378-TCDD	0.46	0.50	0.49	0.42
12378-PCDD	0.96	0.95	1.00	0.92
123478-HxCDD	0.90	1.12	0.98	0.86
123678-HxCDD	0.87	0.89	1.03	0.89
123789-HxCDD	0.90	0.94	1.01	0.87
1234678-HpCDD	1.39	1.63	1.53	1.37
OCDD	3.51	4.74	3.29	3.15
		Furans		
2378-TCDF	0.45	0.46	0.47	0.43
12378-PCDF	0.87	0.83	0.96	0.88
23478-PCDF	0.86	0.83	0.97	0.90
123478-HxCDF	0.88	0.96	0.95	0.88
123678-HxCDF	0.95	0.96	1.06	0.89
123789-HxCDF	0.82	0.81	0.96	0.88
234678-HxCDF	0.91	0.96	0.99	0.84
1234678-HpCDF	1.27	1.41	1.50	1.38
1234789-HpCDF	1.12	1.25	1.45	1.27
OCDF	2.25	2.36	2.46	2.02

[a] Mean of duplicate analysis.

Table 5 Examples of Customized Concentration Variations

PCDD/PCDF congener	Nominal concentrations (ng/ng) 2 parts fortified 1 part unfortified	1 part fortified 1 part unfortified	1 part fortified 2 parts unfortified	1 part fortified 5 parts unfortified
Tetra	0.33	0.25	0.17	0.08
Penta	0.66	0.50	0.33	0.17
Hexa	0.66	0.50	0.33	0.17
Hepta	1.00	0.75	0.50	0.25
Octa	1.65	1.25	0.83	0.42

data. Methods will continue to be developed and improved. Analytical standards are essential for accurate measurement processes. Standards are necessary for calibration as well as for measuring performance of instrumentation. Standards are also necessary for measuring performance of sample preparation procedures. Natural and fortified matrix standards are an important part of determining performance of analytical processes. The availability of matrix standards has been somewhat limited, but as their usefulness is demonstrated, more and more demands will be made for these standards. Methods for preparing high quality matrix standards are being developed, and production of standards will become more straightforward. The overall result from use of matrix standards will be more reliable analytical data.

REFERENCES

1. **M. A. Re, J. C. Bradley, and A. W. Nichols,** Preparation and quality control of polyhalogenated dibenzo-*p*-dioxin and polyhalogenated dibenzofuran analytical standards, Organohalogen Compounds, Vol. 2: Dioxin '90 EPRI Seminar, 1990, 289.
2. **J. C. Bradley, et al.,** Interlaboratory testing study on 2,3,7,8-substituted polychlorinated dibenzo-*p*-dioxins and polychlorinated dibenzofuran isomer standard solutions, *Chemosphere*, 20 (1990) 487.

CHAPTER 5

The Preparation of National Research Council Certified Reference Materials

Scott N. Willie

The development of marine chemistry standards and certified reference materials (CRMs) by the National Research Council of Canada (NRC) began in 1976 with the formation of the Marine Analytical Chemistry Standards Program (MACSP).[1] The MACSP is managed by NRC's Institute for Marine Biosciences (IMB) in Halifax and is operated jointly with NRC's Institute for National Measurement Standards (INMS) in Ottawa. At IMB, analytical methods, reference materials, and standards are developed for organic constituents such as polychlorinated biphenyls (PCBs), polycyclic aromatic hydrocarbons (PAHs), and shellfish toxins. The role of NMS is to develop methods and reference materials for the determination of trace elements and other hazardous constituents in sediments, biological tissue, and water. Chapter 8 in this book describes the preparation and certification of CARP-1, a fish reference material for dioxins, furans, and PCBs distributed by INMS. This program is conducted with the continuing advice of the Canadian Committee on Marine Analytical Chemistry (CMAC), which is composed of Canadian scientists from private laboratories, universities, and government, active in chemical oceanography and in the application of analytical chemistry to marine problems.

HS-1, HS-2, CS-1

Three sediment materials for PCB determinations were collected in 1981. HS-1 and HS-2 were collected from Nova Scotian harbors. CS-1, a carbonaceous clay, was collected with the help of Fisheries and Oceans, Bedford Institute of Oceanography, in the Laurentian Channel midway between Nova Scotia and Newfoundland. The sediments were collected using a 0.25 m^2 Van Veen grab and stored wet in sealed 185 L polypropylene barrels. The samples were freeze dried and tumbled in a modified cement mixer to break the clumps formed during the drying. HS-1 and

Table 1 INRC Certified Reference Materials and Standards

IMB		INMS	
Sediments	**Analyte**	**Sediments**	**Analyte**
CS-1	PCB	MESS-2	Trace elements
HS-1	PCB	BCSS-1	Trace elements
HS-2	PCB	PACS-1	Trace elements and organo tin
HS-3	PAH		
HS-4	PAH		
HS-5	PAH		
HS-6	PAH		
SES-1[a]	PAH		
Biological Tissues			***Biological Tissues***
MUS-1	Domoic acid	TORT-2	Trace elements and methyl mercury
MUS-2	Okadaic acid and DTX-1	DORM-2	Trace elements and methyl mercury
		DOLT-2	Trace elements and methyl mercury
		CARP-1	Dioxins, furans, and PCBs
Calibration Solutions			***Waters***
CLB-1	PCB	NASS-4	Trace elements
CLB-2	^{13}C-PCB	CASS-3	Trace elements
DACS-1	Domoic acid	SLEW-2	Trace elements
PSP-1	STX,NEO,$GTX_{2/3}$	SLRS-3	Trace elements
OACS-1	Okadaic acid		

[a] Noncertified.

HS-2 were sieved through a 100 mesh stainless steel screen; CS-1 was sufficiently fine that this step was not required. The sieved sediments were homogenized further in the cement mixer. The sediments were packaged into quart-sized solvent-rinsed steel cans in 200 g quantities. Determinations were performed using packed column gas chromatography and electron capture detection. Concentrations of ten individual chlorinated biphenyls are reported in HS-1 and HS-2.[1]

CLB-1, CLB-2, DPAC-1, DPAC-2

CLB-1 is a set of four mixtures of pure, individually synthesized chlorinated biphenyl (PCB) congeners. Each mixture covers 14 or 15 individual PCBs for a total of 51 PCBs per set of four ampoules. Crystals of individual PCB congeners were prepared using synthetic procedures described by Mullin et al.[2] The purity was established, and the final CLB-1 solution was prepared by mixing appropriate volumes of the stock solutions. Two liter quantities of each of the four solutions containing selected mixtures were prepared and dispensed in 1.00 mL portions into precleaned glass ampoules, which were immediately flame sealed.

CLB-2 is a set of four ampoules, each containing a 0.5 mL solution of ^{13}C-labeled 2,2′,4,4′,5,5′-hexachlorobiphenyl (IUPAC #153) in isooctane. This material is intended as a calibration solution in the quantitative analysis of polychlorinated biphenyls by combined gas chromatography/mass spectrometry. ^{13}C-labeled 2,2′,4,4′,5,5′-hexachlorobiphenyl was prepared by MSD Isotopes Division of Merk Frosst Canada. The purity of the material was verified, and an appropriate amount was dissolved in distilled isooctane. The solution was stirred for 2 h and subsamples were poured by hand into precleaned glass ampoules. The ampoules were cooled in a dry ice/acetone bath, flushed with nitrogen, and immediately flame sealed.

DPAC-1 and -2 are solutions of deuterated polycyclic aromatic compounds (PACs) in toluene. Prior to the final preparation of these materials an extensive series of checks were performed to ensure that the filling and sealing procedure did not introduce contaminates into the final product. A specially designed dispenser was used to avoid any deposition of the solution onto the neck of the ampoule during filling. The ampoules were washed, using distilled-in-glass hexane, followed by an acetone rinse and overnight heating at 130°C. The filled ampoules were cooled before flame sealing under nitrogen to minimize contact of the vapor with the seal point.

HS-3, HS-4, HS-5, HS-6

These four sediment reference materials for polycyclic aromatic hydrocarbons (PAHs) were collected from four harbors in Nova Scotia. The sites were chosen to reflect varying degrees of commercial and industrial activity. These materials were processed similarly to HS-1, HS-2, and CS-1. Reliable values for the concentrations of sixteen PAHs are reported in the certificate.

MESS-1, MESS-2, BCSS-1, PACS-1

In 1981 NRC issued two sediment CRMs from the Gulf of St. Lawrence, MESS-1 and BCSS-1. MESS-1 is from the slightly contaminated Miramichi River estuary and BCSS-1 from the more pristine Baie des Chaleurs. The material was freeze dried, screened to pass a 120 mesh (125 μm) screen, blended, and bottled in 65 and 80 g aliquot for MESS-1 and BCSS-1, respectively. After bottling the samples were radiation sterilized at 2.5 Mrad. Certified values are reported for 14 trace elements and 12 matrix and minor components.

A contaminated harbor sediment, PACS-1, was collected at Esquimalt harbor in British Columbia and was processed in an identical manner as the previous sediments. The concentrations of most elements in PACS-1 are quite high; in particular, the Cd and Pb concentrations are an order of magnitude higher in PACS-1 than MESS-1 and BCSS-1. The high Sn concentration permitted certified values for tributyl-, dibutyl-, and monobutyltin to be determined, as well as 16 trace elements.

A total of 17 different analytical techniques were used in the certification of MESS-1, BCSS-1, PACS-1 or MESS-2, with 22 different laboratories involved with the collection, processing, or analysis. Certified values are based on at least two independent methods of analysis, although in most cases four or five methods were utilized. MESS-2, an estuarine sediment collected in the Beaufort Sea, has replaced the depleted stocks of MESS-1.

TORT-1, TORT-2

TORT-1, lobster hepatopancreas, is the main ingredient of a gourmet food, lobster tomalley. The starting material was freshly frozen edible grade tomalley containing some red eggs and small pieces of meat and shell. After thawing, the raw material was passed through a Comitrol homogenizer to produce a fine, divided slurry. The material was stored in a stainless steel tank before spray drying. In the spray dryer, an atomizer formed small droplets of uniform size, which were propelled into a circulating hot air stream regulated at 175°C. The outlet temperature of the spray drier was maintained at 85 to 95°C, which dried the product in less than 10 sec. The dried powder was vacuum packed in plastic bags and frozen to await the next processing step.

Three extractions with pesticide grade acetone were required to remove the oil from the spray-dried powder. Powder and acetone were mixed manually in a large stainless steel container. After filtering under vacuum, the defatted powder was vacuum dried at 30°C for 48 h in a stainless steel tumble drier to remove traces of acetone. The product was blended into larger batches, sieved through a 0.058 in. nylon mesh, reblended, and bottled into glass containers. Each bottle contained 35 ± 0.3 g of material. Finally, the bottles were irradiation sterilized at a minimum dose of 2.5 Mrad. These procedures yielded approximately 10% w/w of tomalley powder. Supplies of TORT-1 are depleted; however, a replacement material, TORT-2, has been processed in an identical manner to supersede TORT-1.

Various methods of sample decomposition and instrumental techniques were used by NRC and cooperating laboratories to perform homogeneity testing and obtain certified values for 16 trace elements, 7 major elements, and methylmercury.

LUTS-1

CRMs used to validate analytical procedures are typically defatted spray-dried powders, which do not resemble the type of "real" samples that monitoring laboratories are required to analyze. In response to this, NRC has produced an undefatted CRM from lobster hepatopancreas.[3] The starting material is edible-grade tomalley (similar to TORT); however, the processing was significantly different. The tomalley was purchased from a commercial source, frozen at –30°C, and thawed prior to processing. After the first of four passes through a Comitrol homogenizer, an antioxidant (0.02% ethoxyquin) was added to the mixture to prevent rancidity. The moisture content was increased with distilled water from 69 to 85% to prevent separation of the suspension, when, in a later step, the material was autoclaved. The homogenized slurry was emulsified by four passes through a high pressure milk homogenizer and stored at 2°C in polyethylene bottles previously flushed with N_2. An aliquot of 10.3 ± 0.3 g of slurry was dispensed into 6300 8-mL narrow mouth polypropylene screw-capped bottles. Each bottle was heat sealed with a polyester/aluminum/polypropylene-laminated film closure and autoclaved for 20.5 min at 118.3°C in a steam retort to ensure sterilization. After cooling the polypropylene screw closure was tightened, and each bottle was heat sealed in a trilaminate foil pouch. The moisture and lipid content of LUTS-1 is 85.1 ± 0.07 and 8.2%, respectively. Homogeneity tests and trace metal analysis were performed by NRC and cooperating laboratories. Certified values for 17 trace metals and methylmercury are reported in two manners, "as bottled" and as a dry weight. LUTS-1 can easily be sampled by directly removing an aliquot with a pipet. Alternatively, the entire contents can be transferred to a volumetric flask and diluted with water before subsampling. In either case, it is recommended prior to opening the vial that the sample be placed in an ultrasonic bath for a few minutes. LUTS-1 is distributed in packages of six samples.

DORM-1, DORM-2, DOLT-1, DOLT-2

Dogfish (*Squalus acanthias*) was chosen as a suitable reference material, since the absence of skeletal parts minimized the possibility of sample inhomogeneity in the processed product. In addition, the flesh and liver could easily be separated to produce two different materials, DORM-1 (dogfish flesh) and DOLT-1 (dogfish liver).

Processing of these two CRMs was similar to TORT. Dogfish were collected and stored frozen until ready to be eviscerated and beheaded. The livers were separated, cleaned, and frozen at –35°C. The flesh was stripped from the skin and cartilage and similarly frozen. The thawed dogfish flesh was passed four times through the homogenizer, spray dried, and extracted three times with acetone, similar to the procedure used to prepare TORT. Following blending, screening through a 1.5 mm nylon mesh and reblending, the material was packaged into glass bottles each containing 40 ± 0.5 g of powder.

The thawed dogfish liver slurry was partially cooked at 40°C for 1 h, which permitted a large fraction of the oil to be separated by decantation. Four acetone extractions were required to remove the majority of the remaining fats prior to vacuum stripping at 40°C to remove traces of acetone. After blending, screening, and reblending, the powder was packaged in 25 ± 0.5 g portions in glass bottles.

Both DORM-1 and DOLT-1 were irradiation sterilized and inter- and intrabottle homogeneity tested by NRC and cooperating laboratories. Certified values for 12 trace metals, methylmercury, and four minor elements are reported for both materials. Both of these materials have been superseded by identically processed CRMs, DOLT-2 and DORM-2.

NASS-1, NASS-2, NASS-3, NASS-4

One unique aspect of NRC's CRM program has been the production of natural water samples certified for trace elements. NASS-1, a high salinity open ocean seawater CRM produced in 1982, was collected off the Bermuda coast at a depth of 1300 m, in the vicinity of ocean station "S" (32°10'N, 64°30'W).[6] The water at this site was well characterized by oceanographers, due to the temporal and spatial uniformity of physical conditions in the lower part of the water column.

Twelve liter Go-Flo samplers were deployed from the research vessel, Panulirus II, using an unlubricated stainless steel hydrowire. These samplers, internally coated with Teflon®, had the original O-rings replaced with silicone versions and the air vent and stopcock replaced with parts produced from nylon and Teflon®, respectively. Two 50-L polypropylene carboys, enclosed in plastic bags and immobilized in wooden crates, were used to store and transport the sample. The carboys contained pH 1.6 distilled deionized water (DDW) until just before filling with seawater. The samplers, transfer tubing, and polypropylene carboys had previously been leached with high purity acid. The seawater was transferred under filtered N_2 pressure through the tubing into the carboy. A hole in the wooden crate and a resealable perforation in the plastic bag permitted access to the bottom spigot through which the carboy was filled. Immediately prior to filling, the top of the carboy was slightly loosened, and the carboy was drained of pH 1.6 deionized distilled water. A 250-mL polypropylene bottle filled with diluted high purity nitric acid was inserted into the transfer line through a three way stopcock inserted in the cap. Before the carboy was completely filled with sample, the stopcock was turned to divert the flow through the acid bottle, thus acidifying the carboy to pH 1.6 with an appropriate amount of acid. Each carboy was carefully sampled for trace metal contamination using the laboratory facilities of the Bermuda Biological Station for Research.

The carboys were shipped to Ottawa and homogenized in a 800-L polyethylene tank, under clean room conditions. The seawater was bottled into precleaned 2-L wide mouth, screw cap polyethylene bottles.

At the same time a sufficient amount of seawater was collected to prepare NASS-2 and NASS-3 when stocks of NASS-1 were depleted.

NASS-4 was collected 35 km southeast of Halifax, NS at a depth of 10 m. Because NASS-4 was collected at an entirely different location and depth in the North Atlantic, it possesses a different trace metal profile from the previous NASS series.

In addition, a different procedure was used to collect the sample. The water was peristaltically pumped on board the ship through precleaned silicon tubing. This procedure permitted collection of the sample in less time that the Go-Flo method and also minimized the potential for contamination. Plastic clamps were used to fasten the silicone tubing to a polypropylene line attached to a Teflon®-coated weight. Suspended material was filtered from the water using Gelman Versaflow™ capsules. Each filter capsule incorporates a 0.8 μm acrylic copolymer membrane over a 0.45 μm similar membrane. The pleated design of the filters provides 1390 cm^2 of filtration area. Each filter was precleaned with dilute nitric acid and rinsed free of acid with deionized distilled water. The filter assembly contained three filter capsules connected in parallel with polypropylene Y connectors and plastic hose clamps. One filter capsule was capable of filtering more than 50 L water before a new filter was required. A 250-mL polypropylene bottle filled with diluted high purity nitric acid was inserted into the transfer line after the filter. A three-way stopcock fastened through the cap permitted the water flow to be diverted through the bottle, thus acidifying the sample on-line to the storage container. Similar to NASS-1, the carboys were contained in wooden shipping crates and plastic bags during the entire shipboard operation, and access was made through a hole in the bottom of the crate as each carboy was filled from the bottom through the spigot.

CASS-1, CASS-2, CASS-3

CASS-1, a nearshore seawater, was collected in water 25 to 40 m deep and 400 to 800 m offshore, about 10 km south of the inner Halifax harbor, using a vessel from the Bedford Institute of Oceanography. The salinity of the seawater was 31.8 ppt. The collection procedure was identical to NASS-1 (Go-Flo samplers), with the exception that a precleaned in-line 0.45 μm Millipore filter was added to the transfer line to remove suspended material. CASS-2 and CASS-3 were collected at the same site as CASS-1, except at a depth of 5 to 7 m, using the peristaltic pump procedure described in the previous section.

SLRS-1, SLRS-2, SLRS-3, SLEW-1, SLEW-2

The natural water CRMs in this section were collected using the peristaltic pumping procedure rather than Go-Flo samplers. SLRS-1, a riverine water for trace metals, was collected in the St. Lawrence river with the cooperation of the Department of Fisheries and Oceans. The location, several kilometers upstream from Quebec city, is about 30 to 40 km upriver from the mixing zone.

The samples were shipped to Ottawa and put in storage for a few months. When the sample was to be homogenized, a fungus growth was observed in the carboys. Identification showed it to be a fish parasite, *Exophiala pisciphila*, which was removed by filtration through a 0.2 μm Gelman-pleated capsule. After homogenization and bottling, the sample was irradiation sterilized at 2.5 Mrad. Certified values were reported for 21 elements: Na, Mg, Al, K, Ca, V, Cr, Mn, Fe, Co, Ni, Cu, Zn, As, Sr, Mo, Cd, Sb, Ba, Pb, and U.

SLRS-1 was eventually replaced by SLRS-2 and ultimately SLRS-3. The latter two river water CRMs were collected from an Ontario Hydro Dam on the Ottawa River near Chenaux, Ontario, about 100 km upstream from Ottawa. The water was collected 6 ft from the upriver side of the dam at the 2- to 3-m level. The material was filtered through 0.2 μm filters prior to homogenization and bottling. Following bottling the samples were irradiation sterilized and interbottle homogeniety was tested. Certified values were reported for Al, Sb, As, Ba, Cd, Cr, Co, Cu, Fe, Pb, Mn, Mo, Ni, Sr, U, V, Zn, Ca, Mg, K, and Na. The trace metal profile of these latter Ottawa River CRMs is slightly different from the St. Lawrence River CRM, SLRS-1, most notably for Al, Fe, Mn, and Sr.

SLEW-1, an estuarine water reference material for trace metals, was gathered in the St. Lawrence River estuary at a depth of 5 m, several kilometers upriver from Ile aux Courdes and about 50 km below the mixing zone. A ship from the Department of Fisheries and Oceans was used to collect the sample. The salinity of SLEW-1 is 11.6 ppt, and certified concentrations are reported for 10 trace elements: As, Cd, Cr, Co, Cu, Fe, Pb, Mn, Ni, Zn. SLEW-2, the replacement for SLEW-1, was collected at the same time and place but homogenized and bottled at a later date.

SHELLFISH TOXINS

The materials described in this section are shellfish toxins prepared to aid analysts and researchers in the area of seafood testing and the study of algal blooms. MUS-1 is a reference material prepared following a outbreak of domoic acid (the toxin responsible for amnesic shellfish poisoning (ASP)) in cultured blue mussels (*Mytilus edulis*) from a localized area of Prince Edward Island, Canada.[4] A mixture of precooked contaminated and uncontaminated mussels was processed in a manner identical to LUTS-1. Certified concentrations are reported in the certificate for domoic acid and its diastereomer using the recommended procedure. This material is also useful for the routine monitoring for PSP toxins by mouse bioassay using the AOAC extraction procedure.[5]

DACS-1B is a certified instrument calibration solution prepared to aid the analyst in the determination of domoic acid. Domoic acid was purchased from a commercial source purified by chromatography and crystallization and dried *in vacuo* to the anhydrous form. The purity was checked by several methods revealing the presence of 2 *cis-trans* isomers of domoic acid at less than 1% and some of the C5′-diastereomer at about 1.4%. A portion of the domoic acid was accurately weighed and dissolved in degassed acetonitrile/water (1:9, v/v). The solution was mixed and dispensed into clean argon-filled glass ampoules and immediately flame sealed.

MUS-2 is a homogenized blend of blue mussel (*Mytilus edulis* L.) digestive glands and the microalga *Prorocentrum lima.* It is a CRM for the diarrhetic shellfish poisoning (DSP) toxins, okadaic acid, and the dinophysistoxins, DTX-1. The microalga was added in order to elevate the toxin level to a desired level. MUS-2 is a homogenized liquid slurry utilizing the same technology used to prepared MUS-1 and LUTS-1.

OACS-1 is a certified instrument calibration solution for okadaic acid. The package consists of 4 ampoules with each ampoule containing approximately 0.5 mL of a solution of okadaic acid dissolved in methanol. This CRM was prepared by isolating okadaic acid from a laboratory culture of the dinoflagellate species *Procentrum concavum* and purified by liquid chromatography and crystallization. Following *in vacuo* drying to the anhydrous form, purity analyses was performed by 500 MHZ proton NMR ion-spray mass spectrometry and liquid chromatography with UV detection at 210 nm. A 25-mg portion of the anhydrous okadaic acid was weighed and then dissolved on 1 L of degassed, distilled-in-glass grade methanol. The solution was thoroughly mixed while being cooled in ice. Aliquots were then dispensed into clean argon-filled glass ampoules and immediately flame-sealed.

PSP-1B is a calibration solution for the paralytic shellfish toxins. Saxitoxin (STX-1), neosaxitoxin (NEO-1), a mixture of gonyautoxin 2 and 3 ($GTX_{2/3}$-1) and a mixture of gonyautoxin 1 and 4 ($GTX_{1/4}$-1) are a representative set of toxins associated with paralytic shellfish poisoning (PSP). Approximately 25 mg of each toxin were isolated for cell extracts of the dinoflagellates *Alexandrium excavatum* and *Al lucitanicum.*[7] The cells were extracted twice by sonication at 100 W in 0.1 *M* acetic acid and the supernatant fractions were combined and freeze-dried to concentrate the toxins. The residue was redissolved in 0.1 *M* acetic acid and the toxins were purified using a combination of gel permeation and ion exchange chromatography. The purity of each toxin was checked by several methods prior to dissolving the toxin filtered sterilized 0.1 *M* acetic acid. Each ampoule was sealed under argon to minimize any potential oxidation.

REFERENCES

1. **S. S. Berman, J. W. McLaren, and W. D. Jamieson,** *The Marine Analytical Chemistry Standards Program of the National Research Council of Canada. Quality Assurance for Environmental Measurements,* ASTM STP 867, J. K. Taylor and T.W. Stanley, Eds., American Society for Testing and Materials, Philadelphia, 1995, 334–345.
2. **M. D. Mullin, C. M. Pochini, S. McCrindle, M. Romkes, S. H. Safe, and L. M. Safe,** High resolution PCB analysis: synthesis and chromatographic properties of all 209 PCB congeners, *Environ. Sci. Technol.*, 18 (1984) 468–476.
3. **S. S. Berman and R. E. Sturgeon,** A new approach to the preparation of biological reference materials for trace metals, *Fres. Z. Anal. Chem.*, 332 (1988) 546–548.
4. **M. A. Quilliam, P. G. Sim, A. W. McCulloch, and A. G. McInnes,** High performance liquid chromatography of domoic acid, a marine neurotoxin, with application to shellfish and plankton, *Int. J. Environ. Anal. Chem.*, 36 (1989) 139–154.
5. **J. F. Lawrence, C. F. Charbonneau, C. Menard, M. A. Quilliam, and P. G. Sim,** Liquid chromatographic determination of domoic acid in shellfish products using the AOAC paralytic shellfish poison extraction procedure, *J. Chromatogr.*, 462 (1989) 349–356.
6. **S. S. Berman, R. E. Sturgeon, J. A. H. Desaulniers, and A. P. Mykytiuk,** Preparation of the sea water reference material for trace metals, NASS-1, *Mar. Pollution Bull.*, 14 (2) (1983) 69–73.
7. **M. V. Laucock, P. Thivault, S. W. Ayer, and J. A. Walter,** *Natural Toxins*, 2 (1994) 175.

CHAPTER 6

CCRMP: Preparation and Utilization of Environmental and Geochemical Reference Materials

Maureen E. Leaver and William S. Bowman

For the last 20 years, the Canadian Certified Reference Materials Project (CCRMP) at CANMET has provided reference materials to commercial, industrial, and research laboratories worldwide. These reference materials are used by analytical chemists worldwide to improve the quality of the determinations they perform on environmental and geochemical materials and mineral processing products. The processing, homogeneity and certification assessment, and applications of CCRMP reference materials will be described.

INTRODUCTION

In order to quantify the increase in interest in environmental analysis over the last 40 years, citations in the literature of environmental reference materials have been chosen as an indicator. A compilation of E.S. Gladney and I. Roelandts[1] was used to calculate the data in Table 1. The percentage of citations of certain types of elements in North American-produced reference materials (RMs) from the main chemical journals from throughout the world was calculated over the last 4 decades. These reference materials include those produced by the Canadian Certified Reference Materials Project (CCRMP), the United States Geological Survey, and the National Institute of Standards and Technology (formerly the National Bureau of Standards). This original compilation included some of the earlier CCRMP RMs prepared from soils, syenite and gabbro rocks, and various materials characterized for the radionuclides. It is readily seen that the number of reported analyses of toxic elements in these RMs has increased rapidly, and that of the majors has decreased considerably. Interest in gold and silver peaked in the sixties and interest in the noble metals has remained low and relatively constant since the sixties.

Table 1 Percentage of Total Reported Data

Type of element	1950s	1960s	1970s	1980s
Toxics (As, Cd, Hg, Pb, Sb, Se, Te, Tl)	2.4	4.0	11.0	12.0
Majors	47.0	42.0	24.0	24.0
Au + Ag	0.1	8.9	1.8	1.0
Noble metals	0.1	0.7	0.9	0.5

CCRMP

One ongoing project at CANMET since the early seventies is the Canadian Certified Reference Materials Project (CCRMP). The goal of the CCRMP is to make available samples of the raw materials, products, and by-products of the mineral and energy industries having certified values for constituents of interest.

Presently, some 100 certified RMs are available through the CCRMP and are described in the CCRMP catalogue.[2] All of these are unique in the world. Various types of environmental, geochemical, and radioactive materials, ores, and mineral processing products are available. In addition, it should also be noted that CCRMP has materials that have been characterized for gold, silver, the platinum group elements, and the rare earths. Certified or informational values have been determined for most of the elements in the periodic table in this vast spectrum of materials.

In the following pages, the processing, homogeneity and certification assessment, applications, and state of the art in environmental analysis for RMs produced by CCRMP will be described.

COMMINUTION

Over the years of preparation of RMs at CANMET, the same general procedures have been used. After drying the material on steam beds, if required, as in the case of rocks or agglomerates, a jaw crusher or Denver rollers are used to reduce the size of the material to a coarse particle size. A vibration-energy mill with alumina cylinders as the grinding medium is used to reduce the particle size further. A mechanically agitated sieve is used for screening. Traditionally, a sub-74 μm fraction (–200 mesh) is used to prepare CCRMP RMs. This size fraction has been found to minimize the risk of inhomogeneity due to segregation in the experience of CCRMP. The process of regrinding and screening the oversized material is repeated in order to maximize the yield of the original material. A 570-L conical blender is used to mix the material over several hours. The material is placed in an overhead hopper and fed through an auger to deliver a predetermined unit of sample into each bottle. Materials with a sulfur content of greater than a few percent are sealed under nitrogen in laminated foil pouches to provide long-term protection from oxidation.

Various materials have been processed at the CCRMP facilities for clients, including private companies, Environment Canada, and the National Research Council.

ASSESSMENT OF HOMOGENEITY

In order to assess homogeneity, a statistical analysis is performed on the results of the analysis of a number of randomly chosen units from throughout the entire production. These tests have involved analyzing 3 subsamples from 15 units, or latterly, 2 subsamples from 22 units. In general, the analytical method chosen for homogeneity testing is the most precise method available that is commonly used by analysts.

By using the one-way analysis of variance technique,[3] the ratio of the between-bottle to within-bottle mean squares is compared to the F statistic at the 95% level of confidence to determine if significant evidence of inhomogeneity exists.

THE INTERLABORATORY MEASUREMENT PROGRAM

CCRMP uses the consensus approach to certification. Anywhere from 15 to 40 laboratories volunteer their participation in the interlaboratory measurement program by contributing 5 analyses for the elements of interest for one bottle using an appropriate method of their choosing. These laboratories are carefully chosen from their years of participation in CCRMP interlaboratory measurement programs. The certification report lists all results and methods utilized by individual laboratories.

ANALYSIS OF THE RESULTS FROM THE INTERLABORATORY MEASUREMENT PROGRAM

First, statistical and methodological outliers are removed from further calculations. Next a one-way analysis of variance[3] is used to determine the mean, the 95% confidence intervals, the average within-set and between-set standard deviations. The former deviation is a measure of the average within-laboratory precision for the methodology employed in the certification program. As such, it is a measure of the state of the art for the particular analysis. Generally, the histograms of the results show a distribution similar to the normal distribution. Further elaboration on these values will occur in the section on the state of the art in environmental analyses.

CRITERION FOR CERTIFICATION

For a value obtained by the consensus of interlaboratory measurements to be certifiable by CCRMP criteria, there must be a minimum of ten acceptable, independent sets of measurements. The second criterion places an upper limit of three on the ratio of the between-set to the within-set component standard deviation, with no more than 15% of the sets of results being rejected.

ENVIRONMENTAL APPLICATIONS OF CCRMP RMs

The particular CCRMP RMs that will be described are used in a broad spectrum of applications to environmental analyses. The important factor in the choice of reference material for a particular sample is that the matrix of material and the concentration range of the elements of interest are as close as possible in the samples and the RMs. Given that the number of RMs in the world is limited, some decisions must be made by the analyst regarding the optimum match of RM to sample.

In a joint project with Agriculture Canada, a series of four soil RMs, SO-1–4[4] from different sites in Canada were characterized for a total of 65 major, minor, and trace elements.

More recently, in coordination with the Geological Survey of Canada, a series of four of each of freshwater lake and stream sediments, LKSD-1–4 and STSD-1–4,[5] respectively, were released. These eight materials are totally characterized for majors, minors, traces, and partial extractions with concentrated nitric and hydrochloric acids, and dilute mixtures of these two acids. These extraction techniques are routinely used by geochemists. Each of these two series of materials was specifically prepared so as to span broad concentration ranges and thus can be used as calibration standards.

Presently, as a continuation of this program, the Geological Survey of Canada is collaborating with CCRMP to certify four till samples. Till are a specific type of soil material. These materials are expected to be made available for distribution over the next few months.

Random insertion of these soil and sediment materials in a series of field samples provides a means of verification of the performance of a contract laboratory in the determination of arsenic, antimony, selenium, tellurium, lead, mercury, zinc, and others. These materials are used by companies involved with clearing land sites. Transport of Dangerous Goods regulations require that soils be analyzed before being removed to dump sites. Since these materials are from uncontaminated sites, they can be used as the basis for spike tests, baseline studies, and in the development of new methods.

Acid rain is of common concern, but equally important, though less widely known, is acid drainage. Acid drainage is produced from any site that contains metal sulfides that have been disturbed due to the movement of the earth and/or civil construction. Metal sulfides comprise a significant amount of Canadian mineralogy. The effects of weathering and certain bacteria on the sulfides produce acidic solutions, which leach toxic heavy metals from these sites. The solutions, which are called acid drainage, may then escape into the environment. Acid drainage can damage water supplies and soil, and harm plants and animals. It has been estimated that acid-generating wastes cover 15,000 hectares of land in Canada. Lime is commonly used to neutralize the acid solutions in tailings piles and waste rock, to ensure that the environment is not damaged by acidic drainage. However, the continuous monitoring and maintenance required is not practical for a long-term economic solution to the problem. Several areas within the Canadian government and industry are presently performing research in this area. One aspect of this research was the preparation and certification of four sulfide ore mill tailings materials, RTS-1–4,[6] for major, minor, and trace elemental concentrations. In particular, concentration values were determined for sulfur speciation as well as the elements of environmental interest.

In order to best match the concentration levels in the sample and the RM, consideration may also have to be given to various CCRMP rock RMs, including iron formation, syenite and gabbro types, or one of the fifteen ore RMs. Among the ore materials are those certified for gold, the base metals, iron, tungsten, tantalum, antimony, etc. These are naturally occurring materials in the Canadian environment with certain elements at concentration levels that make mining economically viable.

In Canada about 130 million tonnes of mine wastes have been produced from the mining of uranium ore. In order to monitor and control this contamination as it spreads from the source to the biosphere, CCRMP has prepared RMs from vegetative samples,[7] tailings,[8] and ores[9–11] that are certified for uranium, thorium, radionuclides, heavy metals, and other elements. These RMs have been used at the CANMET Elliot Lake Laboratory in northern Ontario where tests are being conducted to establish national radiation standards for regulating agencies. In addition, this location has the added problem of acid drainage and resultant heavy metals contamination of the groundwater. CCRMP RMs are used here in the evaluation of various technologies for the abatement of these environmental problems.

As a result of a cooperative effort between CCRMP and the Air Pollution Centre of Environment Canada, a nonferrous dust,[12] PD-1, from a base metal smelter was certified for lead, arsenic, and mercury. Informational values for cadmium, silica, zinc, and other elements were also determined. This material is used in a nationwide quality assurance program to monitor the amounts of airborne elements to which industrial workers are exposed.

THE STATE OF THE ART IN ENVIRONMENTAL ANALYSIS

The results from the CCRMP interlaboratory measurement program can offer the analyst some information on expected precisions at particular concentration levels. This information could be used by an analyst when a new method is being developed or an unfamiliar concentration level is being sought. Earlier in the presentation, the concept of the average within-laboratory standard deviation was discussed as being an indication of the state of the art in precision for a particular element. The implication exists therefore that a laboratory using a method of average or better reproducibility should obtain individual results for a given element with a precision that is at least comparable to the average within-set standard deviation obtained in CCRMP interlaboratory programs.

Table 2 shows the results from various interlaboratory measurement programs for arsenic in a lake sediment, a tailings material, and an industrial dust. At the lower levels of arsenic found in the first two RMs, comparable precisions are observed for the very different matrices of the sediment and tailings. At the high level of arsenic found in the dust RM, the relative precision decreases, as would be expected.

Similarily, Table 3 shows the results for lead in various environmental materials. These results show that for soil and sediment materials, at concentration levels of lead between 15 and 80 μg/g, the precisions are comparable. At the high concentration of lead in the dust RM (PD-1), the lowest relative precision is observed.

Table 2 Arsenic Results

Name	Type of material	Certified value (μg/g)	Precision (μg/g)	Relative precision (%)
LKSD-4	Sediment	12.6	1.2	9.5
RTS-3	Tailings	9.1	1.2	13.1
PD-1	Dust	7600.0	200.0	2.6

Table 3 Lead Results

Name	Type of material	Certified value (μg/g)	Precision (μg/g)	Relative precision (%)
SO-4	Soil	15.9	2.0	12.5
SO-2	Soil	20.7	2.6	12.5
STSD-1	Sediment	33.4	1.9	5.7
LKSD-1	Sediment	82.9	3.8	4.6
PD-1	Dust	27,500.0	100.0	0.4

Table 4 Mercury Results

Name	Type of material	Certified value (ng/g)	Precision (ng/g)	Relative precision (%)
SO-3	Soil	15.9	3.0	18.9
SO-2	Soil	80.7	7.1	8.7
PD-1	Dust	389.0	14.0	3.6

Table 5 Zinc Results

Name	Type of material	Certified value (μg/g)	Precision (μg/g)	Relative precision (%)
SO-4	Soil	94.1	3.4	3.6
RTS-2	Tailings	117.0	7.7	6.6
RTS-4	Tailings	158.0	6.4	4.1
LKSD-2	Sediment	200.0	9.0	4.5
STSD-2	Sediment	223.0	10.2	4.6
LKSD-1	Sediment	325.0	12.5	3.8

The results for mercury in Table 4 span the concentration range of 16 to 400 ng/g in two soils and dust RMs. Although these mercury concentrations are in the nanogram concentration range, as compared to the microgram concentration range of the other elements discussed herein (i.e., arsenic, lead, and zinc), relative precisions are comparable.

The results in Table 5 show very small differences in relative precision over the concentration range of 90 to 325 μg/g of zinc, despite different matrices. This is in contrast to the results observed with mercury in the μg/g range.

CONCLUSIONS

CCRMP is continuing to serve analysts in industry, commercial laboratories, and research institutes with a broad range of RMs and information on the state of the art in analysis. The certification of dust, soil, sediment, and tailings RMs have provided analysts with the means to perform soil analysis, environmental assessments, site remediation, method development, and quality control/quality assurance programs. The use of RMs to produce accurate and precise results is essential for good environmental protection.

REFERENCES

1. **E. S. Gladney and I. Roelandts,** *Geostandards Newsl.*, 11 (2) (1987) 133–142.
2. **W. S. Bowman,** Catalogue of certified reference materials, CANMET, Natural Resources Canada, CANMET Report 94-1E (1994).
3. **K. A. Brownlee,** *Statistical Theory and Methodology in Science and Engineering*, 2nd ed., John Wiley and Sons, New York, 1965.
4. **W. S. Bowman, G. H. Faye, R. Sutarno, J. A. McKeague, and H. Kodama,** Soil samples SO-1, SO-2, SO-3 and SO-4-certified reference materials, CANMET, Energy, Mines and Resources Canada, CANMET Report 79-3 (1979).
5. **J. Lynch,** *Geostandards Newsl.*, XIV (1) (1990) 153–167.
6. **C. W. Smith and W. S. Bowman,** RTS-1, RTS-2, RTS-3 and RTS-4: sulphide ore mill tailings reference materials, CANMET, Energy, Mines and Resources Canada, CANMET Report MEND Project 4.3.1 (1990).
7. **J. L. Dalton and W. S. Bowman,** Vegetative Radionuclide Reference Materials, CANMET, Energy, Mines and Resources Canada, National Uranium Tailings Program, NUTP-4E (1986).
8. **C. W. Smith, H. F. Steger, and W. S. Bowman,** Uranium Tailings Reference Materials, CANMET, Energy, Mines and Resources Canada, National Uranium Tailings Program, NUTP-2E (1984).
9. **C. W. Smith and H. F. Steger,** Radium-226 in certified uranium reference ores DL-1a, BL-4a, DH-1a and BL-5, CANMET, Energy, Mines and Resources Canada, CANMET Report 83-9E (1983).
10. **H. F. Steger and W. S. Bowman,** DL-1a: a certified uranium-thorium reference ore, CANMET, Energy, Mines and Resources Canada, CANMET Report 80-10E (1980).
11. **H. F. Steger, W. S. Bowman, and G. Zechanowitsch,** DH-1a: a certified uranium-thorium reference ore, CANMET, Energy, Mines and Resources Canada, CANMET Report 81-11E (1981).
12. **H. F. Steger, W. S. Bowman, B. D. McDonnell, and C. Pupp,** PD-1 a certified non-ferrous reference dust, CANMET, Energy, Mines and Resources Canada, CANMET Report 81-7E (1981).

CHAPTER 7

Development of Fortified and Unfortified Fish CRMs for Chlorinated Dibenzo-*p*-dioxins, Chlorinated Dibenzofurans, and Coplanar PCBs, and Their Use as Reference Materials for Organohalogens

David B. Sergeant and Daniel L. Bolt

Certified reference materials (CRMs) have long been needed for quality assurance and quality control (QA/QC) in the analysis of fish tissues for chlorinated dibenzo-*p*-dioxins (dioxins) and chlorinated dibenzofurans (furans). Reference materials (RMs) for organohalogen pesticides and polychlorinated biphenyls (PCBs) are also needed. Coplanar PCBs are a class of compounds for which analytical methods have only recently been developed. CRMs and RMs for these compounds have not yet been developed. In this chapter the planning, production, and certification process for three fish tissue materials for use as CRMs for dioxins, furans, and coplanar PCBs and as RMs for selected organohalogens are described. The use of these materials to identify and resolve standards and method problems is discussed.

INTRODUCTION

An international need for biological tissue certified reference materials (CRMs) for chlorinated dioxins and dibenzofurans became evident as surveys of the effects of contaminants from kraft pulp and paper mills progressed. In Canada the initial monitoring programs on effluents quickly expanded to full-scale environmental assessment programs with matrices such as pulp, sediments, and biota being monitored upstream and downstream from mills. In May 1988, the Ministers of Fisheries and Oceans, Environment Canada, and Health and Welfare Canada extended monitoring programs to all pulp mills using chlorine bleaching in Canada. The large numbers of samples involved dictated that many different laboratories performed the analyses, and quality assurance/quality control (QA/QC) of these analyses took on increased importance. Contract laboratories were engaged to perform many of the analyses for these projects, as government laboratories did not have the capacity to do all the analyses in the time frames in which the data were needed.

The development of QA and QC guidelines[1] for the analyses was necessary to ensure that data generated by different laboratories for different matrices were comparable and had integrity. Committees of government and private sector scientists met and developed QA/QC protocols to provide benchmarks against which data would be checked. Data failing to meet QA/QC standards resulted in rejection of the data set and required repeat analyses. CRMs were needed to evaluate the performance characteristics of methods used by Canadian laboratories, there having been no such evaluation since the beginning of the studies of pulp mill emissions. Since such materials were not commercially available, planning for the production of a large batch of CRMs (one that would last for 5 to 10 years) was initiated by the Department of Fisheries and Oceans (DFO) in Burlington, Ontario.

Historically, for purposes of comparing data integrity between data from federal, provincial, and contract laboratories, biota QA/QC reference materials were prepared yearly in small batches of 30 to 100 50-g bottles of fish tissue and routinely interspersed among the samples sent to contract laboratories. These QA/QC samples were submitted as both "known" and "blind" reference materials. Quality and integrity of the data set could then be assessed, as data on the same reference material were available from all laboratories participating in each year's analytical program. The problem with this method was that the batches of reference materials quickly ran out, new reference materials had to be implemented, and new QA/QC values had to be tracked and compared between laboratories. No long-term continuity of data sets was therefore possible.

Dioxin and furan analyses have progressed significantly since 1988. High resolution gas chromatography (HRGC)/high resolution mass spectrometry (HRMS) has displaced HRGC/low resolution mass spectrometry (LRMS) for dioxin and furan analyses, due to greater selectivity and sensitivity. Congeners that were not detected by HRGC/LRMS are now routinely determined at detection limits more than 20 times lower and with greater specificity due to the resolution that the HRMS can achieve. With these advances came changes to QA/QC requirements and the development of new criteria for HRGC/HRMS data acceptance.[2–4] Cambridge Isotope Laboratories (CIL) had developed and made available two soil CRMs for dioxins and furans, but CRMs for other matrices had not been developed.

There were reasons for suppliers not making available such CRMs. In some instances toxicity of the dioxins and furans could have caused potential CRM developers to decide against manipulating matrices such as contaminated freeze-dried sediments or tissues. Other reasons could have been problems with spiking and processing a range of concentrations for a matrix, neither experience in the analysis of dioxins and furans nor an appreciation of the QA/QC requirements, no means of characterizing the materials, or budget restraints precluding buying these very costly analyses. In the case of biota, there were additional problems of moisture and lipid content (which can vary from near 0 to 30–40%) making it difficult to store and preserve the tissue homogenates.

It was within these contexts that the decision to make CRMs that would be of use to Canadian and international analysts performing ultra trace dioxin and furan analyses was made. A partnership was formed among Department of Fisheries and Oceans (DFO), Great Lakes Laboratory for Fisheries and Aquatic Sciences (GLLFAS), DFO Institute of Ocean Sciences (IOS), the British Columbia Department of the Environment, and CIL to implement this decision. The analytical capability for dioxins/furans and coplanar PCBs was available at GLLFAS, as well as involvement with evaluating the data of other laboratories, and the experience in making CRMs was available at CIL. Current analytical trends indicated that there would be a future need for coplanar PCB CRMs, and they were therefore included in this project. This chapter discusses the development and certification of three biota (fish) CRMs that provide good reference materials for these organic compounds. In addition some further characterization of the three CRMs and their application as reference materials (RMs) to the QA/QC of other halogenated organic pollutants is described.

PREPARATION OF CRM MATERIALS AND METHODS

Consultation of the scientific literature revealed previous studies in which a trace metal undefatted lobster hepatopancreas CRM[5] and a domoic acid reference material[6] from mussel (*Mytilus edulis*) tissue had been successfully prepared. We adopted this technique for preparation and preservation, but for organic contaminants we opted to use sealed glass ampules instead of polypropylene vials with polyester aluminium foil/polypropylene laminated film closures.[5,6] This in part was due to the concern that coplanar PCBs and selected halogenated organics might be of sufficient volatility to escape (migrate through laminated seals) over time, and the CRMs would become useless.

Three different fish species were selected for use in the project: 1 freshwater, 1 marine, and 1 freshwater/marine species.

Lake Ontario lake trout (*Salvelinus namaycush*) — a naturally contaminated composite of about 65 kg, freshwater species

Sockeye salmon (*Oncorhynchus nerka*) — relatively uncontaminated and suitable for spiking, >65 kg, freshwater/marine species

Pacific herring (*Clupea harengus pallasi*) — representative of a "clean" fish from off the far northern tip of Vancouver Island, >65 kg, marine

Analysis of the fish tissues after processing but prior to ampuling was not possible. Dioxin/furan analyses of replicates from each bulk homogenized fish tissue would have taken so long that the fish would have spoiled in the interim. To ensure that at least one CRM would contain 2,3,7,8-substituted dioxins and furans, the salmon CRM was spiked with a mixture of these compounds. Since coplanar PCBs were becoming increasingly toxicologically significant,[7] it was decided to simultaneously spike three coplanar PCBs into this same CRM. Contamination of tissues was by the normal laboratory spiking technique, but on a much larger scale.

Once the three CRMs were prepared, an international round-robin analysis was conducted to establish the consensus values for the compounds of interest. Certification of the CRMs was based on these consensus values. Questions on homogeneity were addressed by analyses at two DFO laboratories, who were partners in this project, and CIL via Radian Corporation. The development and certification of the three CRMs is described, as well as some additional work at GLLFAS that was aimed at making the materials more useful in the following three phases.

Phase One: Preparation of the Three CRMs

Three different fish species were used in the preparation of the CRMs. Lake trout represented a naturally contaminated high-lipid fish (previous analyses over many years[8] had demonstrated that this species had bioaccumulated significant levels of not only dioxins and furans but also PCBs and organochlorine pesticides). Salmon represented a relatively uncontaminated moderate-lipid fish of high commercial value, for which having a CRM would prove useful for data integrity assessment, and herring was a "clean" low-lipid fish for method verification and development use. The salmon was spiked during processing to the approximate target levels of 25, 50, 75, 100, and 200 ng kg^{-1} (parts per trillion [ppt]) of the tetra- through octa-congeners of dioxins and furans and to 600 ng kg^{-1} for PCBs BZ numbers 77, 126, and 169, as identified by Ballschmiter and Zell.[9] This ensured that at least one CRM would have all the 2,3,7,8-substituted dioxins, furans, and PCBs BZ numbers 77, 126, and 169 in it. The other two tissues were processed without the addition of contaminants.

Since 65 kg or more of each composited fish would be processed, pilot-plant scale equipment would be needed. The Technical University of Nova Scotia (TUNS) in Halifax, Nova Scotia was contracted to process the samples. TUNS had previous experience in performing this type of work, having processed the National Research Council's (NRC) CRMs[5] (see also Chapter 8). All fish were shipped frozen as individual specimens to TUNS. The sockeye salmon (Frazer River) and Pacific herring (Barcley Sound) were whole fish, collected in British Columbia, and the lake trout (frozen, gutted fish from another DFO research project) were from Lake Ontario near Grimbsy, Ontario. Fish were kept frozen at –30°C until just before processing, which took place soon after receipt of the fish.

The fish were processed by the procedures of NRC;[5] however, glass ampules were substituted as the storage device. Since the salmon were spiked, a description of the procedure is given here. The other two materials were processed in the same manner but they were not artificially contaminated with dioxins, furans, or coplanar PCBs.

In all three cases the fish were removed from cold storage and allowed to thaw overnight at room temperature. They were split down the center for processing, and the backbone and a thin layer of skin were removed so that the equipment used for the various homogenization steps would not become plugged. A Bibun SDX16 meat separator (Bibun Co., Hiroshima, Japan) was used to debone and skin the fish prior to processing in a Bibun SUM420 meat strainer. The tissue was then passed through a Comitrol 3600 food cutter (Urschel Laboratories Inc., Valparaiso, IN) equipped with a Type K 28 column 010-020 cutter. Ethoxyquin (0.025% powder) was added as an antioxidant, and the homogenate was passed three more times through the food cutter with a finer 010-010 cutter. Between the second and third passes the slurry was heated to 80°C in a steam-jacketed kettle. The moisture content was adjusted to 85% by addition of distilled water to produce a stable fluid homogenate. Moisture content was determined using a moisture balance determination balance (Model 6010, Ohaus Corporation, Union, NJ).

At this point the salmon was spiked with a nonane solution of dioxins/furans and coplanar PCBs, provided by Radian Corporation (Austin, TX), which had been sonicated for 15 min and diluted in 200 mL methanol. The comminuted tissue was put through the Comitrol 3600 food cutter and divided into two batches; each batch was spiked by stirring in half the spiking solution, and the mixtures were recombined by alternately passing portions of the two halves through the Comitrol 3600.

The slurry was held at 10°C overnight in a coldroom and then processed sequentially four times through a No-Bac Unitherm IV homogenizer (Cherry-Burrell Corporation, Cedar Springs, IA), located at Agriculture Canada Research Station, Kentville, Nova Scotia, at pump homogenizing pressures of 2100/2600, 2500/3000, 2500/4000, and 2500/4000 psi. On the final pass the homogenate was collected in 20-L polyethylene buckets, capped, and stored at 10°C until ampuling.

An ampuling machine (Cozzoli Machine Co., Plainfield, NJ) located at the NRC-IMB laboratories in Halifax was calibrated to dispense a volume weighing 10 g. Some weight variability was observed and was attributed to the varying amount of entrained air in a particular tissue batch. Efforts were made to control this as weights were monitored and the fill volume adjusted when they varied from the set-point. Ampuling consisted of the following sequence: the ampule was flushed with nitrogen, filled with a metered volume of tissue homogenate, and again flushed with nitrogen and sealed with a propane/oxygen flame.

Ampules were returned to TUNS, placed in racks, placed in a steam retort (WSF Industries Inc., Tonawanda, NY), and thermally processed. Salmon were processed for 12 min and herring and trout for 11 min at 118°C. A microprocessor-based system developed by Canadian Institute of Fisheries Technology monitored each batch using a thermocouple-equipped ampule. Target lethalities (F_0) were achieved between 6 and 7 min. Due to the large number of ampules produced, 11 separate batches had to be processed for each of the three CRMs. Once cooled the processed ampules were checked for closure integrity, heat sealed in individual trilaminate retort pouches, identified by a color code ("red" for herring, "green" for salmon, and "black" for lake trout), and packaged in units of 100 in paperboard boxes. Boxes were numbered according to the order in which they were ampuled and retorted. Ultimately 6100 lake trout, 6300 herring, and 6011 salmon ampules were boxed, placed on wooden pallets, wrapped with plastic shipping material, and sent by heated truck to DFO's Burlington warehouse. Shortly after receipt, the materials were divided up and shipped to the other partners, taking care to select boxes across the production run so that each partner received materials across the run.

Phase Two: Consensus Value Determinations

An international round-robin analysis was conducted to obtain values for the selected 2,3,7,8-substituted congeners of dioxins and furans, as well as for coplanar PCBs BZ numbers 77, 126, and 169. Participating laboratories were sent 1 ampule of each of the three CRMs and requested to analyze them using their methodology and analytical/surrogate standards. All laboratories were requested to report data on all 2,3,7,8-substituted dioxins and furans. If coplanar PCBs were analyzed, the PCBs BZ numbers 77, 126, and 169 were to be reported. No special steps were taken to reduce differences between laboratories, since it was desirable to see the real variation in the data based upon different laboratories and methods.

Table 1 lists the participating laboratories. Only those values generated by laboratories using high resolution gas chromatography/high resolution mass spectrometry (HRGC/HRMS) were used to generate the consensus values. The rationale for this was that low resolution mass spectrometry would miss very low-level congeners and bias the consensus values for those congeners. The GLLFAS and IOS laboratories simultaneously performed replicate HRGC/HRMS analyses on each of the three CRMs by selecting ampules from boxes across their portion of the production run to assess homogeneity. Radian Corporation performed similar tests but with HRGC/LRMS analyses. These analyses when coupled with the other interlaboratory analyses gave information on the extent of CRM homogeneity across the batches.

Table 1 List of International Laboratories Participating in Round-Robin for Consensus Value Determination

Laboratory	Country
Triangle Labs, Inc.	USA
Alta Analytical Labs	USA
Zenon Environmental Labs	Canada
AXYS Analytical Labs	Canada
Twin City Testing	USA
Japan Food Research Center	Japan
JMI Institute	Japan
Midwest Research Centre	USA
Pacific Analytical	USA
University of Paderbon	Germany
Dr. Weßling	Germany
Enviro-Test Labs	Canada
Toray Research Center	Japan
Fisheries and Oceans Canada Institute of Ocean Sciences	Canada
Fisheries and Oceans Canada Bayfield Institute (GLLFAS)	Canada
Radian Corporation	USA
RIVO-DLO (Netherlands Institute of Fisheries Research)	Netherlands
Norwegian Institute for Air Research	Norway
Chester/Labnet	USA
Chemwest Labs	USA
ENSCO/CAL Labs	USA
RW-TUV	Germany

Phase Three: Characterization for Other Halogenated Organics

The utility of these CRMs would be increased if they could be used in QA/QC of fish for other halogenated organics. DFO Burlington therefore analyzed replicates (N = 6) of all thee CRMs, using in-house methodology, for a variety of organochlorine pesticides, octachlorostyrene (OCS), total PCBs, PCB congeners, toxaphene, and chlorinated diphenyl ethers (CDPEs).

RESULTS AND DISCUSSION

Analytical quality control is universally accepted as being essential for all analyses, especially environmental analyses. QC is accomplished by analysis of reference materials (RMs), certified reference materials (CRMs), and standard reference materials (SRMs).[10–17] It should be noted that the term SRM is unique to the National Institute of Standards (NIST). In most instances the objective is to document and show that these materials, when analyzed over a short or long period of time, yield comparable results. Graphs are prepared and presented as quality control charts for each analyte.[12,13,15,18] A visual examination of the chart and the error bars representing 1 and 2 standard deviations gives the analyst or QA officer an immediate indication of whether the analyses are under control or out of control. Corrective action can then be taken and sample analyses repeated as required. By producing the three CRMs, it was hoped to make available three useful QA/QC materials for the analysis of dioxins, furans, and coplanar PCBs. These CRMs were expected to be used over the next 5 to 10 years in analytical programs.

This paper does not try to explain the differences between the terminologies for various types of reference materials (CRMs, RMs, etc.). The reader is referred to the literature where there are definitions and discussions of such terms.[11,13,14,19–22] The International Standards Organization (ISO) definition of a certified reference material (CRM) was adopted and is given as: "A reference material one or more of whose property values are certified by a technically valid procedure, accompanied by or to a traceable certificate or other documentation which is issued by a certifying body."[19] The issue of the production of CRMs has for some time and continues to be studied by groups such as the ISO under its ISO9001 program and also by the Association of Official Analytical Chemists (AOAC).[23]

In selecting the method of preservation, the options quickly narrowed to the use of sealed ampules and thermal stabilization. The major considerations were the space and storage conditions that bottled vs. frozen samples would occupy. To put it in context, when the ampules arrived, they occupied three normal size pallets and were piled to about 2 m in height. If this number of ampules had been bottles of frozen fish tissue, the freezer space required would have would been significant and difficult to obtain. Freezing this quantity of material and maintaining its integrity over a number of years would have necessitated buying several freezers and associated alarms, and ensuring that freezer failures were immediately corrected, before the materials thawed and decomposed. Freezing in liquid nitrogen would have necessitated even more freezers, and annual costs of liquid nitrogen would have been an ongoing expense. The ampuled materials are theoretically stable forever, can be stored at room temperature, and should not be affected even by high room temperatures. Protection against freezing is recommended. Although no tests were performed, it is anticipated that the ampules might break when frozen, due to expansion of entrained water (high moisture content). Thus, the key issues of the preservation of a biological material and convenience of the storage method[24,25] were addressed.

No problems were encountered in making the three CRMs. TUNS had performed similar contracts in the past and had established procedures to prevent problems from occurring and to yield a homogeneous product. Spiking of the one CRM presented no problem because of the way it was done.

Quantitative removal of the homogenized fish tissues from the ampules was found to present some problems. The tissues had settled out in the ampules over time, and lipid and tissues were coated on the walls of the ampules. It was a challenge to remove all the tissue through the small hole in the top of the ampule, once the top had been snapped off. The best procedure to ensure quantitative transfer was to first place the ampule in an ultrasonic bath and sonicate it for several minutes before removing the top of the ampule. Weighing by difference was found to be necessary to determine exact sample weights. The tissues were removed with the aid of a small spatula, the ampule was rinsed three times with extraction solvent (added to the remainder of extract), the ampule and top were air-dried to constant weight, final weight measurements were taken, and sample weight was calculated. Rinsing not only removes fine tissue particules but also any lipids coating the ampule walls.

Homogeneity testing prior to ampuling as recommended in ISO Guide 35[26] was not feasible since dioxin/furan analyses require several days of preparation before HRGC/HRMS analysis. Homogeneity of the production run was assessed post ampuling. Each of the two DFO laboratories and the Radian laboratory analyzed ampules selected at random from across the entire production batch. The two DFO laboratories used HRGC/HRMS for analyses, while Radian used HRGC/LRMS for their determinations. Examination of the three data sets showed there were no significant differences for the major congeners in the data across the materials.

As a further check on homogeneity the distribution of all individual congener data from the interlaboratory study, including all the data from the DFO laboratories, was graphed. Figures 1 through 5 show several of these distributions. One tetracongener for each of the three materials (Figures 1, 3, 5), a pentacongener for the lake trout CRM (Figure 2), and a representative octacongener for the sockeye salmon CRM (Figure 4) were plotted. Kolmogorov-Smirnov goodness of fit testing and data plots indicated that most data distributions were normal about a mean for all three CRMs and for most congeners. We therefore concluded that the materials must be homogeneous; otherwise, the graphs would have been badly skewed one way or the other. The coplanar PCBs had insufficient data points ($N = 13$) to obtain a good histogram and are therefore not shown.

An interlaboratory round-robin study was conducted in 1992–1993 with the data to be used to certify the CRMs. Samples of all three CRMs were sent to the participating laboratories (Table 1), who were requested to perform the dioxin, furan, and coplanar PCB analyses using their methods, ^{13}C-surrogates and analytical

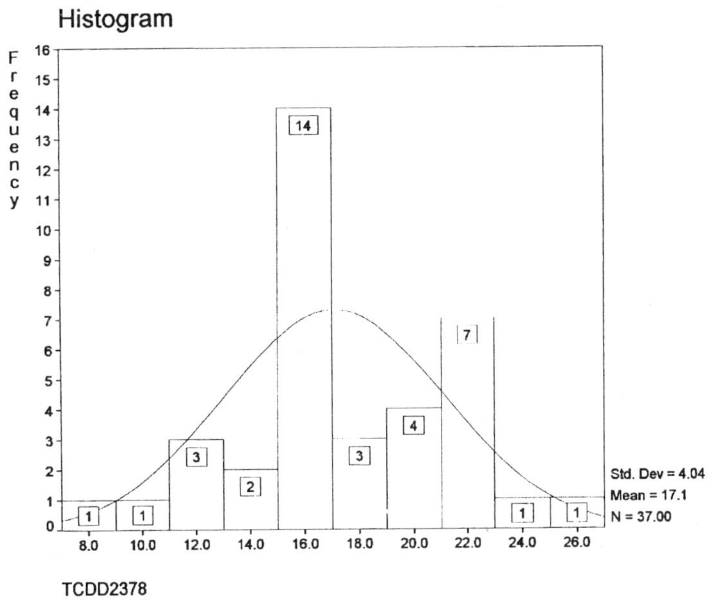

Figure 1 Histogram plot of 2,3,7,8-TCDD, with normal curve superimposed, for EDF-2525 (lake trout) CRM. Concentration is in ng kg^{-1}.

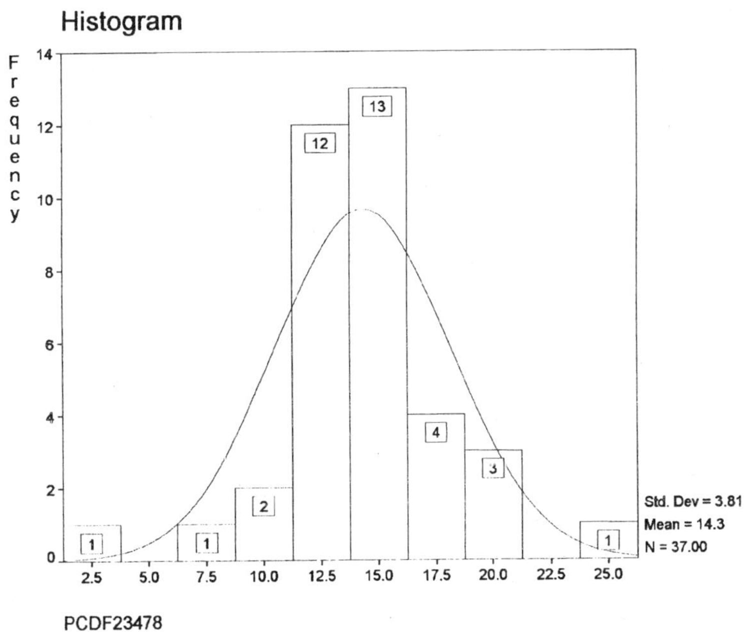

Figure 2 Histogram plot of 2,3,4,7,8-PCDF, with normal curve superimposed, for EDF-2525 (lake trout) CRM. Concentration is in ng kg^{-1}.

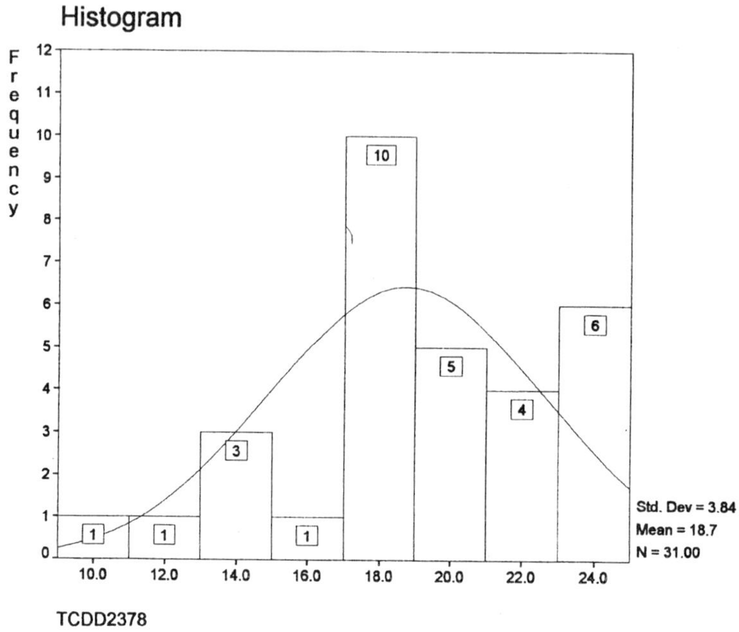

Figure 3 Histogram plot of 2,3,7,8-TCDD, with normal curve superimposed, for EDF-2526 (salmon) CRM. Concentration is in ng kg^{-1}.

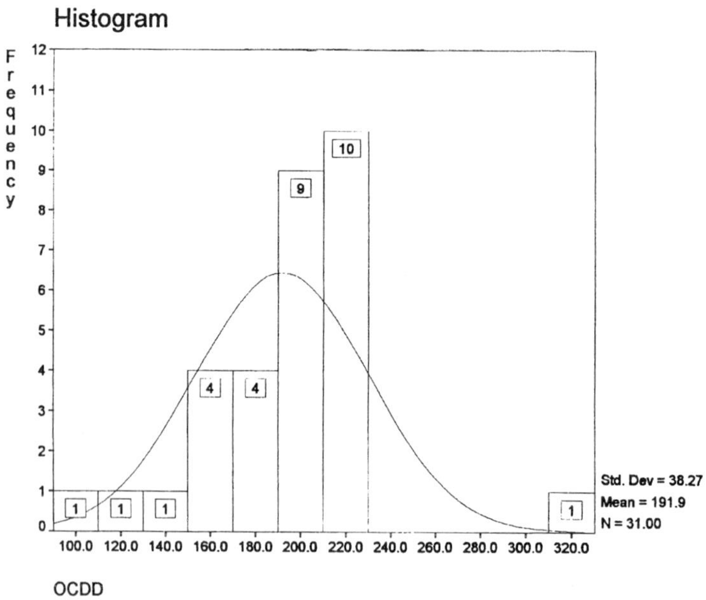

Figure 4 Histogram plot of OCDD, with normal curve superimposed, for EDF-2526 (salmon) CRM. Concentration is in ng kg^{-1}.

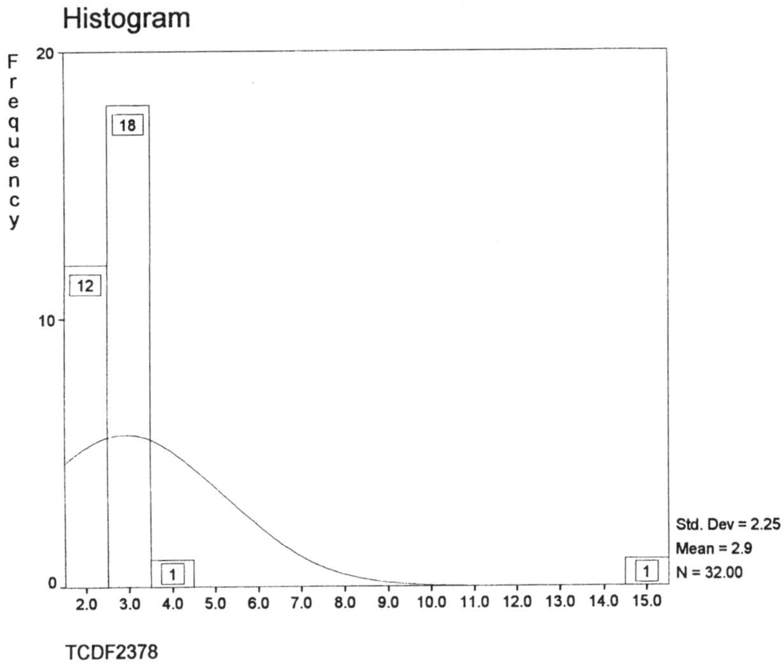

Figure 5 Histogram plot of 2,3,7,8-TCDF, with normal curve superimposed, for EDF-2524 (herring) CRM. Concentration is in ng kg^{-1}.

standards. Results were returned to the authors for evaluation. Detection limits for dioxins and furans were, for the most part, less than 0.5 pg g^{-1} for all congeners, with some octacongeners ranging to 6.0 pg g^{-1}. Coplanar PCB detection limits ranged between 1 and 30 pg g^{-1}, with some outliers.

Target or consensus values were determined by statistical evaluation of the data from the international round-robin study using SPSS V 5.0 for windows software. All data points were counted as one determination; that is to say, the DFO laboratory individual results were each treated as coming from another distinct laboratory. The consensus values were also examined for their 95% confidence intervals. These results are shown in Tables 2 and 3. For the lake trout, N was 37 (more data were available), for the salmon 32, and for the herring 31. When the difficulty of parts per trillion (ppt) analyses of these congeners was considered, the tightness of the data for the dioxins and furans was judged to be good.

Examination of the data for the spiked salmon showed that although the material was homogeneous, the target values for expected concentration were not achieved. Water had been added to the material during processing, and slightly more fish tissue than originally planned was used, which diluted the contaminant concentrations. As shown by Table 2, the target values were approximately 75 to 80% of the planned nominal concentrations, while OCDD and OCDF were almost 100% of the planned value. The reason for this is unknown. That the 2,3,7,8-TCDD and TCDF values in the salmon were almost identical to those in the lake trout was purely coincidental.

Table 2 Dioxin and Furan Consensus Values[a] (ng kg^{-1}) for CRMs

Compound	EDF-2524 (herring)	EDF-2525 (lake trout)	EDF-2526 (salmon)
2378-TCDD	ND[b]	17 ± 1.4	19 ± 1.4
12378-PCDD	ND	4.0 ± 0.57	40 ± 3.0
123478-HXCDD	ND	0.77 ± 0.27	60 ± 4.8
123678-HXCDD	ND	3.0 ± 1.2	56 ± 4.8
123789-HXCDD	ND	0.79 ± 0.26	60 ± 4.4
1234678-HPCDD	ND	1.4 ± 0.53	76 ± 5.9
OCDD	ND	7.2 ± 3.7	192 ± 14
2378-TCDF	2.5 ± 0.16	22 ± 1.6	17 ± 1.5
12378-PCDF	ND	4.9 ± 0.56	40 ± 3.7
23478-PCDF	ND	14 ± 1.3	38 ± 3.5
123478-HXCDF	ND	8.2 ± 3.7	80 ± 8.4
123678-HXCDF	ND	2.7 ± 1.2	63 ± 5.5
123789-HXCDF	ND	0.76 ± 0.35	58 ± 7.0
234678-HXCDF	ND	2.3 ± 1.9	60 ± 5.5
1234678-HPCDF	ND	4.4 ± 6.0	83 ± 9.2
1234789-HPCDF	ND	0.63 ± 0.23	73 ± 7.7
OCDF	ND	2.6 ± 1.3	190 ± 22
PCB-77	24 ± 16	2376 ± 672	619 ± 107
PCB-126	6.5 ± 2.8	884 ± 277	1140 ± 465
PCB-169	2.5 ± 1.2	181 ± 264	1416 ± 553

[a] Concentration is shown as mean of values of interlaboratory study followed by 95% confidence interval.
[b] ND = Not Detected.

The wide standard deviation and 95% confidence interval for the coplanar PCBs was of concern. The cause of this variability can only be speculated on at present. Method bias may be the cause, as the analysis has not been routine for anywhere near as long as that for dioxins/furans. Eight laboratories contributed data, and N was 13 during statistical analysis of the data. A second and more probable cause is suspected to be analytical standard concentration variability and/or quality. This points to the need for "neat" certified reference standards of these PCBs so that problems of this nature do not occur in the future.

Due to the small number of laboratories contributing to the consensus value determination for coplanar PCBs and the suspected problem with standards (one laboratory reported data 2 orders of magnitude greater than data from the others, and it was not used) from different sources, it was decided to see if there was interest in a further round-robin on coplanars in these CRMs. Since there was interest, currently an interlaboratory study is being conducted by DFO and Environment Canada, Burlington, Ontario. ^{13}C-surrogates and analytical standards are being supplied to rule out any variations due to standards. It should then be possible to determine whether observed variations were due to method differences. Data assessment and a report are planned for 1995/1996.

Values for other organochlorine compounds, analyzed at GLLFAS, are given in Tables 3 and 4. Detection limits were 0.002 μg g^{-1} for all compounds except

Table 3 Organochlorine Pesticides and Total PCB Concentrations ($mg\ kg^{-1}$) in CRMs

Compound	EDF-2524 (herring)	EDF-2525 (lake trout)	EDF-2526 (salmon)
α-BHC	ND	0.005 ± 0.001	0.002 ± 0.000
HCB	0.003 ± 0.001	0.031 ± 0.001	0.003 ± 0.001
β-BHC	ND	ND	ND
γ-BHC	ND	ND	ND
δ-BHC	ND	ND	ND
Heptachlor	ND	ND	ND
Aldrin	ND	ND	ND
OCS	ND	0.030 ± 0.002	ND
Hept. epox.	ND	0.020 ± 0.002	ND
γ-Chlordane	ND	0.022 ± 0.001	ND
α-Chlordane	0.002 ± 0.000	0.058 ± 0.001	0.002 ± 0.000
pp′-DDE	0.022 ± 0.003	0.597 ± 0.035	0.006 ± 0.001
Dieldrin	0.002 ± 0.000	0.086 ± 0.002	ND
Endrin	ND	0.009 ± 0.001	ND
β-Endosulfan	ND	ND	ND
pp′-TDE	0.004 ± 0.001	0.181 ± 0.007	0.003 ± 0.000
op′-DDT	ND	0.014 ± 0.004	ND
pp′-DDT	ND	0.007 ± 0.003	ND
Photomirex	ND	0.156 ± 0.011	ND
pp′-Methoxychlor	ND	ND	ND
Mirex	ND	0.239 ± 0.024	ND
Toxaphene	ND	0.132 ± 0.027	ND
Total PCBs	0.134 ± 0.188	2.563 ± 0.141	0.034 ± 0.006

toxaphene, which was 0.030 $\mu g\ g^{-1}$. The lake trout CRM is the only one of the three that contains appreciable amounts of organochlorine pesticides, toxaphene, CDPEs, and octachlorostyrene. Even though the other two CRMs are not highly contaminated, they remain useful for these analyses, as they can be used as matrix blanks, or spiked and used as method evaluation materials.

CONCLUSIONS

The project has produced three CRMs that have been certified by international round-robin study. These materials are commercially available through CIL (Andover, Massacchusetts, U.S. and Burlington, Ontario, Canada) and should prove to be of value for QA/QC programs of laboratories engaged in dioxin/furan and coplanar PCB analyses in biota. The CRMs will be invaluable to laboratories wishing to bring these analyses on-line as they provide consensus values which the laboratory must achieve within 2 standard deviations of the mean before they can say that they have demonstrated their capability to perform these analyses. This type of method validation or proficiency testing in the analytical laboratory is extremely important and should be fully documented for new methods.[27] The materials can also be used to identify problems with calibration standards. The herring CRM especially should prove to be of value for method development purposes, since it is very clean with

Table 4 PCB Congener and CDPE Concentrations (mg kg^{-1}) in CRMs

Compound	EDF-2524 (herring)	EDF-2525 (lake trout)	EDF-2526 (salmon)
PCB28	0.002 ± 0.003	0.009 ± 0.000	ND
PCB52	0.004 ± 0.005	0.036 ± 0.001	ND
PCB101	0.006 ± 0.008	.110 ± 0.009	ND
PCB118	0.002 ± 0.002	0.044 ± .003	ND
PCB153	0.005 ± 0.007	0.147 ± 0.005	ND
PCB137	ND	0.010 ± 0.001	ND
PCB138	0.006 ± 0.009	0.186 ± 0.010	ND
PCB180	0.005 ± 0.009	0.114 ± 0.005	ND
PCB170	0.003 ± 0.004	0.067 ± 0.003	ND
22′44′CDPE	ND	ND	ND
244′5CDPE	ND	ND	ND
33′44′CDPE	ND	ND	ND
22′44′6CDPE	ND	ND	ND
22′44′5CDPE	ND	0.005 ± 0.002	ND
33′44′6′CDPE	ND	ND	ND
22′44′56′CDPE	ND	0.008 ± 0.001	ND
22′344′CDPE	ND	ND	ND
22′44′55′CDPE	ND	0.007 ± 0.005	ND
22′344′6CDPE	ND	0.005 ± 0.004	ND
22′344′6′CDPE	ND	0.004 ± 0.002	ND
22′344′5CDPE	ND	ND	ND
22′344′5′CDPE	ND	0.004 ± 0.002	ND
22′344′66′CDPE	ND	ND	ND
22′344′5′6CDPE	ND	0.005 ± 0.003	ND
22′344′56′CDPE	ND	ND	ND
22′344′55′CDPE	ND	ND	ND

respect to contaminants and can therefore be spiked with compounds of interest for recovery studies and method reproducibility testing.

This study has demonstrated that it is possible to make a CRM by spiking a suitable matrix, mixing to homogeneity, and subdividing into appropriate storage containers for preservation. This would seem to indicate that any miscible contaminant can be turned into a CRM on a custom basis, providing that it can be made water- or lipid-soluble. In the future there might be a need for nonexotic halogenated organic CRMs, and these should be produced to meet forecast market needs.

A second round-robin to further verify the coplanar PCB consensus values is already underway as described. The quality of coplanar PCB standards needs to be examined and appropriate SRMs, CRMs, or RMs developed to assist analysts.

These three CRMs can also be used as RMs for control charting other halogenated organics (Tables 3 and 4) in fish tissue over time. The lake trout sample has environmentally relevant levels of many of these contaminants, making it a useful RM. The other two fish species had not experienced significant enough exposure to accumulate the same profile of contaminants, but are also useful as RMs.

ACKNOWLEDGMENTS

The authors wish to thank Fisheries and Oceans Canada, Institute of Ocean Sciences, and the British Columbia Department of the Environment for their financial support, without which this project would not have been possible. A special thanks goes out to all the laboratories that participated in the interlaboratory study. We also thank Dr. B.K. Afghan of Environment Canada for his scientific advice and encouragement during the planning of this project. Also thanked are Don Barclay and Mark McKinnon for their assistance in statistical data analysis and Dr. M.R. Servos and S.Y. Huestis for their review of the manuscript.

REFERENCES

1. **D. B. Sergeant,** Quality assurance in analysis of environmental matrices for PCDDs and PCDFs, *Chemosphere*, 20(10–12) (1990) 1263–1268.
2. Internal Quality Assurance Requirements for the Analysis of Dioxins in Environmental Samples, Environment Canada, Environmental Protection Series, Report EPS 1/RM/23 1992, 26 pp.
3. Reference Method for the Determination of Polychlorinated Dibenzo-para-dioxins (PCDDs) and Polychlorinated dibenzofurans (PCDFs) in Pulp and Paper Mill Effluents, Environment Canada, Environmental Protection Series, Report EPS 1/RM/19 1992, 45 pp.
4. Method 1613: Tetra- through Octa-Chlorinated Dioxins and Furans by Isotope dilution HRGC/HRMS, Revision A, U.S. E.P.A., 1990.
5. **S. S. Berman and R. E. Sturgeon,** A new approach to the preparation of biological reference materials for trace metals, *Fresenius Z. Anal. Chem.*, 332 (1988) 546–548.
6. **W. R. Hardstaff, W. D. Jamieson, J. E. Milley, M. A. Quilliam, and P. G. Sim,** Reference materials for domoic acid, a marine neurotoxin, *Fresenius Z. Anal. Chem.*, 338 (4) (1990) 520–525.
7. **D. E. Wells,** Current developments in the analysis of polychlorinated biphenyls (PCBs) including planar and other toxic metabolites in environmental matrices, In: *Environmental Analysis: Techniques, Applications and Quality Assurance*, D. Barcelo, Ed., Elsevier Science Publisher, New York, 1993, 113–148.
8. **U. Borgmann and D. M. Whittle,** Contaminant concentration trends in Lake Ontario lake trout (*Salvelinus namaycush*): 1977 to 1988, *J. Great Lakes Res.*, 17(3) (1991) 368–381.
9. **K. Ballschmiter and M. Zell,** Analysis of polychlorinated biphenyls (PCB) by glass capillary gas chromatography, *Fresenius Z. Anal. Chem.*, 302(20) (1980) 20–31.
10. *Monitoring Environmental Materials and Specimen Banking*, Proceedings of the International Workshop, Berlin (West), 23-28 October, 1978, Martinus Nijhoff Publishers, The Hague, 1979, 591 pp.
11. **J. P. Cali, T. W. Mears, R. E. Michaelis, W. P. Reed, R. W. Seward, C. L. Stanley, H. T. Yolken, and H. H. Ku,** The Role of Standard Reference Materials in Measurement Systems, NBS Monograph 148, U.S. Department of Commerce, National Bureau of Standards, Washington, D.C., 1975.
12. **J. K. Taylor,** Quality assurance of chemical measurements, *Anal. Chem.*, 53(14) (1981) 1588A–1596A.

13. **J. K. Taylor,** Standard Reference Materials: Handbook for SRM Users, NBS Special Publication 260-100, U.S. Department of Commerce, National Bureau of Standards, Gaithersburg, MD, 1985, 131 pp.
14. **R. J. Mesley, W. D. Pocklington, and R. F. Walker,** Analytical quality assurance: a review, *Analyst*, 116 (1991) 975–990.
15. **F. M. Garfield,** Quality Assurance Principles for Analytical Laboratories. Association of Official Analytical Chemists (ISBN 0-935584-46-3), U.S., 1991, 196 pp.
16. **W. P. Cofino,** Quality assurance in environmental analysis, In: *Environmental Analysis: Techniques, Applications and Quality Assurance*, D. Barcelo, Ed., Elsevier Science, New York, 1993, 359–381.
17. **E. A. Maier,** Certified reference materials for the quality control of measurements in environmental monitoring, In: *Environmental Analysis: Techniques, Applications and Quality Assurance*, D. Barcelo, Ed., Elsevier Science, New York, 1993, 383–401.
18. **J. P. Dux,** *Handbook of Quality Assurance for the Analytical Chemistry Laboratory*, Van Nostrand Reinhold Company, New York, 1986, 123 pp.
19. General Requirements for the Competence of Calibration and Testing Laboratories, ISO Guide 25, International Organization for Standardization, Geneva, Switzerland, 1990, 6 pp.
20. **H. T. Yolken,** The National Standard Reference Materials Program in the U.S.A., In: Standard Reference Materials and Meaningful Measurements. NBS Special Publication 408, U.S. Department of Commerce, National Bureau of Standards, Washington, D.C., 1975, 237–245.
21. **R. E. Michaelis,** The SRM story at NBS, In: Standard Reference Materials and Meaningful Measurements. NBS Special Publication 408, U.S. Department of Commerce, National Bureau of Standards, Washington, D.C., 1975, 245–257.
22. **J. D. Cox,** Reference materials in the United Kingdom, In: Standard Reference Materials and Meaningful Measurements. NBS Special Publication 408, U.S. Department of Commerce, National Bureau of Standards, Washington, D.C., 1975, 224–235.
23. AOAC Board approves formation of technical division on reference materials, *The Referee*, 17(5) (1993) 9 pp.
24. **T. D. Spittler and J. B. Bourke,** Considerations for preserving biotic samples, In: *Principles of Environmental Sampling*, L. H. Keith, Ed., ACS Professional Reference Book, American Chemical Society, Maple Press Company, York, PA, 1988, 375–381.
25. **M. P. Maskarinec and R. L. Moody,** Storage and preservation of environmental samples, In: *Principles of Environmental Sampling*, L. H. Keith, Ed., ACS Professional Reference Book, American Chemical Society, Maple Press Company, York, PA, 1988, 145–155.
26. Certification of Reference Materials — General and Statistical Principles, ISO Guide 35, International Standards Organization, Geneva, Switzerland, 1989, 32 pp.
27. **M. Thompson and R. Wood,** International Harmonized Protocol for Proficiency Testing of (Chemical) Analytical Laboratories, *JAOAC Int.*, 76(4) (1993) 926–940.

CHAPTER 8

Preparation and Certification of a Biological Reference Material (CARP-1) for Polychlorinated Dibenzo-*p*-dioxin, Dibenzofuran, and Biphenyl Congeners

Catherine A. Fraser, Graeme J. Gardner, Paulette S. Maxwell, Cariton Kubwabo, Roger Guevremont, K.W. Michael Siu, and Shier S. Berman

A biological certified reference material for nine 2,3,7,8-chlorine substituted dibenzo-*p*-dioxin and dibenzofuran congeners and nine common polychlorinated biphenyl congeners is available for the first time. This material (CARP-1) is in the form of a slurry comprising approximately 9 g of carp homogenate sealed in a glass ampule. The only foreign materials in the slurry are an antioxidant added to prevent rancidity and water to facilitate homogenization and ampuling. CARP-1 was found to be homogeneous and stable with respect to its organochlorine concentrations under room as well as elevated temperature storage conditions. The minimum projected shelf life for this material is 5 years.

INTRODUCTION

Organochlorine compounds have aroused much interest among environmental and analytical chemists in the last two decades. Of these, three classes of compounds, polychlorinated biphenyls (PCBs), polychlorinated dibenzo-*p*-dioxins (PCDDs), and polychlorinated dibenzofurans (PCDFs) have received the most attention because of their ubiquitous presence in the environment, their anthropogenic origin, and their assumed toxicity. One of the PCDD congeners, 2,3,7,8-tetrachlorodibenzo-*p*-dioxin, has been labeled as "the most toxic man-made chemical."[1] Although the appropriateness of this description is questionable, the perception of toxicity has helped to push the determination of PCDDs and PCDFs to the foreground of organic contaminant monitoring.

Laboratory analyses and their associated quality assurance procedures pertinent to the monitoring of pollutants depend on the availability and use of certified reference materials (CRMs). The essential requirements[2] of a CRM are that it be: (1) representative of the real sample in that it presents a similar matrix and comparable contaminant load; (2) stable with respect to the analytes and the matrix; (3) homogeneous; and (4) accurately certified. The need for CRMs is, perhaps, best reflected by the general lack of comparability exhibited in results of interlaboratory comparisons (see Reference 3). Despite the important role that CRMs play and the immense environmental impact of organochlorine compounds, very few CRMs for organochlorine pollutants, and none for PCDDs and PCDFs, exist. From a CRM developer's point of view, this absence is understandable considering the relative difficulties involved in producing a CRM and in certifying the concentrations of individual congeners of these compounds.

This report summarizes our efforts in the production and certification of a biological reference material, CARP-1, for priority polychlorinated dibenzo-*p*-dioxin and dibenzofuran congeners.

PREPARATION OF CARP-1

It was decided that a biological material would be the most appropriate choice since most environmental monitoring laboratories performing organochlorine contaminant determinations analyze biological tissues. The particular biological material appropriate as a first CRM must satisfy the following criteria: (1) it must contain relatively high concentrations of PCBs, PCDDs, and PCDFs to render it useful to as many laboratories as possible; (2) it must be available in kilogram quantities; and (3) similar materials and quantities must be available in years to come. After much consultation, the carp from Saginaw Bay in Lake Huron was deemed to be the most appropriate candidate. The Saginaw Bay area is highly contaminated with respect to a variety of organochlorine compounds whose concentrations are not expected to change significantly over the next few years. Further, the carp population is large and stable, and should allow future repetitive sampling in tens of kilograms of material.

An approximately 30-kg sample of carp was made available to us courtesy of the U.S. Fish and Wildlife Service in East Lansing, Michigan. The carp was harvested near the warm-water discharge of the Consumer's Power Plant, Saginaw Bay on December 2, 1988, ground whole, and stored at –20°C. The sample was shipped frozen to the Canadian Institute of Fisheries Technology, Technical University of Nova Scotia, which was contracted to process the material and render it into a stabilized homogeneous slurry suitable for certification.

The following is a summary of the production procedures. The thawed carp was first comminuted four times in a food cutter (Comitrol, Model 3600, Urschel Laboratories Inc., Valparaiso, IN) equipped with a fine Type K 28 column 010-020 cutter on the first pass and a finer 010-010 cutter on subsequent passes. After the first pass, the weight of the raw material was reduced to approximately 16 kg. An antioxidant, 9.5 g of ethoxyquin powder, was added to the material. Distilled water was then added to raise the moisture content of the slurry from 62 to 85%, after which the resulting slurry was processed four times in a homogenizer (No-Bac Unitherm IV, Cherry-Burrell Corp., Cedar Rapids, IA). The slurry was divided and bottled by means of an ampuling machine. Each ampule was filled with approximately 10 mL (about 9 g) of slurry in an atmosphere of nitrogen and then sealed. To stabilize the homogenate, the ampules were heated in a steam retort (WSF Industries Inc., Tonawanda, NY) at 118°C for 11 min, which achieved lethalities, F_0, between 6.0 and 6.6 min. For packaging, the ampules were individually heat-sealed in trilaminate retort pouches and then packed six per box. The material is available as CARP-1 from the Institute for National Measurement Standards, National Research Council of Canada; each unit contains six ampules.

CHARACTERIZATION OF CARP-1

The physical characteristics of CARP-1 are listed in Table 1. Since only water and an antioxidant were added, the material is as similar as possible to a fish homogenate that an analytical chemist would prepare for analysis and satisfies one of the more often overlooked requirements of CRMs.

The >5-year shelf life at ambient temperature is a projection based on the stability of a similarly processed material, LUTS-1, and is a result of the addition of the antioxidant to retard rancidity plus heating to ensure sterilization. The lipid contents were determined gravimetrically after solvent evaporation of hexane extracts of CARP-1 (see later sections).

To ensure homogeneity, random samples of CARP-1 were pulled and analyzed for nine major PCDD and PCDF congeners, and an equal number of PCB congeners, over a period of weeks. The standard deviations of the determinations (using approximately 9 g of the entire vial contents) were better than ± 15% of the means and were considered to be in line with the expected precisions of repetitive determinations of a single homogeneous sample. To test the material's stability, random samples were analyzed on a regular basis over a period of 13 months. No change in the PCDD, PCDF, and PCB congener concentrations monitored was evident. As an example, Figure 1 shows the results of the stability study for 2,3,7,8-tetrachlorodibenzo-*p*-dioxin (2,3,7,8-TCDD) and 1,2,3,7,8-pentachlorodibenzo-*p*-dioxin (1,2,3,7,8-PCDD) that spanned from August 1992 to September 1993. The concentrations of these two congeners (and those of other congeners monitored) remained unchanged over the study period. A few samples of CARP-1 were stored for periods of months at 73°C to simulate extreme storage conditions; there were some signs of physical and chemical alterations to the matrix; the congener concentrations, however, were unchanged.

Table 1 Physical Characteristics of CARP-1

Stable shelf life	>5 years
Content per vial	8.89 ± 0.19 g
Moisture	85%
Lipids	3.92 ± 0.17%

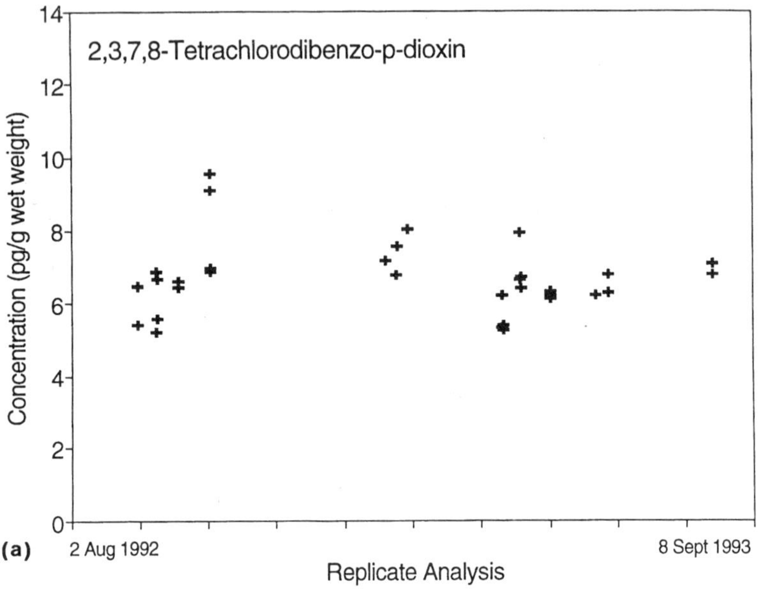

(a)

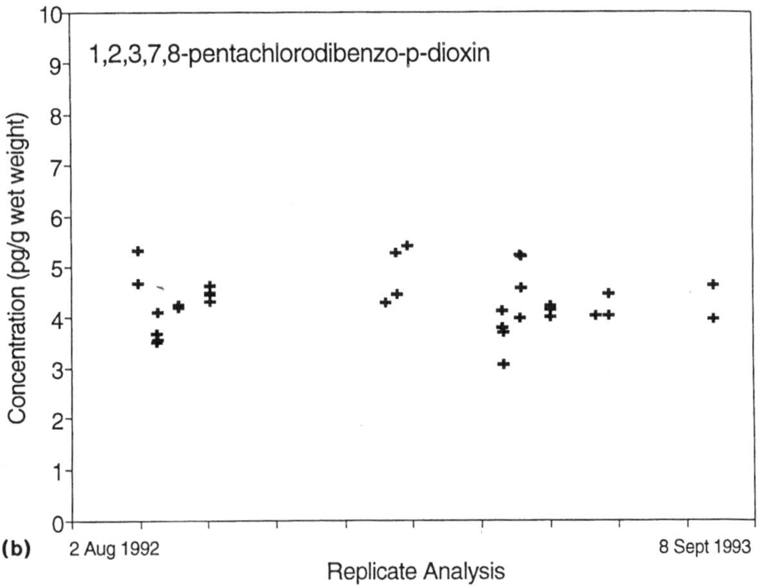

(b)

Figure 1 Stability study of PCDD and PCDF congeners in CARP-1: (a) 2,3,7,8-tetrachlorodibenzo-*p*-dioxin and (b) 1,2,3,7,8-pentachlorodibenzo-*p*-dioxin.

Table 2 Certified Concentrations of PCDF and PCDD Congeners in CARP-1

Congener	Certified conc. (ng/kg)
2,3,7,8-TCDF	11.9 ± 2.7[a]
1,2,3,7,8-PCDF	5.0 ± 2.0
2,3,7,8-TCDD	6.6 ± 0.6
1,2,3,7,8-PCDD	4.4 ± 1.1
1,2,3,4,7,8-HxCDD	1.9 ± 0.7
1,2,3,6,7,8-HxCDD	5.6 ± 1.3
1,2,3,7,8,9-HxCDD	0.7 ± 0.4
1,2,3,4,6,7,8-HpCDD	6.5 ± 1.8
OCDD	6.3 ± 1.9

[a] 95% confidence intervals; n = 25–50, results from five laboratories and six sets of data. Abbreviations of the congeners: T, tetra; P, penta; Hx, hexa; Hp, hepta; and O, octa.

POLYCHLORINATED DIBENZO-*P*-DIOXIN AND DIBENZOFURAN CONGENERS

Of the major PCDD and PCDF congeners present in CARP-1, most have the 2,3,7,8-chlorine substitution pattern. Table 2 shows the nine 2,3,7,8-chlorine-containing PCDD and PCDF congeners' concentrations certified in CARP-1. The uncertainties shown are the 95% confidence intervals; the number of repetitive analyses (n) varied from congener to congener and ranged from 25 to 50. Five expert laboratories, each using their preferred methodologies, participated in the certification exercise. Each of these methods differed from our in-house procedure in details but not in the chemistry employed. The following is a summary of the analytical procedure practiced in our laboratory. All chemicals were ACS grade or better, and all solvents were either Distilled-in-glass or Pesticide grade. PCDD and PCDF standards were purchased from Wellington Laboratories (Guelph, Ontario). Chromatographic adsorbents and gas chromatographic supplies were from Chromatographic Specialties Inc. (Brockville, Ontario).

CARP-1 Extraction

The ampule of CARP-1 was first sonicated in an ultrasonic bath for 15 min, after which the seal was broken and as much as possible of the material was transferred into a mortar and pestle. A sufficient quantity (about 46 g) of anhydrous sodium sulfate (previously heated to 450°C overnight) was added until the mixture attained a free-flowing state. The material was loaded into a Pyrex thimble for Soxhlet extraction. After the thimble was placed inside the extractor, the ampule plus the mortar and pestle were rinsed with a small quantity of the extraction solvent. The wash was transferred quantitatively onto the thimble contents, followed by spiked solutions containing the appropriate amounts of ^{13}C-labeled PCDD and PCDF recovery standards, including 2,3,7,8-tetra-, 1,2,3,7,8-penta-, 1,2,3,4,7,8-hexa-, 1,2,3,6,7,8-hexa-, 1,2,3,7,8,9-hexa-,

1,2,3,4,6,7,8-hepta-, and octachlorodibenzo-*p*-dioxin, as well as 2,3,7,8-tetra- and 1,2,3,7,8-pentachlorodibenzofuran. Extraction was carried out with 180 mL of 2/1 (v/v) acetone/hexane for 7 h. After extraction, acetone was removed by means of partitioning the extract with 40 mL of water. The hexane phase was removed, and the aqueous phase was washed with 50 mL of hexane. The hexane solutions were then combined and washed for the last time with 30 mL of water. To dry the hexane solution, it was passed through a filter paper containing a little anhydrous sodium sulfate on top. The filtrate was collected in a 100-mL volumetric flask. The filter paper was then washed with hexane to bring the final filtrate volume to 100 mL.

Bulk Lipid Removal

If lipid determination was required, a 10-mL portion of this solution was set aside. The remaining portion was transferred to a 250-mL separatory funnel, followed by 60 mL of hexane. Concentrated sulfuric acid (20 mL) was added. After shaking and phase separation, the aqueous phase was removed. This process was repeated five to six times until the aqueous phase became clear and pale yellow in color. The defatted hexane solution was then washed successively with 20 mL of water, 20 mL of 1% aqueous potassium hydroxide, and again 20 mL of water. It was then filtered through a Whatman number 1 paper containing a thin layer of anhydrous sodium sulfate, and reduced in volume to about 1 mL via rotary evaporation.

Sample Cleanup

A modified Ryan[4] multicolumn clean-up was performed using three columns in the following order, acid- and base-impregnated silica gel, Florisil, and carbon. The silica gel column and the Florisil column cleanup were run in tandem. Cesium hydroxide-impregnated silica was prepared as follows: 30 g of cesium hydroxide was dissolved in 100 mL of methanol. This solution was poured into 100 mL of methanol, followed by 50 g of activated silica gel, 40/60 mesh. This mixture was allowed to remain at 55°C for 90 min, after which it was filtered. The silica gel was washed with 200 mL of methanol and then with 200 mL of dichloromethane. Sulfuric acid-impregnated silica gel was synthesized by shaking 2 parts of concentrated sulfuric acid with 3 parts of activated silica gel, 40 to 60 mesh, until the mixture was free of lumps and of a uniform consistency (about 15 min). Both silica materials were prepared ahead of time and stored in closed jars. Since these materials deteriorate slowly under laboratory conditions, column packing was performed just prior to use. One gram of the sulfuric acid-impregnated silica gel, 0.5 g of the cesium hydroxide-impregnated silica gel, and 0.5 cm of anhydrous sodium sulfate were packed, in that order, into a 5-mL serological pipet. The column was first washed with 10 mL of hexane and then mounted on top of a Florisil column.

For the Florisil column, 1.5 g of Florisil, 60 to 100 mesh, previously extracted with dichloromethane, was packed into a 5-mL pipet followed by a 1-cm layer of anhydrous sodium sulphate. This column was activated at 130°C for 24 h. Before use, the column was washed with 10 mL dichloromethane, followed by 10 mL 2% dichloromethane in hexane. The Florisil column was then coupled to the silica gel column.

The concentrated and defatted hexane solution was applied to the silica gel and Florisil column in tandem. The flask and the columns were then washed with a total of about 15 mL of hexane. The silica gel column was removed. A 20-mL solution of 2% dichloromethane in hexane was used to elute PCBs from the Florisil column. PCDDs and PCDFs were eluted with 50 mL of dichloromethane. The volume of this eluate was reduced. Dichloromethane was replaced with hexane; this solution was ready for carbon cleanup.

An adsorbent of 18% Carbopack C, 80 to 100 mesh, on Celite 545 was prepared by means of shaking 3.6 g of the carbon with 16.4 g of the diatomaceous earth until a uniform consistency as well as color was observed. A 0.25-g portion of this material was packed into a 5-mL serological pipet that had previously been packed with 0.2 g of anhydrous sodium sulfate. The column was then washed with 2 mL of toluene, 1 mL of 1/1 cyclohexane/dichloromethane, and 2 mL of hexane, in that order, prior to use. The sample was added to the column, followed by 2 mL of hexane and then 1 mL of 1/1 cyclohexane/dichloromethane to elute remaining PCBs and pesticides. Fifteen milliliters of toluene were used to elute PCDDs and PCDFs.

The toluene solution was reduced to 0.5 mL on a rotary evaporator, and further reduced to approximately 100 µL with the aid of a stream of nitrogen. An appropriate amount of the internal standard ^{13}C-labeled 1,2,3,4-tetrachlorodibenzo-*p*-dioxin was added. The solution was evaporated down and reduced to 20 µL. This solution was ready for gas chromatography/high resolution mass spectrometric (GC/HRMS) analysis.

Instrumentation

Determinations were conducted on a VG Autospec-Q hybrid mass spectrometer having an EBEqQ geometry. This instrument is capable of unattended automatic runs for a period of days. All measurements were carried out solely by means of the high resolution mode with a resolution of 10,000. Sample injections were performed in the splitless mode on a 60-m × 0.25-mm ID DB-5 column with a film thickness of 0.1 µm (J & W Scientific). A DB-1701 column of identical dimensions was also used as an alternative column for verification. A typical temperature program run on the DB-5 column has the following steps: initial temperature, 100°C for 2 min; ramp #1, 20°C/min; hold temperature, 180°C for 0 min; ramp #2, 2.5°C/min; final temperature, 280°C for 14 min. Figure 2 shows chromatographic windows of selected ion monitoring runs for (1) tetrachlorodibenzo-*p*-dioxins, (2) pentachlorodibenzofurans, and (3) hexachlorodibenzo-*p*-dioxins in a CARP-1 extract. Quantifications were performed using standard solutions of PCDDs and PCDFs containing appropriate amounts of ^{13}C-labeled recovery and internal standards as recommended by the U.S. EPA.[5]

POLYCHLORINATED BIPHENYL CONGENERS

Table 3 summarizes the concentrations of the nine PCB congeners certified in CARP-1. Note that the units are in micrograms per kilogram, which means that the

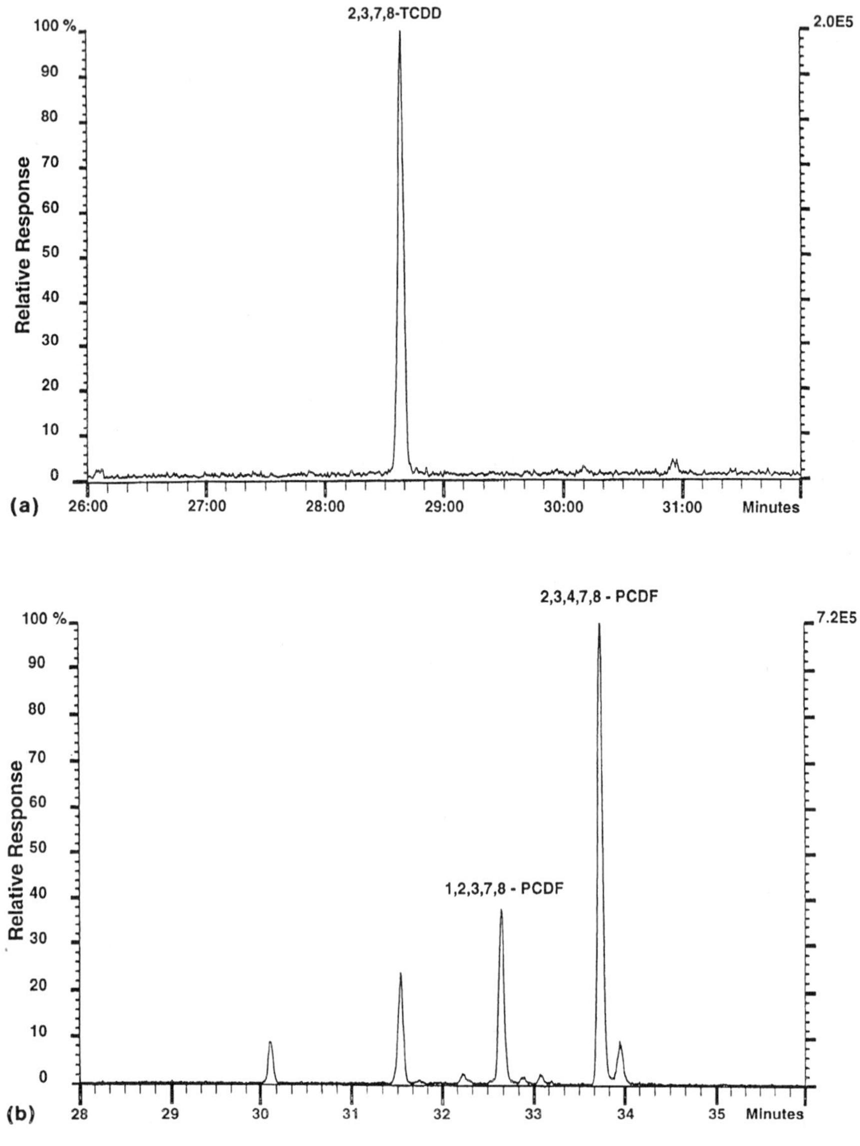

Figure 2 Selected ion chromatograms of PCDD and PCDF congeners in CARP-1: (a) m/z 319.8965 for tetrachlorodibenzo-*p*-dioxin congeners, (b) m/z 339.8598 for pentachlorodibenzofuran congeners, and (c) m/z 389.8157 for hexachlorodibenzo-*p*-dioxin congeners. See Table 2 for explanations of the labels.

PCB congener concentrations are about 4 orders of magnitude higher than the PCDD and PCDF congener concentrations shown in Table 2. Again the uncertainties shown are the 95% confidence intervals. A comparably larger number of expert laboratories (15) participated in the certification of PCBs, resulting in 18 sets of data. The number of repetitive analyses for each congener (n) used for certification was typically

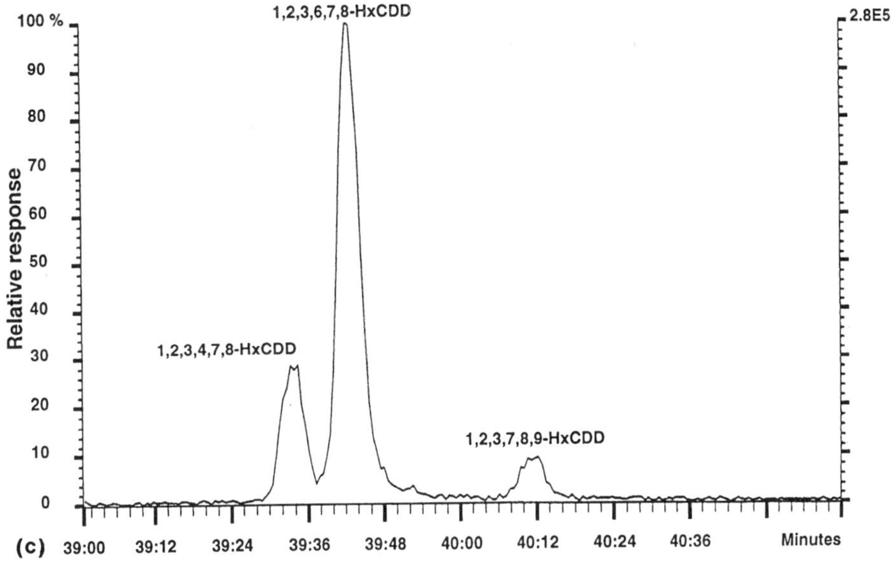

Figure 2 (continued).

Table 3 Certified Concentrations of PCB Congeners in CARP-1

Congener (IUPAC No.)	Certified conc. (μg/kg)
52	124 ± 32[a]
101/90	124 ± 37
105	54 ± 24
118	132 ± 60
138/163/164	102 ± 23
153	83 ± 39
170/190	22 ± 8
180	46 ± 14
187/182	36 ± 16

[a] 95% confidence interval; n ≈ 100, results from 15 laboratories and 18 sets of data.

around 100. As in the case of the PCDD and PCDF exercise, the participating laboratories used their preferred PCB methodologies, which differed from one another in details but not in chemistry. The analytical method used in our laboratory is summarized below. Polychlorinated biphenyl standard solutions were either prepared from the calibration solution CRM, CLB-1 (National Research Council of Canada) or prepared from chemicals or solutions available commercially (Ultra Scientific, North Kingstown, RI, and AccuStandard, Chromatographic Specialties Ltd.) and verified against CLB-1. The ^{13}C-labeled PCB congener, IUPAC no. 153, was available from the National Research Council of Canada as CLB-2; other ^{13}C-labeled congeners were purchased from Cambridge Isotope Laboratories (Andover, MA).

CARP-1 Extraction

An ampule of CARP-1 was first sonicated, transferred, mixed with anhydrous sodium sulfate, and loaded into a Soxhlet apparatus as detailed above. The thimble's contents were spiked with the appropriate amounts of recovery standards: IUPAC no. 78, 188, and 204 for electron capture detection, and ^{13}C-labeled congeners, including IUPAC no. 52, 101, 105, 118, 138, 153, and 180, for mass spectrometric detection. Extraction was performed with 175 mL of 1/1 (v/v) acetone/hexane for 7 h. The extract was then reduced to a few mL by means of rotary evaporation.

Sample Cleanup

Twenty g of Florisil, 60 to 100 mesh, previously deactivated with 2% water, was packed into a 50-mL buret. The extract was transferred to the Florisil column. The PCBs were eluted with 300 mL of hexane. The hexane solution was evaporated down to 1 mL for cleanup on an alumina column. This was prepared by packing a 3.7-g sample of activated alumina F1, 40/60 mesh, into a 10-mL serological pipet, followed by a thin layer of anhydrous sodium sulfate, and activated at 190°C for 2 d. The concentrated hexane extract was applied to the alumina column. The column was then washed with 10 mL of hexane. The PCBs were eluted with 50 mL of 15% ethyl ether in hexane. The extract was reduced to 0.5 mL by means of rotary evaporation and transferred quantitatively into a 2-mL volumetric flask. An appropriate amount of IUPAC no. 194 or 205 was added as an internal standard. Isooctane was added to the mark. This solution was diluted ten times prior to gas chromatography with electron capture detection (GC-ECD) or gas chromatography/high resolution mass spectrometric analysis.

Instrumentation

A Varian, Model 3300, gas chromatograph equipped with a split/splitless injector, a 60-m or 30-m × 0.25 mm ID DB-5 column with a film thickness of 0.25 μm (J & W Scientific), and a constant current-variable frequency electron capture detector was used for the bulk of the gas chromatographic work. For priority congeners that were known to have potential coeluting congeners, determination was also carried out on a two-dimensional gas chromatograph, Siemens, Model Sichromat 2-8, equipped with a 30-m × 0.25 mm ID DB-5 column, 0.25 μm film thickness, as the first column and a 30-m × 0.25 mm ID DB-23, again 0.25 μm film thickness (both from J & W Scientific) as the second column. Injections were made via the splitless mode; chromatography on each column was monitored separately by its own constant current-variable frequency electron capture detector. Typically a temperature program was used to separate the PCB congeners on the DB-5 column; the peak containing the congener of interest was heart-cut onto the DB-23 column and resolved isothermally. Quantification was performed using standard solution calibration; the final concentrations were adjusted using the percent recovery determined via the recovery standards. Gas chromatography/high resolution mass spectrometry was performed on a VG Autospec-Q as previously described. A quantification pro-

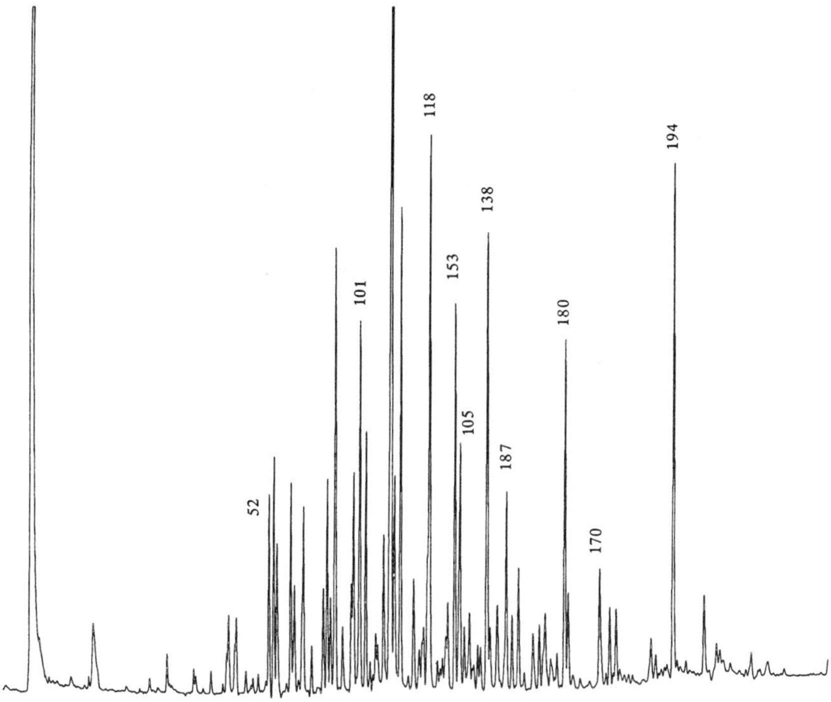

Figure 3 PCB congeners in CARP-1 as separated on a DB-5 column. Electron capture detection; the labels are the IUPAC numbers of the congeners. IUPAC no. 194 was an internal standard added.

cedure identical to the one used for PCDD and PCDF congeners, which involved ^{13}C-labeled recovery standards, was used to calculate the PCB congener concentrations. Figure 3 shows a typical chromatogram of a CARP-1 extract separated on the DB-5 column and monitored with electron capture detection.

CONCLUSIONS

A CRM certified for nine priority PCDD and PCDF congeners as well as for nine common PCB congeners is available for the first time to the general analytical community. We are projecting a shelf life of >5 years although we expect our stock to be depleted before then. Work is progressing well toward the certification of a replacement material, CARP-2.

ACKNOWLEDGMENTS

The cooperation of the following individuals and organizations in the preparation and analysis of CARP-1 is gratefully acknowledged: T. Kubiak and L. Williams,

U.S. Fish and Wildlife Service, East Lansing, MI; M. A. Tung, I. J. Britt, and C. Hotton, Technical University of Nova Scotia, Halifax, Nova Scotia; R. J. Norstrom, M. Simon, and J. Zhu, Canadian Wildlife Service, Environment Canada, Hull, Quebec; R. E. Clement, E. Reiner, S. Cussion, and A. Johnson, Ontario Ministry of the Environment and Energy; J. J. Ryan, L. G. Panopio, and B. P.-Y. Lau, Health and Welfare Canada, Ottawa, Ontario; B. G. Chittim, Wellington Laboratories, Guelph, Ontario; L. Phillips, Axys Analytical Services, Sidney, British Columbia; B. J. Ross, Eli Eco Logic International Inc., Rockwood, Ontario; D. Muir, Freshwater Research Institute, Fisheries and Oceans, Winnipeg, Manitoba; S. A. Wise, R. M. Harris, M. M. Schantz, and W. E. May, National Institute of Standards and Technology, Gaithersburg, MD; J. M. Curtis and M. D. LeBlanc, Institute for Marine Biosciences, National Research Council of Canada, Halifax, Nova Scotia; and D. E. Wells and C. McKenzie, The Scottish Office, Agriculture and Fisheries Department Marine Laboratory, Aberdeen, Scotland. A feasibility study for the production of CARP-1 was funded in part by the Research and Technology Branch, Ontario Ministry of the Environment.

REFERENCES

1. **G. L. Sparschu, F. L. Dunn, and F. L. Rowe,** *Food Cosmet. Toxicol.*, 9 (1971) 405.
2. **P. J. Wagstaffe,** In: W. R. Wolf, Ed., *Biological Reference Materials: Availability, Uses and Need for Validation of Nutrient Measurement*, John Wiley & Sons, New York, 1985.
3. **J. de Boer, J. C. Duinker, J. Calder, and J. van der Meer,** Draft Report on the ICES/IOC/JMG Intercomparison Exercise on the Exploration of the Sea, 1990.
4. **J. J. Ryan, R. Lizotte, L. G. Panopio, B. P.-Y. Lau, D. Lewis, and W.-F. Sun,** Department Report, Food Research Division, Health Protection Branch, Health and Welfare Canada, Ottawa, Ontario, 1989.
5. Polychlorinated Dibenzodioxins and Polychlorinated Dibenzofurans by High-Resolution Gas Chromatography/High Resolution Mass Spectrometry, U.S. Environmental Protection Agency Method 8290, 1990.

CHAPTER 9

Development of Certified Aqueous Reference Materials Using Natural Unpreserved Lake, River, and Rainwaters

Keijo I. Aspila, Harry Alkema, Nabil Arafat, Karen Jankovich, Ellie Kokotich, and Christian Vogel

At the National Water Research Institute, several hundred water samples have been collected from across Canada and have been stored unpreserved at 4°C in 100- to 900-L units. To enhance stability, these waters are first clarified by centrifugation, and then filtered through 0.2-μm cartridge filters to remove vegetative bacteria. Some samples are also pasteurized by heating at 90°C. Characterization of the constituent concentrations are made from the data obtained from reference laboratories or from the consensus values obtained from large intercomparison studies. This report overviews the treatment procedures and the successes or failures in obtaining stable certified reference waters.

BACKGROUND

For over 25 years the Environmental Standards and Statistics Project in the Aquatic Ecosystem Protection Branch, National Water Research Institute, has prepared and delivered performance evaluation studies (round-robins) to a wide variety of environmental agencies in Canada and the United States.

The inorganic performance assessment studies are comprised of ten different water samples that are distributed to between 40 and 70 different laboratories. Most of these studies are completed within 8 to 12 weeks and are analyzed for about 20 different constituents. Each assessment focuses on the interlaboratory measurement bias[1–3] with overall performance[1] based on the frequency of biased parameters and flagged results.[1]

These round-robins are viewed as an excellent mechanism to define the performance of laboratories in the measurement process, as well as providing a powerful stimulant for improving and maintaining precision and accuracy. In these situations, where the laboratory performance was seen as poor, the intercomparison studies provide positive and constructive information to such laboratories with clear indication on how to improve.[1]

Samples used for the NWRI major ions and nutrients studies have always been unpreserved reference waters. Each laboratory receives approximately 500 mL of each test sample. Since these QA studies are frequent (about 8 to 12 studies per year), the volume of stock reference waters required is quite sizable (400- to 900-L units). These routine and frequent round-robins have provided an excellent mechanism for compiling raw data to verify the long-term stability of the reference waters.

The following is a brief description of how these aqueous reference waters have been prepared at the National Water Research Institute for use in the performance evaluation studies and later as certified reference materials.

PREPARING STABLE AQUEOUS REFERENCE SAMPLES

General

For over 25 years, the approach taken for the Environment Canada interlaboratory performance evaluation studies has been the utilization of natural unpreserved reference waters for the major ions, nutrients, and physical parameters. The use of unpreserved natural samples has been preferred since such samples would best reflect the program-specific situations. There have been occasional failures, such as the precipitation of calcium in hard waters and the influence of a high bacterial content on the ammonia and nitrate concentrations in other waters. Nevertheless, there has been overwhelming support for the use of natural waters that originated from pristine lakes and rivers. The stability of these waters is quite adequate to meet the requirements of the quality assurance programs. This will become evident to the reader after examining the tables and figures that follow.

The area of primary concern to the producer of reference samples is whether the nutrients will be sufficiently stable during the time period when participants would make their analyses. This time period is normally 4 to 8 weeks. The processes that create instability are well documented in the literature.[4,5] The major element is believed to be microbial. The microbe-driven nitrogen cycle is illustrated in Figure 1. To suppress the bacterial conversion of nitrate, nitrite, and ammonia, several options are available for the treatment of water samples. One could irradiate with UV or gamma rays or add preservatives such as mercury or phenols to kill bacteria. These procedures have not been used as they are cumbersome or impractical.

We have used three different procedures to stabilize water samples, namely (1) centrifugation only, (2) centrifugation with heat treatment at 90°C to pasteurize, and (3) centrifugation followed by removal of the particulates and bacteria with membrane filtration. These procedures are described in detail below and are followed by a discussion of the merits and disadvantages of each procedure.

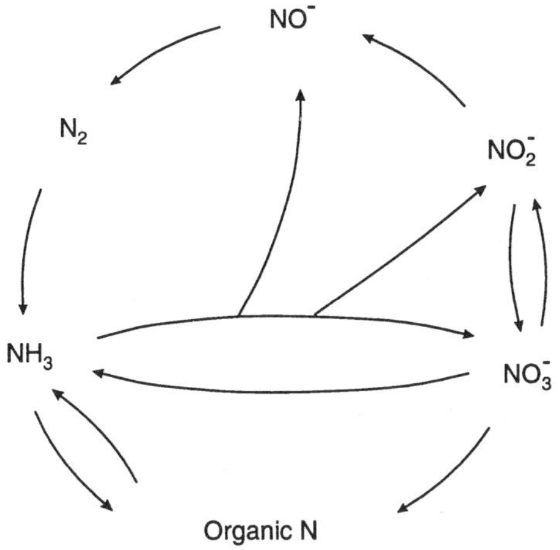

Figure 1 The nitrogen cycle (1). (From *The Nitrogen and Sulphur Cycles*, J. A. Cole and S. J. Ferguson, Eds., Cambridge University Press, 1988. With permission.)

Bottle Washing Procedures

Large 100-, 200-, 300-, 400- or 900-L polyethylene barrels are prepared by scrubbing with a stiff brush and tap water from a garden hose. Containers are then rinsed with a 10% solution of normal household bleach (sodium hypochlorite) followed by a thorough rinse using sterile water. The polyethylene closures for all barrels are treated in a similar manner. Although the containers may appear to be sterile initially, these conditions cannot be maintained as these large vessels are moved about when being filled and when the reference water is being subsampled over the duration of its containment (often many years).

The sample bottles used for transporting the unpreserved natural waters are linear polyethylene (Nalge). These bottles are simply rinsed, filled with deionized water, and stored at room temperature for a period of 4 to 6 months before being filled with the reference samples. Repeated analysis of these bottles for the major ions and nutrients confirm the absence of measurable contaminants. Our current practice is to purchase several thousand bottles at a time, specifying that the order be the same lot number from the manufacturer.

Collection of Bulk Water Samples

Large lake and river reference waters are collected using submersible pumps to transfer 200-L aliquots of water into prewashed polyethylene barrels. Two to five barrels of water are normally collected from each station, and these are transported back to the Institute by either field vehicles or by commercial carrier.

Large rainwater samples are collected off a very large greenhouse roof (Alkema Greenhouses) in Grimsby, Ontario. The greenhouse roof was previously modified to allow the drainage of rainwater into a manifold of five 200-L polyethylene barrels. A "transfer valve" is placed on the main pipeline to allow the collection to take place when a rainfall sample is required. No special care is taken to prewash the roof of insects or debris since rain samples collected are centrifuged within a day or so after collection.

Clarification by Centrifugation

The removal of particulates from these water samples is necessary in order to (1) allow for safe operation of precision pumps (see filtration, below), (2) improve the homogeneity of subsamples, and (3) reduce the need for extensive replacement of coarse prefilters in the filtration process.

The instrument used for clarifying large volume water samples (100 L and larger) is the Westphalia Clarifier (Model KA-02-06-75). This unit has a four-chamber stainless steel bowl and is normally operated at a bowl speed of 9400 to 10,000 rpm. A flow rate of about 4 to 6 L/min has been adopted, as this flow rate has been shown to remove approximately 90% of the particulate load (sediments) from most waters.[6,7] It is also known that as much as 70% of the particulates captured by this clarifier are finer than 0.2 microns. This process is a major step in removing the vegetative bacteria affecting the nitrogen species especially since many bacteria are associated with the suspended particulates.

Clarified waters are temporarily stored in clean polyethylene barrels (100- to 900-L capacity) and these are transferred and maintained in coldrooms at 4°C. These clarified water samples, if so required, are then processed further by filtration or by heat treatment (pasteurization).

Pasteurization

To pasteurize a large reference water sample (900 L), two 40-L units of the centrifuged water are transferred into a large autoclave, and the partially sealed containers are flooded with steam until the water reaches a temperature of 90°C. A thermocouple in the water samples is required to track the water temperature to 90°C, at which time the autoclave is disengaged. Normally, a temperature of 90°C is reached in the water sample after a period of about 4 h.

The heat-treated aliquots are then removed, and two new samples are introduced to continue the process until the large 900-L units are all processed. The heat-treated samples, after a brief cool-down period, are then poured into the sterile 900-L containment vessels (polyethylene).

It must be emphasized that heat treatment is acceptable only for soft waters of low alkalinity. Processing hard waters in this way means a major risk of carbonate precipitation. Filtration is an important alternative process and indeed one of preference since it is simple and much less labor intensive.

Membrane Filtration

Of the four major types of contaminants found in water (organics, inorganics, particles, and microorganisms), the membrane filtration unit eliminates two of them, namely particles and microorganisms.[8] It is an accepted procedure to use membrane filters to achieve sterility and to remove particulates.[9] Thus, to ensure that a liquid is sterile and particle free and will pass rigid final inspection, membrane filters are used in developing stable reference waters.

In many industrial and medical processes, membrane filtration is used for final product filtration as well as to prepare raw materials, such as environmental water samples, sterile gases, and air. Samples that are heavily contaminated with particulate matter and bacteria are usually subjected to serial filtration procedures[10] where a depth prefiltration stage is placed upstream of a final 0.2 μm sterilizing filter unit.[11]

The filtration system used for reference waters consists of a rotary vane pump, a Millipore prefilter (130-mm, 5-μm glass fiber filter) followed by the cartridge filter. The water pump (Model #900-576-46M) has carbon graphite vanes. It is direct drive, self-priming, and has pulseless flow up to 4 gpm at 100 psi. Two pressure meters are placed in the plumbing line, one ahead of the prefilter and the other before the cartridge filter in order to monitor any build-up of pressure caused by a filter saturation. These meters are only precautionary to prevent damage and to allow recognition of the need for filter replacement. The pump assembly also has a built-in pressure relief valve that is designed to cut in should the internal pressure exceed 70 psi.

The filter assembly is a Gelman 0.2 μm Supor® membrane filter. The filter cartridge (10 in.), is contained in a Gelman stainless steel housing unit.

All pipes and fittings to connect the prefilter, the pump, and the filter are custom made using high-grade stainless steel. This plumbing is assembled on a rigid but very mobile platform.

Prior to the filtration of a designated reference water, the filtration system is sterilized with a 25% (v/v) solution of household bleach (6% hypochlorite). This solution is left in the system for about 5 min, after which time the system is carefully flushed with at least 50 L of clean water. In a routine operation, the cartridge filters are replaced after approximately 1500 L of sample are processed or when the pressure meters indicate a 15 to 20 psi pressure build-up.

Prior to the use of filtration for reference water samples, a study was undertaken to confirm the effectiveness of the filter to remove bacteria. The water for this test, which was taken from Hamilton Harbour, was known to be chemically unstable and heavily contaminated with bacteria. Results are discussed below.

Storage of the Bulk Samples and the Subsampling Procedures

All large unpreserved water samples are maintained in large walk-in cold rooms at 4°C. The storage vessels are barrels of 100-, 200-, or 300-L capacity. Each of these units has a screw cap closure. The 400- and 900-L units have loose fitting flat plastic lids with lips as closure. None of these closures can be considered as air tight.

Subsamples of the large reference water are removed by employing a simple syphon prepared from a glass tube inserted into the middle section of the water tank and syphoning water into bottles through surgical latex tubing. Normally these subsamples are half-liter units.

DISCUSSION OF RESULTS

General

A study was undertaken to assess the capability of the filtration process to produce sterile and stable reference water. For this study, Hamilton Harbour water was collected and subdivided into two groups. One aliquot was unprocessed and had copious particulates. The other sample was the same water but clarified by centrifugation only. Both of these test samples were passed through the filtration process, after which both input and clarified waters were checked for bacterial activity. The bacteriological purity was tested using a spread-plate technique,[12] nutrient agar, 1 mL volume of sample, and a 5 d incubation period at 20 to 22°C.

The bacterial tests on the filtered waters were negative, whereas the results for waters only centrifuged were positive but low. In contrast, the unprocessed harbor water indicated very high bacterial contamination. These results were encouraging, and additional tests were made using other large water samples. In all cases, filtrates from the filtration module were free of bacteria. However, it must be noted that if the subsamples from the filtration module were simply placed into the half-liter polyethylene Nalgene sample bottles, the bacterial tests on these samples revealed trace levels of bacteria. These follow-up tests were not too surprising since no major efforts were made to ensure sterility of these sample bottles. Similar bacteriological tests of the large volume samples, contained in 200- to 400-L units (after filtration or pasteurization), revealed that some bacteria were present. This nonsterility of the large volume samples was attributed to the continuous exposure to the coldroom air and the inadvertent introduction of bacteria from the sampling devices.

The overall assessment of the various processes indicates that filtration would be the more suitable technique in preparing reference water samples. The long-term stability of these different reference waters is obtained from round-robins and is discussed below.

Stability of Natural Lake Water Reference Samples

The analytical data illustrating the long-term stability of some typical reference samples produced by the above procedures are given below in Tables 1 to 10 and in Figures 2 to 5.

Samples Treated by Centrifugation Only

Only one example is given for which the pretreatment was clarification by centrifugation. This reference water originated from Lake Superior. This lake is very large and oligotrophic[13,14] with very low bacterial counts.[15] The reference sample, ION-91, was taken from a depth of several meters in the open waters many kilometers from the shoreline. The unique stability of Lake Superior water is evident from the long-term data illustrated in Figure 2. The stability also applies to the major ions (Ca, Mg, Na, K, SO_4, and Cl), as well as to silica and alkalinity, although this is not illustrated in Figure 2. In our experience, even raw, unprocessed waters derived from Lake Superior have long-term stability for almost all of the constituents listed.

Table 1 Comparison of Interlaboratory Median Values for AUD-O1, over Several Studies (LRTAP Data), Plastic Lake

Parameter	Units of measurement	Study no.: / Sample no.: / Date of study:	0019 / 8 / Aug 88	0021 / 9 / April 89	0022 / 6 / June 89	0031 / 4 / Sept 92	0035 / 5 / Jan 94
Color	Hazen units		6.000	5.000	4.500	4.500	3.550
Specific conductance	µmho/cm		22.200	22.100	22.100	22.200	21.900
Acidity to pH 8.3	mg/L $CaCO_3$		1.630	1.799	2.018	2.650	1.440
pH	pH units		6.050	6.000	6.030	6.069	6.085
Dissolved organic carbon	mg C/L		2.055	2.200	2.100	2.370	2.060
Alkalinity-fixed endpt pH4.5	mg/L $CaCO_3$		2.000	2.000	2.195	2.400	2.400
Alkalinity-gran, inflec extrap	mg/L $CaCO_3$		0.835	0.890	0.790	0.918	0.805
Alkalinity, gran titration	mg/L $CaCO_3$		0.800	0.810	0.810	0.965	0.905
Dissolved inorg carbon	mg C/L		0.409	0.485	0.580	0.578	0.550
Nitrate + nitrite	mg N/L		0.043	0.042	0.040	0.009	0.002
Nitrate-IC	mg N/L		—	—	—	0.008	0.001
Ammonia	mg N/L		0.030	0.010	0.004	0.006	0.004
Total Kjeldahl nitrogen	mg N/L		0.174	0.158	0.135	0.130	0.120
Sodium	mg/L		0.492	0.500	0.491	0.499	0.497
Magnesium	mg/L		0.477	0.487	0.480	0.489	0.487
Reactive silica	mg Si/L		0.310	0.310	0.310	0.320	0.310
Sulfate, IC methods	mg/L		6.320	6.350	6.320	6.340	6.334
Sulfate, non-IC methods	mg/L		6.360	6.300	6.365	6.400	6.431
Chloride, IC methods	mg/L		0.384	0.380	0.381	0.380	0.400
Chloride, non-IC methods	mg/L		0.400	0.400	0.360	0.417	0.416
Potassium	mg/L		0.212	0.210	0.213	0.210	0.212
Calcium	mg/L		2.035	2.030	2.050	2.060	2.050

Note: Centrifuged and pasteurized; 65 months of monitoring.

Table 2 Comparison of Interlaboratory Median Values for AUD-O2, over Several Studies (LRTAP Data), Harp Lake

Parameter	Units of measurement	Study no.: Sample no.: Date of study:	0019 9 Aug 88	0022 7 Jun 89	0030 5 May 92	0035 6 Jan 94
Color	Hazen units		14.000	11.500	11.200	10.500
Specific conductance	µmho/cm		34.300	34.300	34.700	34.400
Acidity to pH 8.3	mg/L $CaCO_3$		1.680	1.557	1.814	1.539
pH	pH units		6.700	6.660	6.550	6.550
Dissolved organic carbon	mg C/L		3.900	4.200	6.190	5.620
Alkalinity-fixed endpt pH4.5	mg/L $CaCO_3$		4.830	5.500	5.190	5.400
Alkalinity-gran, inflec extrap	mg/L $CaCO_3$		3.815	4.060	3.531	3.700
Alkalinity, gran titration	mg/L $CaCO_3$		3.780	4.135	3.840	4.090
Dissolved inorganic carbon	mg C/L		1.000	1.205	0.725	1.120
Nitrate + nitrite	mg N/L		0.020	0.009	0.005	0.002
Nitrate-IC	mg N/L		—	—	0.007	0.001
Ammonia	mg N/L		0.014	0.009	0.006	0.003
Total Kjeldahl nitrogen	mg N/L		0.220	0.210	0.212	0.220
Sodium	mg/L		1.240	1.220	1.240	1.230
Magnesium	mg/L		0.915	0.915	0.930	0.933
Reactive silica	mg Si/L		1.032	1.028	1.070	1.050
Sulfate, IC methods	mg/L		7.400	7.406	7.500	7.468
Sulfate, non-IC methods	mg/L		7.650	7.700	7.595	7.650
Chloride, IC methods	mg/L		1.250	1.230	1.250	1.264
Chloride, non-IC methods	mg/L		1.255	1.200	1.300	1.330
Potassium	mg/L		0.544	0.540	0.543	0.550
Calcium	mg/L		2.988	2.995	3.005	3.030

Note: Centrifuged and pasteurized; 65 months of monitoring.

Table 3 Comparison of Interlaboratory Median Values for AUD-O3, over Several Studies (LRTAP Data), Twelve Mile Lake

Parameter	Units of measurement	Study no.: Sample no.: Date of study:	0019 10 Aug 88	0021 10 April 89	0022 5 June 89	0030 7 May 92
Color	Hazen units		6.500	7.000	6.250	5.000
Specific conductance	µmho/cm		54.400	54.000	54.000	54.200
Acidity to pH 8.3	mg/L $CaCO_3$		1.520	1.499	2.180	1.675
pH	pH units		7.150	7.100	6.950	7.221
Dissolved organic carbon	mg C/L		3.665	3.800	4.275	2.590
Alkalinity-fixed endpt pH 4.5	mg/L $CaCO_3$		12.000	12.300	12.270	12.900
Alkalinity-gran, inflec extrap	mg/L $CaCO_3$		11.040	11.230	11.190	11.080
Alkalinity, gran titration	mg/L $CaCO_3$		11.000	11.290	11.200	11.360
Dissolved inorganic carbon	mg C/L		2.785	2.750	2.910	2.760
Nitrate + nitrite	mg N/L		0.073	0.015	0.010	0.080
Nitrate-IC	mg N/L		—	—	—	0.080
Ammonia	mg N/L		0.008	0.005	0.010	0.034
Total Kjeldahl nitrogen	mg N/L		0.212	0.218	0.200	0.160
Sodium	mg/L		1.565	1.570	1.550	1.560
Magnesium	mg/L		1.460	1.480	1.470	1.460
Reactive silica	mg Si/L		1.010	1.010	1.001	1.020
Sulfate, IC methods	mg/L		8.800	8.810	8.795	8.808
Sulfate, non-IC methods	mg/L		8.950	8.970	8.800	9.100
Chloride, IC methods	mg/L		2.100	2.080	2.082	2.080
Chloride, non-IC methods	mg/L		2.100	2.100	2.100	2.100
Potassium	mg/L		0.736	0.730	0.730	0.730
Calcium	mg/L		5.585	5.600	5.585	5.570

Note: Centrifuged and pasteurized; 44 months of monitoring.

Table 4 Comparison of Interlaboratory Median Values for AUD-O5, over Several Studies (LRTAP Data), Halls Lake

Parameter	Units of measurement	Study no.:	0020	0022	0024	0025	0027	0029	0030	0032
		Sample no.:	9	3	6	2	4	6	3	10
		Study date:	Jan 89	June 89	May 90	Sept 90	May 91	Feb 92	May 92	Jan 93
Color	Hazen units		5.000	5.000	3.000	6.750	4.750	3.800	3.400	4.000
Spec. cond.	µmho/cm		34.250	34.100	33.650	36.000	33.700	34.000	33.600	34.000
Acidity (pH 8.3)	mg/L $CaCO_3$		2.018	1.700	1.665	2.100	1.650	1.670	1.660	2.050
pH	pH units		6.580	6.640	6.545	6.590	6.575	6.600	6.625	6.725
DOC	mg C/L		2.370	2.590	2.200	2.860	2.435	2.070	2.000	1.930
Alka (pH 4.5)	mg/L $CaCO_3$		4.200	4.315	4.000	4.260	4.300	4.615	4.450	4.595
Alka (g,i,ex)	mg/L $CaCO_3$		2.790	2.850	2.880	3.300	2.920	2.980	3.015	2.924
Alka (gran)	mg/L $CaCO_3$		2.880	3.090	2.980	3.330	2.930	3.010	3.025	3.085
DIC	mg C/L		0.956	0.939	0.800	0.900	0.839	0.900	0.840	0.839
Nitrate + nitrite	mg N/L		0.120	0.124	0.123	0.131	0.120	0.120	0.126	0.127
Nitrate-IC	mg N/L		—	—	—	—	0.126	0.124	0.123	—
Ammonia	mg N/L		0.019	0.010	0.014	0.086	0.025	0.023	0.030	0.037
TKN	mg N/L		0.192	0.150	0.143	0.260	0.135	0.128	0.160	0.133
Sodium	mg/L		1.031	1.040	1.010	1.080	1.010	1.050	1.020	1.020
Magnesium	mg/L		0.750	0.760	0.740	0.790	0.740	0.758	0.749	0.750
Aluminum	mg/L		—	—	0.010	0.006	0.006	0.006	0.006	—
Reactive silica	mg Si/L		0.770	0.764	0.750	0.810	0.752	0.770	0.770	0.770
Sulfate, IC	mg/L		7.975	7.915	7.813	8.240	7.741	7.945	7.769	7.778
Sulfate, non-IC	mg/L		7.975	8.000	7.850	8.400	7.450	7.960	7.650	7.880
Chloride, IC	mg/L		1.260	1.250	1.250	1.321	1.214	1.270	1.239	1.230
Chloride, non-IC	mg/L		1.205	1.260	1.230	1.300	1.215	1.290	1.300	1.235
Potassium	mg/L		0.550	0.549	0.531	0.580	0.540	0.540	0.540	0.530
Calcium	mg/L		3.155	3.165	3.070	3.260	3.110	3.180	3.070	3.105

Note: Centrifuged and pasteurized; 48 months of monitoring.

Table 5 Comparison of Interlaboratory Median Values for AUD-06, over Several Studies (LRTAP Data), Eagle Lake

Parameter	Units of measurement	Study no.: Sample no.: Study date:	0020 10 Jan 89	0022 2 June 89	0053 2 May 91	0062 1 Jan 93	0036 10 May 94
Color	Hazen units		8.000	7.500	6.310	5.500	6.200
Specific conductance	µmho/cm		53.100	52.600	53.600	53.950	53.100
Acidity to pH 8.3	mg/L $CaCO_3$		2.230	1.590	—	—	1.410
pH	pH units		7.189	7.160	7.110	7.140	7.340
Dissolved organic carbon	mg C/L		3.400	3.480	3.000	2.700	2.800
Alkalinity - fixed end point 4.5	mg/L $CaCO_3$		13.800	14.190	—	—	14.550
Alkalinity - gran, inflec, extrap	mg/L $CaCO_3$		12.880	12.910	—	—	13.100
Alkalinity - gran titration	mg/L $CaCO_3$		12.880	12.865	—	—	13.390
Dissolved inorganic carbon	mg C/L		3.295	3.340	3.200	3.050	3.270
Nitrate + nitrite	mg N/L		0.083	0.086	0.082	0.087	0.087
Ammonia	mg N/L		0.022	0.012	0.019	0.040	0.050
Total Kjedahl nitrogen	mg N/L		0.232	0.200	0.184	0.220	0.189
Sodium	mg/L		0.990	0.997	1.000	1.010	1.000
Magnesium	mg/L		1.613	1.620	1.620	1.630	1.630
Reactive silica	mg Si/L		1.050	1.040	—	—	1.055
Sulfate, IC methods	mg/L		8.420	8.305	—	—	8.344
Sulfate, non-IC methods	mg/L		8.600	8.415	8.350	8.400	8.710
Chloride, IC methods	mg/L		0.940	0.942	—	—	0.960
Chloride, non-IC methods	mg/L		0.905	0.990	0.910	0.994	1.000
Potassium	mg/L		0.730	0.731	0.730	0.720	0.736
Calcium	mg/L		5.770	5.800	5.770	5.850	5.840

Note: Centrifuged and pasteurized; 52 months of monitoring.

Table 6 Comparison of Interlaboratory Median Values for EU-O8, over Several Studies (LRTAP Data), CCIW Rainwater

Parameter	Units of measurement	Study no.: Sample no.: Study date:	0006 8 Dec 88	0007 10 Feb 89	0010 7 June 89	0012 8 Sept 89	0016 8 Feb 90	020L 6 Jan 89
Specific conductance	µmho/cm		10.780	10.640	10.400	10.490	10.490	10.400
Gran acidity	mg/L $CaCO_3$		0.000	0.000	0.000	0.000	0.000	0.000
pH	pH units		5.515	5.495	5.540	5.548	5.540	5.530
Dissolved organic carbon	mg C/L		0.200	0.100	0.116	0.192	0.200	0.100
Alkalinity, gran titration	mg/L $CaCO_3$		0.000	0.000	–.100	0.100	0.200	0.000
Dissolved inorganic carbon	mg C/L		0.273	0.288	0.291	0.380	0.000	0.274
Nitrate + nitrite	mg N/L		0.243	0.252	0.261	0.247	0.240	0.245
Ammonia	mg N/L		0.236	0.230	0.237	0.240	0.233	0.235
Nitrate (IC method)	mg N/L		0.242	0.245	0.246	0.245	0.240	—
Sodium	mg/L		0.149	0.151	0.150	0.150	0.160	0.147
Magnesium	mg/L		0.087	0.089	0.089	0.090	0.089	0.090
Reactive silica	mg Si/L		0.034	0.036	0.037	0.033	0.040	0.040
Sulfate, IC methods	mg/L		1.918	1.896	1.950	1.920	1.941	1.943
Chloride, IC methods	mg/L		0.270	0.280	0.280	0.273	0.277	0.284
Potassium	mg/L		0.030	0.029	0.030	0.034	0.037	0.028
Calcium	mg/L		0.696	0.710	0.706	0.708	0.706	0.713

Note: Centrifuged and pasteurized; 14 months of stability.

Table 7 Comparison of Interlaboratory Median Values for GRM-01, over Several Studies (LRTAP Data), Grimsby Rainwater

Parameter	Units of measurement	Study no.: Sample no.: Study date:	0025 10 Sept 90	0028 6 Dec 91	0029 7 Feb 92	0032 2 Jan 93	0034 2 Sept 93	0037 6 Sept 94
Color	Hazen units		10.000	3.000	3.000	2.500	3.000	2.500
Specific conductance	µmho/cm		96.200	97.450	97.600	98.500	111.10	118.000
Acidity to pH 8.3	mg/L $CaCO_3$		4.100	2.500	2.550	2.980	5.700	6.510
pH	pH UNITS		6.060	6.485	6.442	6.480	4.370	4.250
Diss organic carbon	mg C/L		3.190	2.200	1.880	1.800	1.610	1.619
Alka. (edpt pH 4.5)	mg/L $CaCO_3$		2.800	3.600	3.800	3.850	0.000	—
Alka. gran, inf, extrap	mg/L $CaCO_3$		1.700	2.100	2.100	2.300	–2.190	–3.580
Alka, gran titration	mg/L $CaCO_3$		1.650	2.190	2.300	2.400	–2.460	–3.195
Diss inorg carbon	mg C/L		.625	.707	.707	.700	.256	.300
Nitrate + nitrite	mg N/L		3.355	3.400	3.400	3.410	4.149	4.220
Nitrate-IC	mg N/L		—	2.946	2.925	2.957	4.200	4.280
Ammonia	mg N/L		1.549	1.665	1.689	1.715	.996	.900
Total Kjeldahl nitrogen	mg N/L		1.970	1.880	1.885	1.900	1.190	1.006
Sodium	mg/L		.200	.200	.206	.204	.202	.200
Magnesium	mg/L		2.605	2.650	2.640	2.670	2.680	2.689
Reactive silica	mg Si/L		.100	.110	.100	.108	.110	.110
Sulfate,IC method	mg/L		22.494	22.80	22.740	22.900	23.30	22.897
Sulfate non-IC method	mg/L		22.250	22.65	22.500	22.600	22.30	22.050
Cholride IC	mg/L		.865	.845	.835	.835	.860	.920
Chloride non-IC	mg/L		.875	.890	.900	.872	.800	.929
Potassium	mg/L		.320	.322	.326	.326	.330	.322
Calcium	mg/L		8.665	8.650	8.678	8.685	8.700	8.800

Note: Centrifuged and pasteurized; 48 months of monitoring.

Table 8 Comparison of Interlaboratory Median Values for GRM-02, over Several Studies (LRTAP Data), Grimsby Rainwater

Parameter	Units of measurement	Study no.: Sample no.: Study date:	0026 10 Jan 91	0027 8 May 91	0029 9 Feb 92	0030 2 May 92	0031 8 Sept 92	0033 3 May 93	0034 3 Sept 93
Color	Hazen units		5.000	3.000	2.400	1.900	2.500	3.000	3.850
Specific conductance	μmho/CM		43.300	43.500	49.200	49.000	49.700	49.200	48.450
Acidity to pH 8.3	mg/L $CaCO_3$		3.780	2.800	3.900	3.837	3.700	3.270	3.590
pH	pH units		6.360	6.465	4.600	4.600	4.600	4.640	4.740
Dissolved organic carbon	mg C/L		1.695	1.520	1.000	1.005	1.000	1.180	0.930
Alka fixed endpt pH 4.5	mg/L $CaCO_3$		4.000	5.200	0.385	0.615	0.390	0.345	0.600
Alka-gran, inflex, extrap	mg/L $CaCO_3$		3.620	3.790	−1.270	−1.095	−1.200	−1.180	−0.880
Gran titra alka	mg/L $CACO_3$		3.720	3.830	−1.375	−1.370	−1.320	−1.250	−1.040
Dissolved inorg carbon	mg C/L		1.100	1.250	0.300	0.270	0.308	0.200	0.260
Nitrate + nitrite	mg N/L		1.610	1.620	2.380	2.393	2.420	2.455	2.444
Nitrate-IC	mg N/L		—	—	2.433	2.410	2.410	2.420	2.450
Ammonia	mg N/L		1.130	1.150	0.460	0.450	0.466	0.514	0.589
Total Kjeldahl nitrogen	mg N/L		1.300	1.250	0.540	0.525	0.505	0.585	0.686
Sodium	mg/L		0.204	0.200	0.207	0.210	0.205	0.200	0.200
Magnesium	mg/L		1.070	1.060	1.080	1.080	1.080	1.080	1.100
Aluminum	mg/L		0.016	0.014	0.028	0.028	0.030	0.027	0.029
Reactive silica	mg Si/L		0.080	0.070	0.070	0.072	0.080	0.080	0.080
Sulfate, IC methods	mg/L		6.660	6.659	6.808	6.820	6.810	6.895	6.926
Sulfate, non-IC	mg/L		6.750	6.750	6.700	6.685	6.785	6.600	6.700
Chloride IC	mg/L		0.526	0.514	0.530	0.522	0.519	0.545	0.547
Chloride non-IC	mg/L		0.600	0.565	0.520	0.540	0.500	0.500	0.435
Potassium	mg/L		0.190	0.190	0.190	0.190	0.196	0.198	0.200
Calcium	mg/L		3.255	3.287	3.300	3.285	3.330	3.295	3.345

Note: Centrifuged and pasteurized; 34 months of monitoring.

Table 9 Comparison of Interlaboratory Median Values for ION-96, over Several Studies (FP Data), Grand River

Parameter	Units of measurement	Study no.: Sample no.: Study date:	0051 4 Jan 91	0058 4 Mar 92	0062 10 Jan 93	0063 10 July 93
Color	Hazen units		18.00	15.00	19.00	20.00
Specific conductance	µmho/cm		811.00	817.90	805.00	818.00
pH	pH units		8.03	8.25	8.34	8.34
Boron	mg/L		0.04	0.05	0.042	0.05
Dissolved organic carbon	mg C/L		6.55	5.50	5.31	5.90
Total alkalinity	mg/L $CaCO_3$		201.00	201.00	203.00	201.00
Dissolved inorganic carbon	mg C/L		45.70	48.90	46.50	46.70
Nitrate + nitrite	mg N/L		4.73	4.67	4.76	4.64
Ammonia	mg N/L		—	0.007	0.013	0.006
Total N	mg N/L		5.14	4.85	5.44	4.85
Total Kjeldahl nitrogen	mg N/L		0.54	0.47	0.51	0.54
Fluoride	mg/L		0.18	0.167	0.17	0.16
Total hardness	mg/L		329.00	329.00	331.00	332.00
Sodium	mg/L		43.05	43.40	42.60	43.00
Magnesium	mg/L		22.60	22.50	22.60	22.80
Silicates	mg SiO_2/L		6.63	6.54	6.60	6.64
Sulfate, all methods	mg/L		101.50	102.00	101.00	103.08
Chloride, all methods	mg/L		71.50	72.00	71.85	72.40
Potassium	mg/L		3.90	3.90	3.97	3.91
Calcium	mg/L		94.00	93.90	94.65	95.00

Note: Centrifuged and filtered; 18 months of monitoring.

Table 10 Comparison of Interlaboratory Median Values for BMOOS-01, over Several Studies (LRTAP Data), Big Moose Lake

		Study no.:	0031	0033	0035	
	Units of	Sample no.:	1	2	1	Average
Parameter	measurement	Study date:	Sept 92	May 93	Jan 94	Median
Color	Hazen units		10.000	8.395	9.950	9.448
Specific conductance	µmho/cm		23.000	23.200	23.300	23.167
Acidity to pH 8.3	mg/L $CaCO_3$		3.890	3.490	3.480	3.620
pH	pH units		5.385	5.370	5.330	5.362
Dissolved organic carbon	mg C/L		3.100	3.800	4.005	3.635
Alkalinity (pH = 4.5)	mg/L $CaCO_3$		1.740	1.680	1.800	1.740
Alka.- gran, inflec, extrap	mg/L $CaCO_3$		0.206	0.255	0.100	0.187
Alkalinity, gran titration	mg/L $CaCO_3$		0.290	0.223	0.195	0.236
Dissolved inorganic carbon	mg C/L		0.300	0.215	0.335	0.283
Nitrate + nitrite	mg N/L		0.410	0.410	0.410	0.410
Nitrate-IC	mg N/L		0.417	0.410	0.411	0.412
Ammonia	mg N/L		0.029	0.027	0.024	0.027
Total Kjeldahl nitrogen	mg N/L		0.170	0.190	0.173	0.178
Sodium	mg/L		0.600	0.590	0.600	0.597
Magnesium	mg/L		0.310	0.310	0.310	0.310
Aluminum	mg/L		0.226	0.228	0.230	0.228
Reactive silica	mg Si/L		1.925	1.940	1.940	1.935
Sulfate, IC methods	mg/L		5.380	5.400	5.400	5.393
Sulfate, non-IC methods	mg/L		5.640	5.600	5.500	5.580
Chloride, IC methods	mg/L		0.420	0.430	0.440	0.430
Chloride, non-IC methods	mg/L		0.490	0.480	0.500	0.490
Potassium	mg/L		0.404	0.408	0.410	0.407
Calcium	mg/L		1.935	1.925	1.930	1.930

Note: Centrifuged and filtered; 16 months of monitoring.

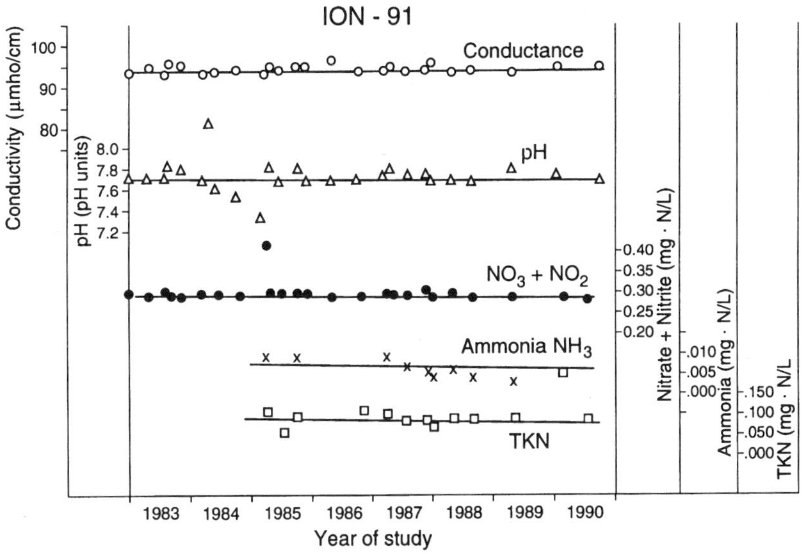

Figure 2 An illustration of the long-term stability of reference sample ION-91. Data points are the interlaboratory median values obtained from many different performance evaluation studies.

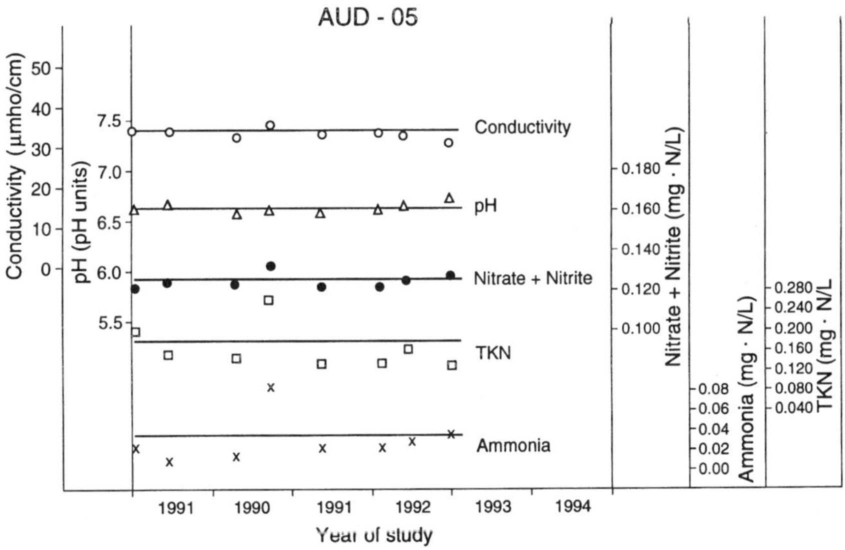

Figure 3 An illustration of the long-term stability of reference sample AUD-05. Data points are the interlaboratory median values obtained from many different performance evaluation studies.

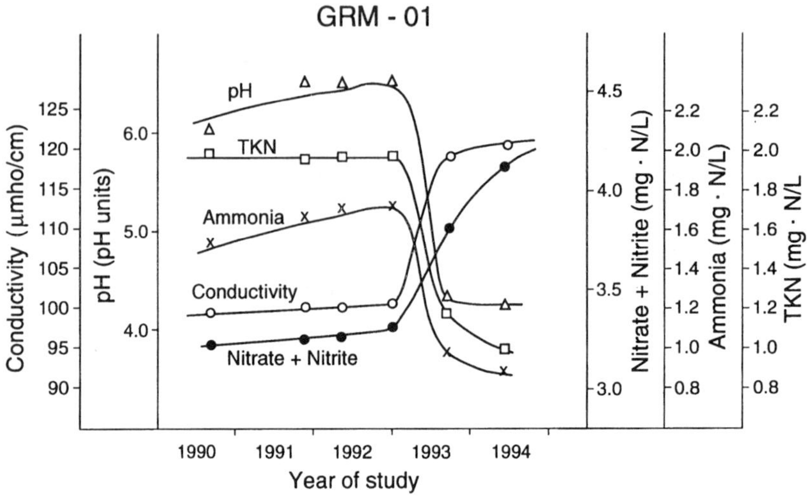

Figure 4 An illustration of the long-term stability of reference sample GRM-01. Data points are the interlaboratory median values obtained from many different performance evaluation studies.

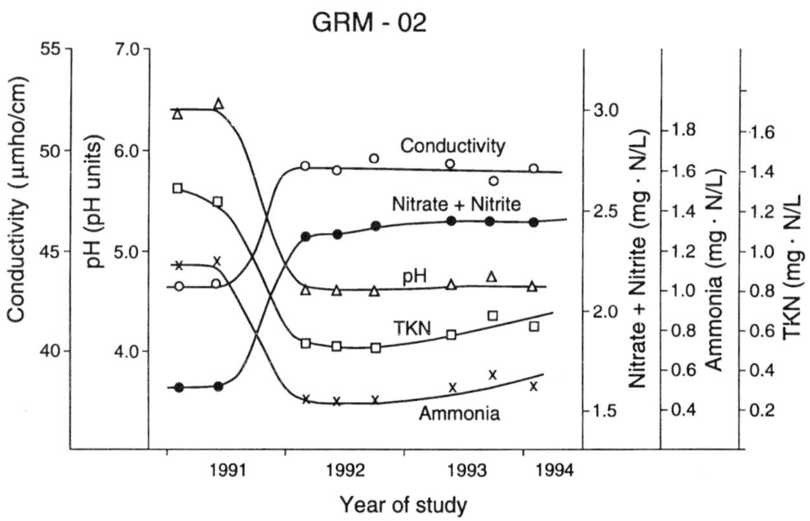

Figure 5 An illustration of the long-term stability of reference sample GRM-02. Data points are the interlaboratory median values obtained from many different performance evaluation studies.

Samples Treated by Centrifugation and Pasteurization

The samples listed in Tables 1 through 8 were both centrifuged and pasteurized. The first five tables illustrate the stability of several soft waters derived from acid rain-impacted lakes in the Dorset Ontario watershed. Tables 6, 7, and 8 reflect the stability of three rainwaters.

All data given in these tables are the interlaboratory median values obtained from many different round-robins. These medians are the middle results obtained from a population of between 40 and 70 laboratories. For some less frequently analyzed constituents, such as acidity, alkalinity, and aluminium, the population of data is only five to twenty. The date for each round-robin is identified in the tables.

The stability for the lake waters is high, which may be rather surprising. Nutrients, especially ammonia and nitrate, are often analyzed as immediates since the holding periods are normally rather brief (hours or weeks).[16,17] A graphical layout to illustrate the stability of the water from Halls Lake is given in Figure 3. The pH, conductivity, and nitrate are quite constant, but TKN, ammonia, and the DOC (not illustrated) show a very slow change over several years. The data for study 25 (Sept. 1990) are anomalous and are interpreted as secondary contamination when a 40-L unit was syphoned and subdivided into 80 aliquots.

An extended holding period for nutrients such as ammonia and nitrate should not be interpreted from data in Tables 1 through 10. These reference waters were highly processed and were never intended to reflect the chemical content of the original body of water from which they were collected.

Samples Treated by Centrifugation and Filtration

The results for two samples are provided to illustrate the stability of samples both centrifuged and filtered. These data are given in Tables 9 and 10. One sample (Table 9) is a hard river water with conductivity of 810 μmhos/cm. The other one is a soft water obtained from Big Moose Lake, New York. Both of these show stability over 1.5 years of monitoring. Several other water samples, data not provided here, have similar stability.

Stability of Rainwater

The stability of three rainwater samples that were centrifuged and pasteurized is described in Tables 6, 7, and 8. Sample EU-08 was a small volume rainwater (40 L) pasteurized and used in six special interlaboratory studies involving only ten laboratories. The data were obtained over a 14 month interval. This sample was very stable for all constituents (refer to Table 6). This sample differs from the Grimsby rainwaters in that it was contained within the carboy in which it was pasteurized. Subsamples for EU-08 were taken by spigot, and round-robin data were obtained within 1 to 2 weeks of study preparation.

The other two rainwater reference materials, GRM-01 and GRM-02, were very large samples (500 to 1000 L) collected off a greenhouse roof. Both of these samples underwent dramatic changes in the nitrogen components, which are illustrated in Figures 4 and 5. Ammonia is oxidized to nitrate, releasing the hydronium ion, with a consequent decrease in pH and an increase in conductivity. The energy for this bacterial process is assumed to be derived from the consumption of DOC, which declined accordingly.

It is very possible that the instability of nutrients for samples in Figures 4 and 5 reflects the same type of microbial activity that was mentioned for Figure 1. These rainwater samples may have been sterile initially but lost sterility and revealed small bacterial levels in all of the containment vessels in later tests. The contamination may have been introduced by syphons or by airborne transport.

Our experience with rainwater samples causes us to raise a note of caution. If all our rainwater were influenced by the same bacterial processes as seen in GRM-01 and GRM-02, then the same caution is raised for those atmospheric networks that allow insects to reside in natural samples for extended periods of time. Most of those samples could be expected to be unstable, particularly in terms of conductivity, pH, ammonia, and nitrate. A subtle but distinct instability of nitrogen constituents was noted by NIST in their development of their simulated rainwater, SRM 2694.[18]

CERTIFIED REFERENCE WATERS AVAILABLE FROM THE NATIONAL WATER RESEARCH INSTITUTE

The use of CRMs plays a large role in establishing the accuracy and traceability of analytical measurements. In the area of natural waters, the Institute has played a unique role in providing reliable measurement standards. Waters, like fingerprints, have differing characteristics; hence, as listed in Table 11, the Institute has developed a variety of certified waters.

As mentioned before, waters intended for CRMs are carefully processed and stored. The prospective samples are then analyzed and monitored. A sizable part of the bulk samples is sent out in interlaboratory studies to "characterize" them. It is our experience that with clean samples the mineral constituents show excellent stability. This indicates that the waters are processed and handled in a consistent and proper manner and that the analyses by the many different laboratories, with a variety of different methodologies, provide reliable results. Realizing that the analytical results are reliable, the results for the nutrient constituents can be carefully monitored. Usually, lake and pristine waters are the most dependable.

To establish the certified values, the analytical data for each constituent is traced through each study. The mean values are compared to the medians as a check that no outliers are unduly affecting the means. Outliers, as determined by the flagging procedures used in the round-robin programs are excluded.[1,19] Each certified value is authenticated by a 95% confidence interval (2 standard deviations).

The nutrient concentrations are checked from time to time. We have two ways to check these values. One method is a reanalysis by a reference laboratory and the other is a reanalysis by reinserting the sample into another interlaboratory study.

Table 11 List of Available NWRI Certified Reference Waters

Sample Name	Source	Parameters	Conductance (μS/cm) or conc. range (mg/L)
ION-911	Lake Superior	Major ions & nutrients	96
ION-92	Distilled water	Major ions & nutrients[a]	444
ION-94	Soft water	SO_4, major ions & nutrients	37
ION-95	Lake Ontario	Major ions & nutrients	291
ION-96	Grand River	Major ions & nutrients	807
ION-20	Lake Ontario	Major ions & nutrients[a]	172
AUD-06	Eagle Lake	Major ions & nutrients	53
GRM-02	Grimsby rain	Major ions & nutrients	49
SOUR-01	Souris River	Major ions & nutrients	485
Huron-03	Lake Huron	Major ions & nutrients	210
TM-23	Lake Ontario	Trace metals[a]	0.002–0.008
TM-24	Lake Ontario	Trace metals[a]	0.003–0.015
TM-26	Lake Ontario	Trace metals[a]	0.010–0.025
TM-27	Lake Ontario	Trace metals[a]	0.003–0.004
TM-28	Lake Ontario	Trace metals[a]	0.001–0.003
TMDA-51	Lake Ontario	Trace metals[a]	0.03–0.1
TMDA-52	Lake Ontario	Trace metals[a]	0.1–0.3
TMDA-52	Lake Ontario	Trace metals[a]	0.2–0.5

[a] A fortified sample (constituents are listed below)

Major Ions & Nutrients

Color	Conductance	Turbidity	Nitrate (+ nitrite)[b]
Fluoride	Total alkalinity	pH	Total hardness
Sodium	Magnesium	Silicates	Sulfate
Chloride	Potassium	Calcium	Boron
DOC	DIC	TKN	Ammonia
Total nitrogen			

[b] The samples are stabilized; there should be no nitrite present.

Trace Metals

Aluminium	Vanadium	Chromium	Manganese
Iron	Cobalt	Nickel	Copper
Zinc	Strontium	Molybdenum	Cadmium
Barium	Lead	Arsenic	Selenium
Antimony	Silver	Lithium	Beryllium
Uranium			

Note: Not all constituents are characterized for each reference sample.

The latter is preferred because the verification is made by many different laboratories. If warranted, the certificate for a CRM is updated to include the additional new data from the interlaboratory study, which also provides data to help verify sample stability. In every case to date, every CRM, except for one proposed CRM, has proven to be stable. Client feedback for sample concerns is always welcomed and an important part of sample evaluation.

SUMMARY

Overall, the results clearly indicate that for the major ions and the physical parameters, the processing (centrifugation, pasteurization, or filtration) of natural waters led to a high level of stability. For some water samples, such as those derived from Lake Superior, nothing needs to be done to yield a stable reference water. Lake Superior is unique, with very low ammonia levels and a very low bacteria count.

Centrifugation as a means of clarifying is an efficient process to remove particulates and to reduce the bacterial content when particulates are the host for the living organisms.

Pasteurization (heat treatment at 90°C) for soft lake and river waters is effective and produces samples with considerable stability for the three major nutrients (nitrate, TKN, and ammonia). For a few types of water, there was a slight increase in ammonia over several years.

The rainwater samples presented a unique situation, even after pasteurization. A complete collapse of the equilibrium between ammonia, nitrate, and TKN was observed. This collapse, as identified by decreased pH, ammonia, and TKN, along with the parallel increase in nitrate and conductivity, is viewed as possibly microbial in origin. This process terminates at the lower pH (approximately 4.8) and a new long-term stability appears to have been achieved.

Filtration of large water samples through 0.2-μm membranes is a cost effective means of creating sterile reference waters. Those water samples that have been processed have stability confirmed. It remains to be seen just how long these waters will be stable. Current data indicates some samples have been stable for almost 3 years. This is encouraging, but continued vigilance appears prudent.

ACKNOWLEDGMENTS

A sincere acknowledgment is extended to the Project Chief, Dr. Klaus L.E. Kaiser, for his constructive review and input to this report, to Barney Dutka and Dr. Salem Rao for their timely advice on the microbiology of water, and to the staff of the NWRI Research Support Branch, Technical Operations Section, in acquiring the many reference waters from remote and often difficult locations.

REFERENCES

1. **K. I. Aspila,** A manual for effective interlaboratory quality assurance, In: *Quality Assurance in Water Quality Monitoring*, J. E. Gaskin, Ed., Canada Communication Group — Publishing, Ottawa, Canada, 1993.
2. **K. I. Aspila, R. E. White, and J. Clark,** Quality assurance aspects of the International Joint Commission Great Lakes Monitoring Program, In: Quality Assurance for Environmental Measurements, J. Taylor and T. W. Stanley, Eds., ASTM Special Technical Publication 867, 1985.

3. **W. J. Youden and E. H. Steiner,** Statistical Manual of the Association of Official Analytical Chemists, AOAC, Washington, D.C., 1975.
4. Forty-Second Symposium of the Society for General Microbiology, University of Southhampton, In: *The Nitrogen and Sulphur Cycles*, J. A. Cole and S. J. Ferguson, Eds., Cambridge University Press, 1988.
5. **P. L. Brezonik,** Nitrogen, sources and transformation in natural waters, In: *Nutrients in Natural Water*, H. E. Allan and J. Kramer, Eds., Wiley-Interscience, Toronto, 1972.
6. **D. Burrus, R. L. Thomas, J. Dominik, and J.-P. Vernet,** Recovery and concentration of suspended solids in the Upper Rhone River by continuous flow centrifugation, In: *Hydrological Processes*, John Wiley & Sons, New York, Vol. 3, 1989, 65–74.
7. **K. Kuntz,** In: Niagara River Sampling Protocol for Upstream/Downstream Niagara River Monitoring Data, A joint publication by Environment Canada, U.S. Environmental Protection Agency, Ontario Ministry of Environment and Energy and the New York State Department of Environmental Conservation, August 1994. Available from the author, Ecosystem Conservation Branch, P.O. Box 5050, 867 Lakeshore Road, Burlington, Ontario, L7R 4A6.
8. **C. F. Frith,** Sterile Water for the Use in Microelectronics Production, Circuits Manual, 9(9), 1969.
9. **B. J. Dutka,** *Membrane Filtration. Applications, Techniques, and Problems*, Marcel Dekker, New York, 1981, 612 pp.
10. **C. F. Frith,** Membrane filters in industry, In: *Membrane Filtration. Applications, Techniques, and Problems*, B. J. Dutka, Ed., Marcel Dekker, New York, 1981.
11. Millipore Corp., Filtration of Serum Products, AB719, Bedford, MA, 1975.
12. **B. J. Dutka,** Methods for Microbiological and Toxicological Analysis of Waters, Wastewaters and Sediments, Rivers Research Branch, National Water Research Institute, Burlington, Ontario, Canada, 1989.
13. **S. S. Rao,** Observations on Bacteriological Conditions in the Upper Great Lakes 1964-1974, Inland Waters Directorate, CCIW Branch, Burlington, Ontario, Scientific Series No. 64, 1976.
14. The Great Lakes, an Environmental Atlas and Resource Book, 1988, Available from Environment Canada, 25 St. Clair Avenue East, Toronto, Ontario, Canada, M4T 1M2. Catalog No. EN40-349/1987E.
15. **S. S. Rao and A. A. Jurkovic,** Differentiation of the trophic status of the Great Lakes by means of bacterial index ratio, *J. Great Lakes Res.*, 3(3-4) (1977) 323–326.
16. Handbook for Sampling and Sample Preservation of Water and Wastewater: U.S. Environmental Protection Agency, U.S. Government Printing Office; Washington DC, 1982, EPA-600/4-82-029.
17. **L. H. Keith,** *Principles of Environmental Sampling*, ACS Professional Reference Book, 1988.
18. Methods and Procedures Used at the National Bureau of Standards to Prepare, Analyze and Certify SRM 2694, Simulated Rainwater, and Recommendations for Use, NBS Special Publication 260-106, W.F. Koch, Ed., July 1986.
19. **H. Alkema,** Interlaboratory Study for Major Ions and Nutreints, FP/GLAP 65, NWRI Report No.AEPB-TN-94-06, National Water Research Institute, 1994.

CHAPTER 10

The Preparation and Validation of Generator Columns for Highly Hydrophobic Materials

Glenys Foster Roberts and Barry G. Oliver

INTRODUCTION

This chapter is intended to provide information to assist the reader to develop generator columns within their own laboratory for organic analytes of interest. The intent is to indicate, where relevant, the reasons for the technical approach recommended. This information, if applied widely, can assist in improving the quality of data generated by environmental laboratories.

Standard reference materials are an integral part of the generation of high quality data. They are essential to the establishment of accuracy of the standards used to quantify target analytes and can also be used as an effective tool in the assessment of the success of the total analytical train for a specific matrix. Many standard reference materials are available in the field of inorganic environmental analysis, and target analytes are available in a wide variety of matrices to assess methodology and to provide the laboratory with confidence in their data. This has come about as a result of the widespread presence of inorganic contaminants in a wide variety of matrices either as a result of natural abundance of the elements or as a result of industrial contamination. These samples are also, in general, very stable and unlikely to encounter losses from volatilization.

The field of trace organics analysis in environmental samples is lagging seriously behind that of their inorganic counterparts in the availability of organic standard reference materials. There is a crucial need to develop new reference materials for use in the assessment of the accuracy of standards being used for quantification and in the assessment of method performance overall for organic analytes. The lack of such materials has arisen as a result of the routine absence of many environmental contaminants in most matrices, the low solubility of many organic compounds in water, their perishability as a result of biological or other degradation, and the relative volatility of many organic compounds of interest. Preparation of reference materials by spiking complex matrices with target organic analyte poses a major challenge to the chemist as a result of the difficulty of preparing a large, homogeneous, and stable sample that is rugged enough to be used over a long period of time. This is especially difficult in an aqueous matrix because of the low concentrations of organics in water and the consequently large volumes of sample that one needs to use in order to detect the contaminants of interest.

Standards for many of the contaminants of interest are available from a variety of reputable suppliers. These standards, from secondary sources, can be used to check the accuracy of the standards used for quantification. In order to assess the overall performance of a method, including the sample preservation, extraction, concentration, cleanup, and instrumental analysis, spiking studies need to be performed. It is, however, difficult to obtain a true value for a spiked sample. Sources of error can include the errors in the preparation of the concentrated stock standard in solution, in the dilution of the stock to prepare intermediate or working standards, and in the measurement of the standard to prepare the spike. These can all be compounded with the normal sources of error inherent in the analysis of trace level organics in environmental samples, which can include the incomplete extraction of the native material. An additional weakness in this approach is its inability to provide information regarding the stability of the analyte in the aqueous sample during the collection stage, its transportation, and storage.

GENERATOR COLUMNS

Development

Generator columns have been used to produce saturated aqueous solutions of hydrophobic organics. As a result of early work by May and several co-workers,[1,2] these generator columns have been used to develop aqueous solutions which could be used as reference materials. Pioneering work in the area was done at the National Bureau of Standards (NBS) (currently known as National Institute of Standards and Technology (NIST)). May, Wasik, and Freeman used generator columns of glass beads coated with various polycyclic aromatic hydrocarbons (PAHs) to measure the solubility of the PAHs at several temperatures. PAHs were dissolved in dichloromethane (DCM), the solution placed in a flask with the glass beads, and the solvent removed by rotary evaporation to provide a loading of the glass beads in the 0.5 to 1% range. The coated beads were then packed into stainless steel columns about 60 cm in length by 0.6 cm in diameter, and water was pumped through the tubing at flow rates between 0.1 and 5 mL/min. The concentration of the PAHs in these saturated aqueous solutions, in these early experiments, was measured directly using high performance liquid chromatography (HPLC) with an ultraviolet detector. More recently, the generator column technique has been used to estimate the aqueous solubility of polychlorinated biphenyls(PCBs)[3–6] and of chlorinated dioxins.[7–9] The method has also been used for the determination of octanol-water partition coefficients.[10,11]

The only recent modification in the technique has been the substitution of Chromosorb W for the glass beads in the solid phase used in the columns. This change was made because Chromosorb W has a higher surface area, approximately 380 m^2/g as compared to 26 m^2/g for similar mesh glass beads. The higher surface area provides better contact between the water and the solute, so equilibrium is more quickly established. This approach has been used in the determination of physical constants such as the measurement of the aqueous solubility, octanol–water partition coefficient, and vapor pressure of six pesticides — alaclor, atrazine, bromocil, dimethoate, diuron and chlorpyrofos.[12]

Applications

The value of the generator column approach was realized by these early workers in that the concentration of the eluent solution from a generator column operated under standardized conditions was constant and could be relied upon for a defined volume of eluent. This fact was used in the establishment of physical constants, including the solubilities and octanol-water partition coefficients of many hydrophobic compounds, and it is still used as a standard approach to this exercise.[13] To an analytical chemist looking for a method to establish accurate and precise analytical measurement systems, the utilization of generator columns can be broadened. Their value can also be seen in their application in support of quality assurance programs for the sampling and analysis of trace level organics in drinking, ground, surface water or effluent monitoring programs. The solutions developed from generator columns can be applied to testing of an entire sample collection, transportation, storage, and analytical scheme and also as a tool for performance evaluation in interlaboratory studies. The means of accomplishing this would be by treating the solutions generated from the columns as a true sample in the analytical train and subjecting the aqueous solution to all of the aspects of the analytical protocol, including, if required, preservation and transportation. The results obtained by this approach can be compared to the results obtained from the approach used to determine the accurate concentration of organic component in the solution and provide information regarding the accuracy of the analytical process. This can be supplemented by the repetition of the analysis in order to obtain precision data for the method and to obtain estimates of method detection limits. Generator column samples can also be submitted to a laboratory as unknown samples as part of an inter- or intralaboratory study in order to assess either the performance of a particular methodology or of a particular laboratory.

Column Preparation — General

In order to prepare a generator column for application in the development of standard solutions, it is necessary to coat the surface of a solid support uniformly with the compound(s) of interest. The intent is to develop a solid support that is coated with

1% or less by weight of the chemical of interest. Each generator column developed as described here will contain 1 g of coated support. The solid support of choice is Chromosorb W (60/80 mesh) because of its inertness combined with its large surface area. The large surface area enables a rapid development of the equilibrium between the chemical on the surface and the eluent, an improved loading of the chemical of interest on the surface, and an extension to the life span of the final column.

Chromosorb Preparation

The Chromosorb W must first be cleaned with solvent to remove impurities and any fines that are present. One successful means of accomplishing this follows:

- Place several grams of Chromosorb W into a beaker;
- Add at least 100 mL of methylene chloride, swirl, let stand until the Chromosorb has settled; decant the solvent;
- Add at least 100 mL of hexane, swirl, let stand until the Chromosorb has settled; decant the solvent;
- Air dry the sorbent, and if necessary, place the Chromosorb into a glass column with glass wool plugs at either end and pass ultrapure nitrogen through the column.

Standard Solution Preparation

The data presented here were developed using high-purity crystals obtained from U.S. EPA at Research Triangle Park, North Carolina. (Since U.S. EPA no longer provides an extensive service, alternative suppliers now need to be used. Several reputable suppliers currently provide standards which are traceable and of high purity.) Solutions of compounds of interest are prepared as follows:

- Weigh, accurately, at least 20 mg of crystals using a Mettler Model HK60 balance or equivalent that can be read to ±0.01 mg;
- Dissolve the crystals in hexane, and bring the solution to an accurate volume in hexane using a 100mL volumetric flask.

Coated Chromosorb Preparation

To prepare Chromosorb W coated with 1% of the compound of interest:

- Weigh out accurately 2 g Chromosorb W, and place in a round bottom flask (250 mL);
- Add exactly 100 mL of hexane containing 20.0 mg of the study compound, and allow to equilibrate at room temperature for 1 h;
- Remove the hexane completely, by rotary evaporation using a water aspirator to develop the vacuum and holding the water bath at a temperature of 35 to 40°C.

Column Preparation

Each column prepared as follows contains about 1 g Chromosorb W and 10 mg of the chemical of interest (1% loading level):

- Pack the coated Chromosorb W into a stainless steel column of the dimensions, 0.64 cm outside diameter by 25 cm length, and use a vibrator to assure tight packing;
- Seal both ends using stainless steel HPLC filter frits and Swagelok fittings.

A support with a 0.5 or 0.05% loading level can similarly be developed by adjusting the concentration of the hexane solution. Mixed columns can be prepared by weighing the individual components and producing a single solution of the organics in hexane at a known concentration and coating the Chromosorb W using this solution.

Aqueous Solution Development

The production of a consistent concentration of the target compound in an aqueous solution is dependent on the temperature of the column and on the flow rate of the water through the column. Both of these parameters need to be closely controlled. An example of the effect of temperature on the concentration of a typical hydrophobic analyte is presented in Figure 1. The optimal means of ensuring the necessary control is to maintain the column in a temperature-controlled environment and to flow water through the column with a pump designed to ensure constant flow conditions. These criteria can be met using an HPLC system, with a pump that enables flow to be accurately controlled and with temperature control of the column enclosure. An alternate means of temperature control is to immerse the column in a thermostatically controlled water bath and to maintain the system away from draughts. High-purity laboratory grade water (e.g., filtered/carbon absorption/reverse osmosis/ion exchange (RODI)) is used as the source of water to produce the standard solutions.

Various flow rates and temperatures were used for a column containing hexachlorobenzene (HCB) to study the effect of these variables on the column eluent. Subsequently, the following optimum temperature and flow rate program was chosen for testing of additional columns

- Attach the packed column to an HPLC pump with a delivery capacity of up to 15 mL/min;
- Maintain the column in an environment with no draughts and at a temperature of 20°C ± 0.1°C;
- Pass high-purity lab water (RODI) through the column at a flow rate of 10 mL/min;
- Collect the solutions from the column using volumetric flasks (50 or 500 mL).

Analytical Procedures

The success of the development of these generator columns has been demonstrated with a number of compounds of different water solubilities. The analytical system to determine these concentration levels was designed carefully with the intent that the approach used must minimize any losses of the component of interest and must also minimize variability and enhance accuracy of the determination. The eluent

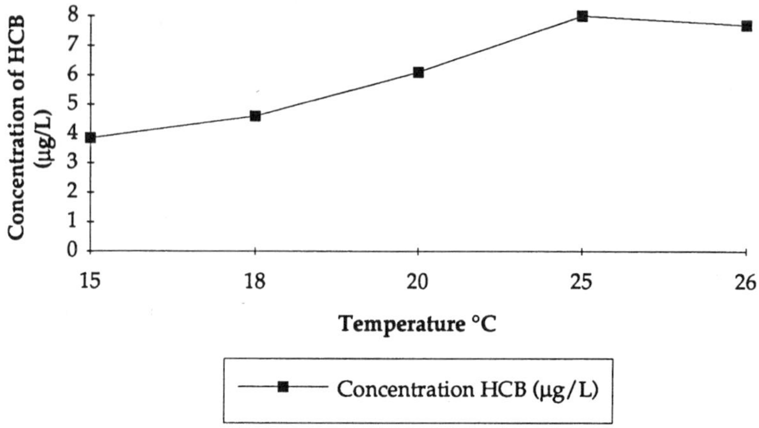

Figure 1 Effect of temperature on HCB concentration at a flow rate of 10 mL/min.

volume used in the extraction was selected based on literature values of solubilities for the compounds studied. The volume of solvent used and the final volume of extract were selected, based on minimizing evaporative and processing losses. Since the chemicals of interest are highly hydrophobic, this approach, in combination with an extremely sensitive and selective separation and detection system, enables highly accurate and reproducible results to be obtained.

The concentration of the organic components in the eluent from the generator column are established by using a liquid/liquid extraction technique followed by gas chromatography/electron capture detector (GC/ECD) analysis. The details for hexachlorobenzene, aldrin, and endosulfan I are as follows:

- Place the aliquots of 50 mL of aqueous eluent from the generator columns into a separatory funnel, and extract with a single portion of 25 mL hexane;
- Dry the hexane layer through a column of anhydrous sodium sulfate, and place in autosampler vials in preparation for instrumental analysis.

The very low water/solvent ratio assures that greater than 99% of the chemicals are recovered in the extraction process. In addition, minimal analyte losses are expected, since no cleanup or concentration of the extract needs to be performed.

In the case of mirex, the water solubility is much lower, and additional preparation steps need to be incorporated into the process in order to enable adequate detection of the levels present at equilibrium in the eluent:

- Extract aliquots of 500 mL of generator column eluent solutions with a single portion of 25 mL of hexane;
- Dry the hexane extract through a column of anhydrous sodium sulfate;
- Concentrate extract to 5 mL by rotary evaporation, prior to analysis.

The very high octanol/water partition coefficient and low vapor pressure of mirex should result in minimal analyte losses.

In the development of the data presented here, the analyses were performed using a Hewlett Packard Model 5880A gas chromatograph equipped with an electron capture detector. Because only single components or well-separated multicomponent systems were studied, a packed column (3.2 mm OD by 2 m long OV 101/210) was used for the analysis. The GC temperature program was 185°C (0.5 min hold), 6.5°C/min to 250°C (final hold 10 min). The injector temperature was 225°C, and the ECD was at 325°C. Later studies were carried out using a 30 m × 0.25 mm id capillary column coated with DB5 (0.25 μm film thickness) and a splitless injection system.

Some variability is seen within the data sets generated by the techniques described here. In order to assess the source of the variability, the standard deviation was calculated for repeat analyses of the same extract at different times as well as for the analysis of different aliquots of the eluent solution, thereby assessing the standard deviation of the total analytical procedure. The following table indicates that the greatest source of variation is derived from the sample development and preparation steps.

Analyte	Total %RSD	Instrumental %RSD
HCB	4.53	1.26
Aldrin	5.87	3.50
Endosulfan 1	2.76	0.47
Mirex (packed column)	9.45	5.86
Mirex (capillary column)	7.2	2.10

The effect of the additional sample preparation and analysis steps is clearly shown in the case of mirex. It has been possible to improve the precision of the measurement by converting the analysis from a packed column to a capillary column, but other sources of variation are still inherent in the method.

APPLICATIONS

A number of organochlorine compounds have been studied in detail. These include hexachlorobenzene (HCB), aldrin, endosulfan I, and mirex. The technique has also been expanded to include a series of three PCB isomers, 2,2′,5,5′-tetrachlorobiphenyl, 2,2′,4,5,5′-pentachlorobiphenyl, 2,2′,4,4′,5,5′-hexachlorobiphenyl. For these compounds, the solubilities established using the generator column technique were consistent with published literature values. These three compounds have solubilities in the low μg/L range and were selected as prominent isomers of the PCB family and representative of a group of compounds of intense environmental concern. In addition, initial experiments with lindane indicated that the generator column approach would be unsuccessful in developing stable reference solutions since the columns were quickly exhausted because of the relatively high water solubility of this compound(mg/L range). This technique, therefore, has good applicability to organic compounds with solubilities in the low μg/L and lower range, while it does not appear to be applicable to chemicals that have water solubilities in the mg/L range.

Elution Conditions

The conditions of the elution of analyte from the column can make a significant difference to the concentration level of the analyte developed in the eluent. The most significant impact is shown by changing the temperature of the column and the eluent; the results are shown for HCB in Figure 1. The solubility of HCB increases by about 0.38 μg/L per degree Celcius in the temperature range between 15 and 25°C. This is about 6.8% increase per degree Celcius. Thus it is necessary to control the temperature to ±0.3°C to ensure reproducible aqueous concentrations. The temperature dependence of solubility fits the equation:[11]

$$\log X = -\Delta Hs/RT + \text{constant}$$

where X is the mole fraction solubility, ΔHs is the enthalpy of solution, R is the ideal gas constant, and T is the temperature in °K. The data in Figure 1 yield a ΔHs of 4.8 kcal/mole for HCB. This is close to the value reported by Doucette and Andren[11] for 2,2′,4,4′,5,5′-hexachlorobiphenyl.

The effect of flow rate on the concentration level produced is minimal, as can be seen in the example of HCB in Figure 2. Lower flow rates may actually produce slightly lower HCB values, possibly due to volatilization. The lower the flow rate, the longer the time required to collect the 50 mL sample in the open volumetric flask. This flow rate effect seems to be small at 20°C and lower temperatures, but seems to be greater at 25°C. This would tend to confirm the volatilization hypothesis.

Columns for three additional pesticides were studied in detail, and the following Figures 3, 4, and 5 show the concentration levels developed in the aqueous eluent from the generator columns. It also demonstrates the consistency of the concentration level produced with variable total eluent flow volumes. From these data it is possible to assess the solubilities of the components of interest. They were determined as follows at 20°C:

Component	Solubility
Hexachlorobenzene	6.1 ± 0.27 μg/L
Aldrin	23.5 ± 2.8 μg/L
Endosulfan I	470 ± 13 μg/L
Mirex	0.048 ± 0.004 μg/L

Figure 6 shows the results generated from columns containing both aldrin and HCB. The concentrations produced are in good agreement with the concentrations produced from the individually spiked columns and are reproducible. This result is expected since, theoretically, the solubility of one component should not affect that of the other component for such sparingly soluble solutes.

Figures 7 and 8 show the results obtained for HCB and HCB plus aldrin at much lower column loading — 0.05% compared to 1%. The data agree quite well with results for 1% columns, indicating that it is possible to achieve saturated solutions for these hydrophobic organic compounds using lower column loadings. This result will be important in the application of these techniques to highly toxic

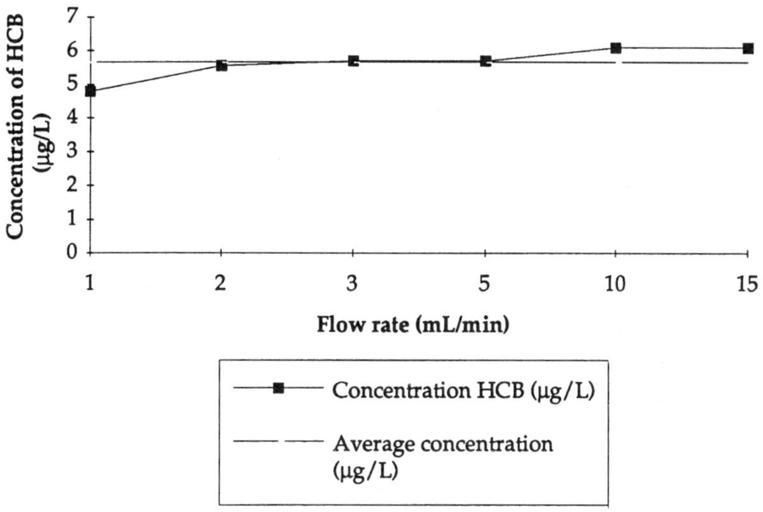

Figure 2 Effect of flow rate on concentration level of HCB from a 1% loading column at 20°C.

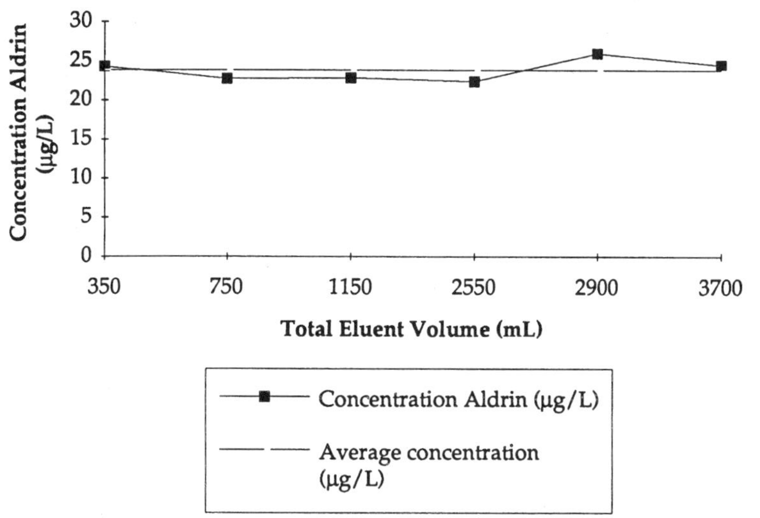

Figure 3 Concentration levels developed from 1% aldrin column (µg/L).

compounds, such as polychlorinated dibenzo-*p*-dioxins. It is likely that saturated solutions can be prepared using far less material, thus minimizing costs and risks to laboratory scientists.

Column Stability

The stability of organic standard solutions has been defined in many analytical protocols, especially those deriving from U.S. EPA. For organochlorine pesticides,

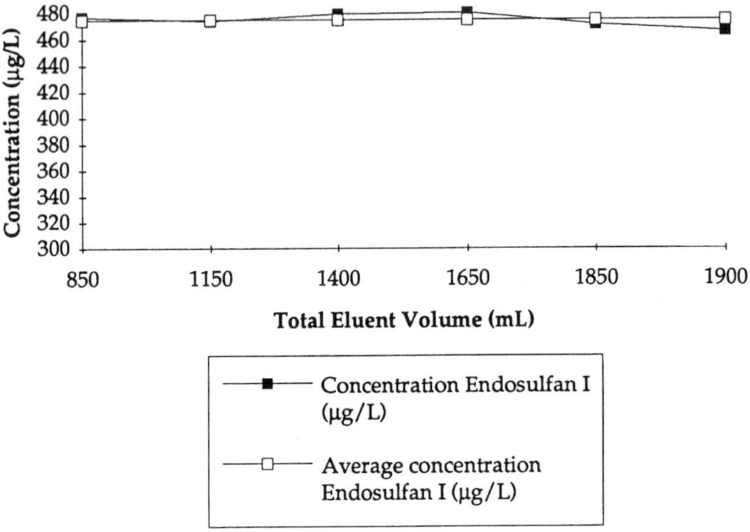

Figure 4 Concentration levels developed from a 1% endosulfan I column (µg/L).

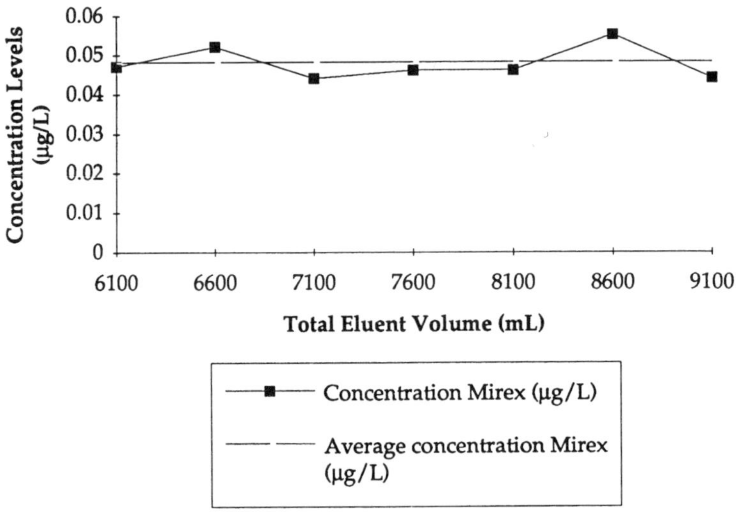

Figure 5 Concentration levels developed from a 1% mirex column (µg/L).

stock standard solutions have expiration times defined to be as short as 2 months (EPA method 508[14]) to as long as one year (EPA method 8080A[15]). Other semivolatile organics have similar expiration times defined within regulated analytical protocols. Standards deteriorate as a result of the evaporative losses, which would affect accuracy of a standard solution. They may also be subject to chemical, physical, or photochemical degradation during storage. Generator columns have a distinct advantage in that the chemical is bound to the surface of the solid support, and it is not

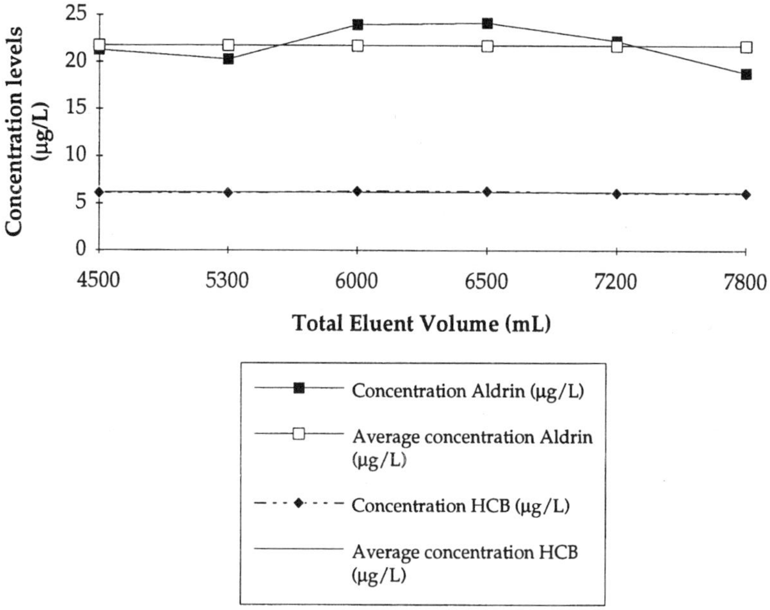

Figure 6 Concentration levels developed from a mixed 1% loading column with aldrin and HCB (µg/L).

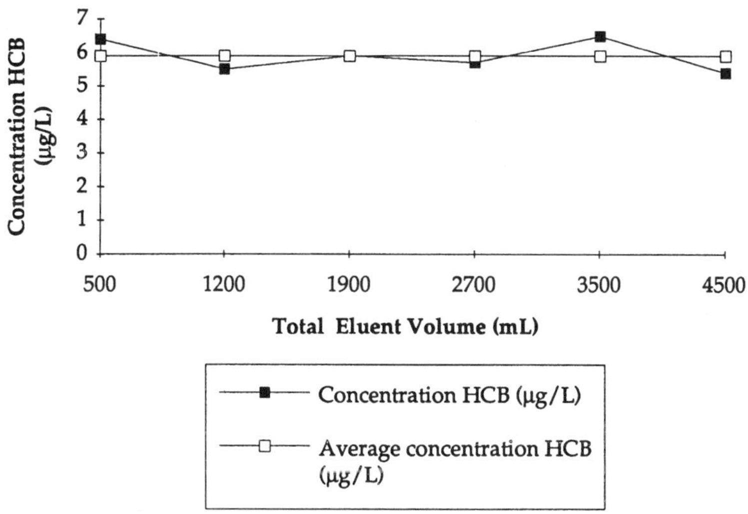

Figure 7 Concentration of HCB developed by a loading of 0.05% (µg/L).

subject to evaporative losses. The chemical is also maintained in a closed system in the absence of light and as such will be anticipated to remain as a stable entity for a period of many months, if not years.

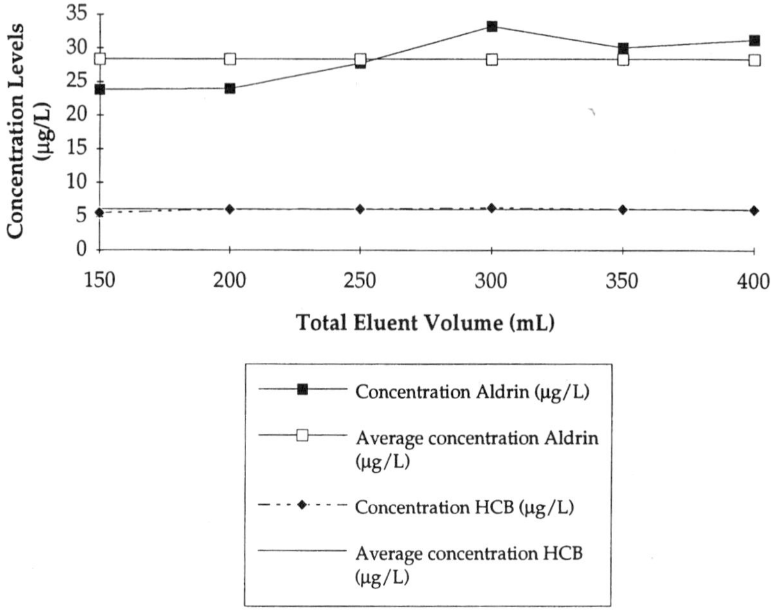

Figure 8 Concentration levels developed from a mixed 0.05% loading column with aldrin and HCB (µg/L).

Column Life

The life span of a generator column will depend on the loading level of the column and also the aqueous solubility of the analyte(s) coated on the surface of the Chromosorb. As mentioned previously, the column coated with lindane presented with a very short column life, while the HCB column, after passing 16 L of water through, had only removed approximately 0.1 mg of the 10 mg of HCB on the column. Theoretically the elution volume of the column would be over 1000 L at a flow rate of 10 mL/min. Endosulfan I has a much higher aqueous solubility than HCB, and Figure 9 shows the running of this column to near exhaustion. The concentration of endosulfan I begins to fall below saturation (470 ± 12) after about 13 L of water passed through the column. At this point about 62% of the endosulfan I on the column has been removed. The rest of the endosulfan must be associated more strongly with the Chromosorb and must dissolve more slowly. Thus, for conservatively estimating the useful life of such generator columns, it is probably safe to assume that they will produce consistent saturated solutions until about 50% of the material in the column is dissolved. The following have been estimated as conservative life spans of some columns developed for analytical purposes:

HCB	Safe water volume = 800 L
HCB/Aldrin	Safe water volume = 200 L
Endosulfan I	Safe water volume = 10 L

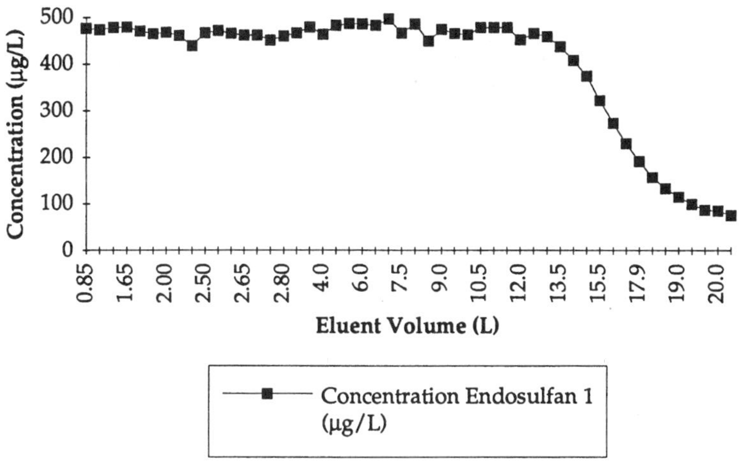

Figure 9 Variation in concentration of endosulfan I with eluent volume (µg/L).

Solution Storage

Since the solutions produced are saturated at 20°C, the solutions should be diluted prior to using for any applications testing or prior to storage. A common storage procedure for samples is to refrigerate at 4°C. This temperature drop would reduce the solubility of HCB by a factor of 3.1. A larger temperature solubility dependence may occur with other compounds. Therefore, if storage of the solutions is required, we suggest that the solutions be diluted by a factor of 10 in order to prevent the chemicals from precipitating. Such dilutions may be performed with natural waters to better simulate actual field conditions if the solutions will be used to monitor sample stability and preservation in the field or during storage.

CONCLUSIONS

Generator columns have shown themselves to be useful adducts to the arsenal of techniques environmental organic analytical chemists have at their disposal to assist in the provision of high quality data. They can be used to generate aqueous solutions of a consistent, known concentration for an extended period of time. Their stability is anticipated to be greater than an equivalent stock solution developed in an organic solvent within the laboratory setting, since the solution is developed *in situ*. The aqueous solutions generated can be used within a laboratory to assess the accuracy and precision of a particular method. They can also be used in interlaboratory studies to either assess individual laboratory performance or to assess the performance of a particular method or modification to a method or in method validation studies. An additional benefit of these solutions is that they can be used to duplicate the conditions experienced by field samples during preservation, storage, transportation, and handling prior to the analysis being initiated.

Methods for the preparation of generator columns for a variety of organic compounds have been detailed, as well as information regarding their behavior under certain conditions. The ability of these columns to measure solubilities of hydrophobic organic compounds has been demonstrated as well as the ability to act as reference solutions for analytical purposes. They are applicable primarily for organic compounds that have solubilities in water in the µg/L range. For compounds with higher solubility, the life span of the column is short; therefore, a consistent reference solution will not be available from such a source. An additional benefit of these columns is that they can be coated with a mixture of organic contaminants, which can be developed for a number of environmental field or laboratory applications.

This chapter has provided detailed technical information to enable a chemist to prepare and use generator columns in the development of high quality data.

ACKNOWLEDGMENTS

The authors would like to thank Anthony Marr, Mike Newman, and Jim Zvonar for technical assistance, John Coburn for his advice during the early stages of the project, and Jane Pagel, Barry Loescher, and Paul Roberts for editorial assistance. The financial assistance of the Ontario Ministry of the Environment and Energy in funding this development work is gratefully acknowledged.

REFERENCES

1. **W. E. May, S. P. Wasik, and D. H. Freeman,** Determination of the aqueous solubility of polynuclear aromatic hydrocarbons by a coupled column liquid chromatographic technique, *Anal. Chem.*, 50 (1978) 175.
2. **W. E. May, S. P. Wasik, and D. H. Freeman,** Determination of the solubility behaviour of some polycyclic aromatic hydrocarbons in water, *Anal. Chem.*, 50 (1978) 997.
3. **R. M. Dickhut, A. W. Andren, and D. E. Armstrong,** Aqueous solubilities of six polychlorinated biphenyl congeners at four temperatures, *Environ. Sci. Technol.*, 20 (1986) 807.
4. **F. M. Dunnivant and A. W. Elzerman,** Aqueous solubility and Henry's Law constant data for PCB congeners for evaluation of quantitative structure-property relationships (QSPR's), *Chemosphere*, 17 (1988) 525.
5. **A. Opperhuizen, F. A. P. C. Gobas, J. M. D. Van der Steen, and O. Hutzinger,** Aqueous solubility of polychlorinated biphenyls related to molecular structure, *Environ. Sci. Technol.*, 22 (1988) 638.
6. **J. W. Billington, G. L. Huang, F. Szeto, W. Y. Shire, and D. Mackay,** Preparation of aqueous solutions of sparingly soluble organic substances: I. Single component systems, *Environ. Toxicol. Chem.*, 7 (1988) 117.
7. **K. J. Friesen, L. P. Sarna, and G. R. B. Webster,** Aqueous solubility of polychlorinated dibenzo-p-dioxins determined by high pressure liquid chromatography, *Chemosphere*, 14 (1985) 1267.

8. **W. Y. Shiu, W. Doucette, F. A. P. C. Gobas, A. Andren, and D. Mackay,** Physical-chemical properties of chlorinated dibenzo-*p*-dioxins, *Environ. Sci. Technol.*, 22 (1988) 651.
9. **W. J. Doucette and A. W. Andren,** Aqueous solubility of selected biphenyl, furan and dioxin congeners, *Chemosphere*, 17 (1988) 243.
10. **K. B. Woodburn, W. J. Doucette, and A. W. Andren,** Generator column determination of octanol/water partition coefficients for selected polychlorinated biphenyl congeners, *Environ. Sci. Technol.*, 22 (1988) 382.
11. **W. J. Doucette and A. W. Andren,** Estimation of octanol/water partition coefficients evaluation of six methods for highly hydrophobic aromatic hydrocarbons, *Chemosphere*, 17 (1988) 345.
12. **W. E. May and M. M. Schantz,** Personal communication, 1990.
13. OECD Guidelines for Testing of Chemicals, Method 105, adopted 12 May 1981.
14. U.S. EPA Method 508, Determination of chlorinated pesticides in water by gas chromatography with an electron capture detector, from Methods for the Determination of Organic Compounds in Drinking Water, ORD Publications, CERI, EPA/600/4-88/039, December 1988.
15. U.S. EPA Method 8080A, Test Methods for Evaluating Solid Waste Physical/Chemical, 3rd ed., Rev. 1, November 1990.

Part II
Using Environmental Reference Materials

CHAPTER 11

Establishing the Credibility of Environmental Laboratories

Donald E. King

Across Canada, there is increasing use of commercial and private laboratories to monitor and evaluate a wide range of environmental conditions and to ensure compliance to regulatory requirements. Proper resolution of the issues surrounding site remediation, water, and waste water treatment, effluent control, or long-term environmental management depends on the reliability of the data provided and the credibility of the laboratories involved. The cost of analytical services tends to limit the amount and type of confirmatory testing performed. Yet the overriding concern for all parties — industry, government, and the general public — is the quality of the data generated. Therefore, laboratory managers must dedicate some of their resources, in a visible way, to enhance and validate the credibility of their facilities, procedures, and analysts.

This discussion provides an overview of the types of information needed to establish credibility in the environmental laboratory. Each has its own list of pros and cons. Each can be misused. The topics covered include good laboratory practice, good analytical practice, analytical method and performance documentation, the use of reference materials, participation in interlaboratory method and performance studies, and external audit for the purposes of certification/accreditation.

INTRODUCTION

The selection of a laboratory is the most difficult task confronting those involved in the study of environmental and health effects or those subjected to one or more of the environmental protection regulations of government. Even experienced laboratory personnel suffer qualms when faced with this task. Despite the emphasis placed on laboratory data quality management over the past 20 years, and while standards have been developed for good laboratory practice, and while most competent laboratories work diligently to implement and advertise their quality management (QM) programs, innumerable interlaboratory performance evaluations (round-robins) continue to document deficiencies in data compatibility.

Most environmental analytical services are extremely expensive. The analytical market place is competitive, highly technological, and very image conscious. Laboratories advertise their sophisticated equipment, their modern facilities, the qualifications of their scientific and technical staff, their comprehensive Quality Management (QM) program, and their professional sales/promotion staff. They tend to depend on the technical ignorance of their potential clients. There is a perception that they are selective in revealing the outcome of their participation in interlaboratory studies and in providing references from previous clients. When challenged, analysts tend to attack the credibility of opposing analysts, rather than providing positive evidence of their own analytical capability. In the following discussion we will consider the perspectives of both the client and the analyst.

THE CLIENT'S PERSPECTIVE

The laboratory client usually has a problem that requires more than just an analytical measurement. Resolution of this problem represents a project that often requires expert assistance. In other words, the client asks, "What do I need?" and "How can I be sure I will I get what I need?" In the case of environmental investigations, several factors beyond the laboratory operation must be addressed. Laboratories may not have all of the expertise required to initiate an appropriate investigation. Issues affecting program/project credibility include the following:

- Rationale:
  - What are we trying to prove, or disprove?
  - What type and quality of information do we need?
- Resources:
  - Does field and laboratory capability match our needs?
  - Will we be able to find what we are looking for?
- Sampling:
  - Will the design produce all the facts?
  - Will the samples taken be worth analyzing?
  - Will the samples reflect original conditions?
  - How can we ensure sample integrity and identity?
- Analysis/measurement:
  - What do we look for, how, and why?
  - How low do we go, and what performance is required?
  - Will the data reflect original conditions?
- Report:
  - Can we ensure valid data management, interpretation?
  - Will conclusions be supported by the data?
  - Will data for and against the hypothesis be gathered and reported?
  - Will the analyst have sufficient expertise to explain/interpret findings?

Because many of the above factors are perceived to be the responsibility of non-laboratory personnel, they tend to be minimized or overlooked by the laboratory client in an effort to reduce costs. Unfortunately, when they are not looked after properly, the analyst tends to be left holding the bag. It is difficult to establish data credibility when one is unable to establish the credibility of the sample, the sampling process, the chain of identity/custody, or the sample management processes.

The purchase of commercial analytical services generally includes the purchase of some expertise in the above areas. But laboratory QM programs rarely address field QM issues. As indicated above, many of the factors affecting the analytical result are beyond the direct control of the laboratory analyst. To this extent, the analyst cannot be held responsible if the sample examined does not truly represent the environmental situation under study. Of course, when the analytical result does not in fact represent the real world situation despite the best efforts of the laboratory, the promulgation of quality statements and corresponding documentation becomes a cruel joke that can actually work to destroy the credibility of the organization involved. The resolution of the project issues discussed above will affect the type of laboratory service required. Ultimately, the client needs a good project manager and a good laboratory analyst. If the client finds these and is willing to pay what is needed to get the job done, then we are also looking for the following qualities:

- Reliable: Will do a good job on schedule!
- Knowledgeable: Will find the best way to get the right job done!
- Capable: Has the experience, facilities, materials needed!
- Well-known: Is responsible and accountable!
- Economical: Will do no more than necessary to provide a solution!

Credibility is a big issue in selecting a laboratory. Can the client believe what the laboratory tells him/her? The challenge for the laboratory is to provide evidence to demonstrate a quality product that is

- Comparable:
  developed using accepted standard procedures, practices, methods, equipment.
- Compatible:
  mergeable with similar data from other sources, has controlled precision, accuracy, recovery, specificity, selectivity.
- Valid:
  proper measure of the environmental characteristic, acceptable for determining compliance, trends.

But now we have a new issue:

- Comparability: Who decides which methods/equipment are proper?
- Compatibility: Do I have to hire two labs to check each other?
- Validity: What criteria are used to evaluate data?
- Quality: How good is good enough?

In other words, the question is whether anyone is setting standards.

ESTABLISHING LABORATORY CREDIBILITY

Several years ago, in a private communication, Stan MacBeth, Chief of an Ontario Ministry of the Environment regional laboratory, contributed the following thoughts on establishing credibility.

> We can all cite examples of organizations whose scientific status and pronouncements are seldom questioned. While a cursory examination sometimes reveals that the spokesperson for the group is a charismatic individual of demonstrated sagacity, closer examination reveals a dedicated commitment on the part of all scientific, technical, and support personnel to clearly specified organization goals which include and stress quality. In view of the fact that charismatic and competent leaders are generally in short supply, most laboratories are obliged to initiate a more laborious program of directed quality control (QC) and quality assurance (QA) activity to establish, enhance, and maintain the overall validity and acceptability of their data. The objective is to minimize those occasions when published data assails the client's credulity. Ultimately this requires a philosophy which extends beyond QC and expands QA beyond the arena of bench practices to include overall data evaluation and assessment. A scientific approach to data evaluation must apply the test of reason to each analytical result.

This includes the following:

- Does the result appear reasonable for this sample?
  Could the sample as described contain this substance?
  Is this the most likely number to apply to this substance?
- Will this result be acceptable to the client?
  Will it introduce questions or concerns that might affect the credibility of the data (and the laboratory)?
- If the result might not be acceptable to the client, what information can we provide to substantiate and support the finding or suggest factors that might have affected the outcome?

In other words, laboratory analysts who depend upon rote application of QA practices run the risk of overlooking something crucial. It is critical that the laboratory not create the impression that it supports hiding unpleasant facts under the rug. The same painstaking care applied to obtain the result must be extended to ensure that the result is appropriate for its intended use. Clients are usually looking for solutions beyond the cold facts and hard numbers. Laboratory credibility centers on three issues:

1. Good analytical practices (GAP), including the use of
   a) documented procedures;
   b) accurate, validated standards;
   c) well-maintained, properly calibrated, measurement devices;
   d) quality control and data quality assessment protocols (Note: Procedures and calibration materials should both be traceable to internationally recognized standard procedures and materials).

2. Good laboratory operations (GLO), including the provision of
   a) qualified staff;
   b) appropriate well-maintained laboratory facilities and equipment;
   c) well-documented sample/workload/data/staff management practices;
   d) implementation of stringent quality management practices.
3. Performance evaluation (PE), including
   a) routine determination and documentation of factors such as measurement precision, accuracy;
   b) evaluation against external reference standards/materials to establish analytical recovery, specificity, selectivity;
   c) implementation of "blind sample" self-audits;
   d) participation in external interlaboratory analytical method studies;
   e) participation in external interlaboratory analytical performance studies.

The QA Manual and PE Reports

Tables 1 and 2 provide checklists relevant to establishing GAP and GLO status. The laboratory manager needs to evaluate the relative importance of these issues in establishing a program of quality improvement. Ultimately, the resolution and documentation of GAP and GLO status will and should lead to the preparation of a laboratory quality assurance (QA) manual. The QA manual provides a source of descriptions of all the standard operating practices (SOP) to be followed by staff in all aspects of day-to-day laboratory operation. SOPs should include performance criteria and procedures for evaluating and recording the outcome of quality performance checks. It is critical that they be assembled in a central location, to facilitate document control, distribution, updating, staff training, and external audit. Copies must also be readily accessible to staff. Any request for accreditation will generally not proceed until the accrediting agency is provided with a copy of this assembled package. Of course it is not sufficient to have assembled this information. Laboratory inspectors expect to see proof that the program has been implemented and is being maintained on an ongoing basis. Therefore performance criteria and performance reports are critical components of the QM program.

Reference Methods and Materials

Most laboratories adopt methods that are in some way traceable to established and recognized "standard methods". But the competent analyst must ensure that the method as described is appropriate for the sample matrix being tested. Ongoing advances in technology provide opportunities for improvement. But the analyst has a responsibility to maintain and ensure the compatibility of analytical results across laboratories. Compatibility ensures that data can be interpreted against the standard criteria for precision and accuracy. Reference materials have two distinct functions:

Reference Standard Materials (RSM):

essentially pure (solutions) materials, with defined true values based on direct physical and chemical principles, used to establish the accuracy of in-house standard solutions, to calibrate the stimulus/response factor for measurement devices, and to assess the specificity of analytical detection systems.

Reference Certified Materials (RCM):
: specimens of natural materials, carefully measured by a variety of analytical methods by a range of competent analysts, certified to contain a specified amount of a specified analyte, within an acceptably narrow range, used to evaluate and establish the performance of analytical methods, in terms of selectivity and recovery.

In order to gain professional recognition, the analyst is obliged to establish the traceability of analytical measurements, via a comprehensive paper trail back to the use of standard methods, RSMs, and RCMs.

ACCREDITATION

When the issues of GLO, GAP, and PE have been addressed internally and problems have been properly resolved and documented, the laboratory can be considered ready to establish its credibility. Credibility is a critical component in the art of persuasion. Sales personnel are particularly conscious of the need to establish credibility, and the process of image development depends on the selection of those facts that will promote it. Credibility is the most difficult characteristic to achieve and the easiest to lose. Of course, if you are tall, dark, leanly handsome, with piercing blue eyes, you probably have a head start in establishing an image. But the problems often start the moment you open your mouth.

There is a growing trend toward having a recognized independent authority express its confidence on behalf of (in lieu of) the laboratory. Accreditation of laboratories by appropriately recognized and designated institutions/agencies/associations fills this role. The accrediting body must be perceived to be impartial and to have applied specific criteria to gather facts that support the decision being made. The laboratory client's confidence is then based on confidence in the accrediting body and the accreditation process. Accreditation incorporates two components, inspection and performance assessment.

Laboratory inspection:
: evaluation by audit, against pre-established criteria, of the proper documentation and implementation of all analytical procedures, laboratory operations, and performance evaluation practices.

Performance assessment:
: evaluation by (replicate) blind analysis of standards/samples for specified parameters/analytes in selected matrices at a few different concentration levels.

The selection of assessors is critical. They must be trained in audit procedures and criteria. They must be independent, unbiased, and have recognized status. They must understand the technical and practical aspects of laboratory operation and must apply their tests in a standard and fair manner. But proper performance assessment is even more critical. Because of the nature of accreditation, one must carefully balance the need for demonstrated quality against the need to allow and encourage improvement. When laboratories experience difficulties, how they react should be examined. Therefore, performance criteria should be set that recognize those who have demonstrated better than average performance, while at the same time permitting others to apply appropriate remedial activity without losing status.

The audit process is very expensive. Who will pay? Clients tend to look to government to regulate the practices of industry. But commercial laboratories are well aware of the impact of overregulation. And government does not have the resources to assess compliance. In the end the client will pay for the cost of accreditation. But if laboratory management does not provide value to cover the added expense, they may find that their reputation precedes them into bankruptcy. There is a tendency on the part of laboratory managers to misunderstand the accreditation process. They prepare their QA Manual to look good. They sometimes forget that it is meant to be an instruction manual, not a sales pitch. They use unusual practices to ensure that they do well on performance evaluation samples. Perhaps it is sufficient to suggest that misdirection is the primary tool of magicians and shady sales personnel. Credibility lies in the eye of the beholder (not the generator). Attempts to establish credibility by misdirection may be perceived as "lies" by the client. We are all aware of the poor image suffered by even the best, most conscientious door-to-door and used car salesmen. To many potential laboratory clients, the purchase of a particular laboratory service feels much like buying a used car. Caveat emptor!

LABORATORY PERSPECTIVE

Therefore, for a laboratory to be credible, it must demonstrate that it is

Capable:	has the facilities, equipment, and staff;
Knowledgeable:	advises/provides appropriate analytical service;
Known:	relies on word-of-mouth to attract new clients;
Reliable:	excellent service convinces clients to return;
Right price:	price reflects the cost of providing good service.

The laboratory will attract and convince new clients by applying GAP, GLO, and PE principles in its day-to-day operation. This is a staged, ongoing, cyclical process:

Step 1: Determine capability
- facilities — clean, well-organized
- operations — efficient, documented
- equipment — current, maintained
- methods — documented, referenced, standardized
- staff — trained, professional attitude
- performance — precision, accuracy, control, validity

Step 2: Upgrade to meet standards
- competitors — define laboratory's niche

standards — identify and exceed the minimum
client desires — balance dreams vs. reality
integrity — personal honesty, credibility

Step 3: Verify, measure, and document capability: internal and external
precision — short-term vs. long-term
accuracy — standards/calibration
control — bias: day-to-day remedial action
validity — bias: sample/matrix/method effects
RSMs — verify calibration accuracy
RCMs — evaluate matrix/method interaction
ref. methods — evaluate in-house method bias
interlaboratory studies
— evaluate method ruggedness and proper use
— evaluate peer group comparability/compatibility

Step 4: Implement quality program
quality planning
— establish policy, goals, objectives
— establish accountability, audit process
quality management
— establish good laboratory operations — GLO
— establish good analytical practices — GAP
quality assurance and improvement
— determine and implement control/audit processes
— monitor and evaluate performance
— implement remedial activities

Step 5: Establish recognition
documentation
— QA manual: actual practices/procedures!
— advertising: judicious exaggeration?
— brochures: "rose-colored" generalities?
word-of-mouth
— can the laboratory depend on client's perspective?
accreditation
— third party: is the laboratory ready?

Step 6: Don't "rest on laurels"
continue review and improvement!
laurels are meant to wear.... not to sit on!

SUMMARY

There is no question that participation in accreditation programs can benefit both the laboratory and the client. But accreditation is neither the first nor the last step on the road to credibility. Government regulations, or specific client needs will usually require additional ongoing effort and documentation to show ability to meet specific data quality objectives. The subjective and volatile nature of credibility demands that it be established in an organized and logical fashion. The facts supporting it must be well documented. Each facet of the operation must be examined in minute detail, with particular regard given to those that are most subject to attack. We all tend to attach credibility to facts that support our own beliefs. Contrary facts are incredible.

Analysts believe strongly in their own credibility because they usually have information to support the reproducibility of their data. But they are also well aware of the weaknesses of their operation. Faced by a challenge from another analyst, they tend to assume, with reason, that the same problems exist in the opposing data. In the absence of supporting data, they assume the other analyst's results are very nice but wrong. Of course, the feeling is mutual. No matter how much time one spends in establishing one's own competence, there is no guarantee that others will be convinced unless the facts are well documented and made available to the client. Credibility is established by behavior, not words or desires or fancy gold-framed certificates.

Table 1 Good Laboratory Operations (GLO) Checklist

The following activities should be interpreted as occurring at the upper levels of laboratory operation. They are subject to director approval and reflect management responsibility.

ADMINISTRATION (GENERAL)	Appropriate organizational structure
	Good client services
	Quality management program
	Suitable, adequate staffing
	Sample/workload management
	Adequate laboratory facilities
	Appropriate use of computers/automation
CLIENT SERVICES	Description of services available
	Specified sampling/submission requirements
	Documented performance criteria
	Current performance reports/summaries
	Good communications to ensure:
	Lab sample requirements are known
	Lab requirements are met
	Client needs are known
	Client needs are met
	Problem tracking/resolution
QUALITY MANAGEMENT	Quality management plan staff responsibilities
	Management/supervisory accountability
	Independent quality assurance personnel
	Internal audits
	QM system
	QA/QC records/data
	Analytical procedure
	Performance and data quality
	External audits
STAFF MANAGEMENT	Adequate numbers
	Qualifications, experience, resumes
	Job descriptions and training

Table 1 Good Laboratory Operations (GLO) Checklist (continued)

WORKLOAD MANAGEMENT	Workload planning and management system Timely assignment of tests to lab areas Accurate test assignment process Efficient work-listing system Timely initiation of analysis Control on throughput and work outstanding Problem reports and follow-up system
SAMPLE MANAGEMENT	Sampling, sample handling instructions for containers/preservation/storage Sample submission instructions Sample container supplies Materials, cleaning, quality checks Container handling, distribution Preservatives, suitable quality Sample reception protocols — routine, high priority, legal, etc. Sample suitability checks (as received) Adequate sample storage facilities Sample tracking/security Safe disposal practices
DATA MANAGEMENT	Effective records management & security Accurate data reporting processes Comprehensive data reporting procedure Results and interpretations Data quality reporting system Data qualifiers/remark codes Sign-off by accountable personnel QA/QC data: Collection, interpretation Archiving, summarization Timely remedial action
FACILITIES MANAGEMENT	Cleanliness, ventilation, safety Adequate equipment/instrumentation Appropriate location of work areas Suitable use of space and equipment Adequate storage: samples, reagents Waste management systems Security
EQUIPMENT	Equipment lists: model/serial# Age, condition, availability Maintenance schedules/logs (tags) Operating procedures Servicing and backup Replacement strategy
INSTRUMENTATION	Instrumentation lists: model/serial# Regular maintenance schedule (tags) Physical calibration checks, alignment, etc. Analytical calibration/standardization protocols (tags)
STORES	Adequate stocks, timely replacement of supplies — stock control, back-orders, purchasing, etc. Quality of supplies Safety aspects, ventilation, etc. Degree of automation/computerization

Table 2 Good Analytical Practices (GAP) Checklist

The following activities should be interpreted as occurring at the day-to-day level of laboratory operation (within the laboratory section/unit). They are subject to manager approval and reflect supervisory responsibility.

ANALYTICAL QUALITY MANAGEMENT STRATEGY	Description of section QM strategy
	General section/unit QA/QC protocols
	Staff responsibilities
	QC task verification by supervisors
	Accountability of supervisory staff
	QA evaluation reports
ANALYTICAL QUALITY ASSURANCE PROGRAM	Quality Assurance Manual — principles, protocols, procedures
	Performance evaluation procedures
	Written acceptance/rejection criteria for
	Methods
	Runs (analytical batches)
	Individual sample data
	Remedial action plan
ANALYTICAL QUALITY CONTROL ACTIVITIES	Quality Control Manual to control
	Requirements
	Procedures
	Records for control activities data
	Control charts, control/warning limits
	QC data archive system
SAMPLE MANAGEMENT	Timely initiation/completion of tests
	Defined sample handling/storage specs
	Regular verification of storage facilities — proper conditions, clean, good order
	Sample custody tracking in lab
ANALYTICAL METHODS	Formal detailed methods/procedures
	Traceable to recognized standard methods
	Readily available bench procedures
	Method development protocol
	Method approval protocol
	Protocol for trial modifications
	No changes without notice/approval
ANALYTICAL QUALITY	Performance monitoring and control activity — blank, calibration, precision, recovery, etc.
	Performance criteria
	Repeatability, recovery, specificity, selectivity
	Matrix effects, interference, identification criteria
	Use of appropriate RCMs
	Interlaboratory comparison
	Analytical performance reports
DATA QUALITY	Run quality data archives
	Run rejection criteria
	Sample data rejection criteria
	Data qualification protocols & codes
	Data reporting practices (W, T, etc)
	Detection limits (MDL)
STAFF PROFICIENCY	Performance criteria
	Proficiency testing of new staff and permanent staff
	Training of new staff and for job rotation

Table 2 Good Analytical Practices (GAP) Checklist (continued)

LABORATORY CHEMICALS & SUPPLIES	Lab reagent water quality
	Lists of reagents and quantities
	Reagent quality specifications
	Regular checks of reagents/supplies
	Old reagent replacement/disposal
	Labware lists and specifications
	Labware quality checks
	Sample container lists and specs
	Container quality checks
CALIBRATION STANDARDS	Preparation/verification procedures
	Source materials: known purity/stability
	Verified vs. RSMs
	Storage, handling, "ageing" protocols
	Safe disposal practices

CHAPTER 12

Standard Reference Materials for the Determination of Trace Organic Contaminants in Environmental Samples

Stephen A. Wise and Michele M. Schantz

INTRODUCTION

Since 1980 the National Institute of Standards and Technology (NIST), formerly the National Bureau of Standards, has issued a number of Standard Reference Materials (SRMs) for use in the determination of trace organic constituents in environmental samples. These SRMs have focused primarily on organic contaminants such as polycyclic aromatic hydrocarbons (PAHs), polychlorinated biphenyls (PCBs), and chlorinated pesticides. Several reviews have been published describing the SRM activities at NIST for PAH, PCB, and chlorinated pesticide measurements.[1–4]

The first SRMs developed at NIST for organic contaminant measurements focused on PAH measurements in fossil fuels, i.e., shale oil[5,6] and petroleum crude oil,[7,8] followed by PAHs in air and diesel particulate material.[9] The most recent natural matrix SRMs for PAH measurements have focused on the marine environment with sediment[10–12] and mussel tissue samples.[13] The first SRMs for PCB measurements were natural matrices (oil and human serum) that had been fortified with PCB mixtures and quantified as Aroclor mixtures; however, more recent efforts in PCB determinations have been directed toward the measurement of individual PCB congeners in natural samples such as cod liver oil,[14] marine sediment,[10,12] river sediment,[15] mussel tissue,[13] and whale blubber tissue.[16] The marine sediment and mussel tissue materials were also the first materials to be analyzed for both PAHs and organochlorine compounds. In this chapter the NIST SRMs available for use in the determination of organic contaminants in environmental samples will be described with particular emphasis on the natural environmental matrix materials. The analytical approach used to provide certified concentrations will be described in detail, using as examples two recently issued natural matrix SRMs related to organic contaminants in the marine environment: SRM 1941a, Organics in Marine Sediment and SRM 1945, Organics in Whale Blubber.

NIST SRMs FOR ORGANIC CONTAMINANTS

Reference materials are used primarily for the following purposes: (1) to calibrate the measurement system, (2) to validate the reliability and precision of a new analytical method, and (3) to provide quality control of routine analyses by analyzing the SRM at appropriate, regular time intervals. The NIST SRMs for organic environmental analyses are summarized in Tables 1 and 2. The SRMs in Tables 1 and 2 represent two different levels of analytical difficulty: (1) simple calibration solutions containing a number of analytes and (2) natural matrix materials. The calibration solutions (see Table 1) are useful for validating the chromatographic separation step (e.g., retention times and analyte detector response), whereas the natural matrix materials (see Table 2), which are similar to the actual environmental samples analyzed, can be used to validate the complete analytical procedure, including extraction, isolation/clean-up procedures, and the final chromatographic separation and quantification.

For each of the SRMs listed in Tables 1 and 2, information is provided on the number of analytes for which "certified" and "noncertified" concentration values are provided in the Certificate of Analysis that accompanies each SRM sample. The certified values for each of these SRMs are based on the agreement of results obtained from two or more independent and reliable analytical techniques, whereas the noncertified values are generally results from only one analytical technique. For the calibration solution SRMs in Table 1, the certified values are generally based on gravimetry (i.e., the accurately weighed quantities of the analytes and solvent used to prepare the solution) and on measurements obtained using liquid chromatography (LC) and gas chromatography (GC) with flame ionization detection (FID), electron capture detection (ECD), and/or gas chromatography-mass spectrometry (GC-MS). The natural matrix materials present a more difficult analytical task, which usually requires extraction to remove the organic compounds from the matrix, cleanup procedures to isolate the compounds of interest from the complex mixture of organic compounds, and finally analysis using GC or LC with selective detection to measure the individual compounds in the complex natural mixtures. For the PAH measurements in the natural matrix SRMs, the primary analytical techniques used at NIST have been reversed-phase LC with fluorescence detection (LC-FL), GC-FID, and GC-MS. For PCB and pesticide measurements, GC-ECD has been the primary technique, with GC-MS used recently as the second technique. To satisfy the requirement of independent analytical procedures for the certification of natural matrix SRMs, the extraction and cleanup/isolation steps that precede the final chromatographic measurements should also be based on different separation characteristics to minimize the possibility of similar biases in both measurements. The complete details of the analytical approach will be described in a later section of this chapter.

Table 1 NIST Calibration Solution SRMs for the Determination of Organic Contaminants

SRM no.	Title	Date issued	Certified constituents	Noncertified constituents
1491	Aromatic Hydrocarbons in Hexane/Toluene	1989	PAHs (23)	PAHs (1)
1492	Chlorinated Pesticides in Hexane	1992	Pesticides (15)	
1493	Chlorinated Biphenyl Congeners in 2,2,4-Trimethylpentane	1995	PCBs (18)	PCBs (2)
1494	Aliphatic Hydrocarbons in 2,2,4-Trimethylpentane	*	Hydrocarbons (20)	
1583	Chlorinated Pesticides in 2,2,4-Trimethylpentane	1985	Pesticides (5)	Pesticides (1)
1584	Priority Pollutant Phenols in Methanol	1984	Phenols (10)	Phenols (1)
1585	Chlorinated Biphenyls in 2,2,4-Trimethylpentane	1986	PCBs (8)	
1586	Isotopically Labeled and Unlabeled Priority Pollutants in Methanol	1984	Priority pollutants (10)	
1587	Nitrated PAHs in Methanol	1985	Nitro-PAHs (6)	Nitro-PAHs (1)
1596	Dinitropyrene Isomers and 1-Nitropyrene in Methylene Chloride	1987	Nitro-PAHs (4)	
1614	Dioxin (2,3,7,8-TCDD) in Iso-octane	1985	Dioxins (2)	Dioxins (2)
1639	Halocarbons (in Methanol) for Water Analysis	1983	Halocarbons (7)	
1644	Generator Columns for PAH	1981	PAHs (3)	
1647c	Priority Pollutant PAHs (in Acetonitrile)	1992[a]	PAHs (16)	
2260	Aromatic Hydrocarbons in Toluene (Nominal Concentration 60 µg/mL)	1991	PAHs (23)	PAHs (1)
2261	Chlorinated Pesticides in Hexane (Nominal Concentration 2 µg/mL)	1992	Pesticides (15)	
2262	Chlorinated Biphenyl Congeners in 2,2,4-Trimethylpentane (Nominal Concentration 2 µg/mL)	1995	PCBs (25)	PCBs (3)

[a] SRM 1647 was first issued in 1981 and has been reissued three times.
* Certification in progress.

Table 2 NIST Natural Matrix SRMs for the Determination of Organic Contaminants in the Environmental Samples

SRM no.	Title	Date issued	Certified constituents	Noncertified constituents
1580	Organics in Shale Oil	1980	PAHs (5); Phenols (3); PANH (1)	Phenols (6); PANH (1)
1581	Polychlorinated Biphenyls in Oils	1982	Aroclor 1242/1260	
1582	Petroleum Crude Oil	1984	PAHs (5); PASH (1)	PAHs (5); Phenols (2); PANH (1)
1588	Organics in Cod Liver Oil	1989	PCBs (5); Pesticides (10)	PCDDs/PCDFs (7)
1589	Polychlorinated Biphenyls (as Aroclor 1260) in Human Serum	1985	Aroclor 1260	Dioxins (2)
1597	Complex Mixture of PAHs from Coal Tar	1987	PAHs (12)	PAHs/PASH/PANH (18)
1648	Urban Particulate Matter	1978	Trace Elements (9)	Trace Elements (25); PAH (13)
1649	Urban Dust/Organics	1982	PAHs (5)	PAHs (9)
1650	Diesel Particulate Material	1985	PAHs (5) Nitro-PAHs (1)	PAHs (6); Nitro-PAHs (3); PAQ (1)
1939	Polychlorinated Biphenyls (Congeners) in River Sediment	1990	PCBs (3)	PCBs (14); Pesticides (5); PAHs (5)
1941	Organics in Marine Sediment[a]	1989	PAHs (11)	PAHs (24); PCBs (15); Pesticides (7); Trace Elements (32)
1941a	Organics in Marine Sediment	1994	PAHs (23); PCBs (21) Pesticides (6)	PAHs (14); PCBs (7); Pesticides (3) Aliphatics (17); Trace Elements (27)
1945	Organics in Whale Blubber (Frozen)	1994	PCBs (27); Pesticides (15)	PCBs (2); Pesticides (2)
1974	Organics in Mussel Tissue (*Mytilus edulis*) (Frozen)[a]	1990	PAHs (9)	PAHs (19); PCBs (13) Pesticides (9); Trace Elements (34)
1974a	Organics in Mussel Tissue (*Mytilus edulis*) (Frozen)	*	PAHs; PCBs; Pesticides	PAHs; PCBs; Pesticides; Aliphatics; Trace Elements
1975	Diesel Particulate Extract	*	PAHs; Nitro-PAHs	
2974	Organics in Mussel Tissue (*Mytilus edulis*) (Freeze-dried)	*	PAHs; PCBs; Pesticides	PAHs; PCBs; Pesticides
8045	GESREM II Mussel Tissue	*	PAHs; PCBs; Pesticides	PAHs; PCBs; Pesticides

Notes: PANH = polycyclic aromatic nitrogen heterocycles; PAQ = Polycyclic aromatic quinones; PASH = polycyclic aromatic sulfur heterocycles; PCDD = polychlorinated dibenzodioxins; PCDF = polychlorinated dibenzofurans.

[a] Material no longer available; renewal material has been issued or is in progress.

* Certification in progress.

Calibration Solutions

The calibration solution SRMs are useful for several purposes, including (1) calibration of chromatographic instrumentation for retention times and detector response factors for quantitation, (2) spiking or fortifying samples, (3) analyte recovery studies, and (4) determining method response factors. NIST has issued 17 different calibration solution SRMs since 1981, as shown in Table 1. The most recent emphasis has been on four groups of organic contaminants routinely measured in marine environmental monitoring programs (i.e., PAHs, PCBs, chlorinated pesticides, and aliphatic hydrocarbons). The compounds and concentration ranges of the calibration solution SRMs from these four groups of analytes are summarized in Table 3. The compounds used to prepare these calibration solutions are obtained in the highest purity available (preferably certified reference materials where available); if the purity of the material is unknown, analyses are performed using GC, LC, and/or differential scanning calorimetry to establish the purity.

The first and by far the most popular of the organic calibration solution SRMs is SRM 1647, an acetonitrile solution of the 16 PAHs on the U.S. Environmental Protection Agency's (EPA) priority pollutant list. This SRM was first prepared in 1981 at the request of EPA in support of EPA Method 610, which specifies the use of reversed-phase LC with fluorescence detection for the determination of PAHs. SRM 1647 has found widespread use as a calibration solution to determine retention times and detector response factors in LC. Because of the popularity of SRM 1647, it has been reissued three times since first issued in 1981 and is now available as SRM 1647c. Since SRM 1647c is used primarily for calibration of reversed-phase LC instrumentation and spiking of aqueous-based matrices (i.e., because of the acetonitrile solvent), two similar solutions (SRMs 1491 and 2260) have been prepared in hexane and/or toluene to provide a solvent more compatible with GC and normal-phase LC analyses and for spiking into nonaqueous matrices. In addition to the 16 PAHs included in SRM 1647, SRMs 1491 and 2260 contain eight additional PAH analytes (see Table 3), which were included in the SRMs specifically to meet the needs of a national marine pollution monitoring program in the U.S. (i.e., the National Status and Trends Program) sponsored by the U.S. National Oceanic and Atmospheric Administration (NOAA). SRMs 1491 and 2260 contain the same 24 analytes but at concentrations that differ by approximately a factor of 10. SRMs 1491 and 2260 can be diluted to produce multipoint calibration curves. SRM 1494, Aliphatic Hydrocarbons in 2,2,4-Trimethylpentane, was developed for use in the measurement of aliphatic hydrocarbons, particularly for assessment of petroleum contamination in the marine environment. SRM 1494 contains *n*-alkanes from C_{10} to C_{20}, even number *n*-alkanes from C_{22} to C_{34}, pristane, and phytane. Pristane and phytane are often measured for use as marker compounds to determine whether the input of hydrocarbons in the sample is of petrogenic or natural origin.

In 1985 and 1986 calibration solution SRMs were issued containing six chlorinated pesticides (SRM 1583) and eight PCB congeners (SRM 1585). Recently, these SRMs have been superseded by calibration solution SRMs that contain more analytes (i.e., SRMs 1492 and 2261, which each contain 15 chlorinated pesticides and SRMs 1493 and 2262 which contain 20 and 28 PCB congeners, respectively). The number

Table 3 Analytes in NIST Calibration Solution SRMs for PAHs, Aliphatic Hydrocarbons, PCBs, and Chlorinated Pesticides

Polycyclic aromatic hydrocarbons in SRMs 1491 and 2260 (~4–8 and 50–70 μg/mL)	Aliphatic hydrocarbons in SRM 1494 (~6–130 μg/mL)	Polychlorinated biphenyl congeners in SRMs 1493 and 2262 (~0.2 and 2 μg/mL)		Chlorinated pesticides in SRMs 1492 and 2261 (~0.2 and 2 μg/mL)
Naphthalene[a]	*n*-Decane (*n*-C_{10})	PCB 1	2-Chlorobiphenyl[c]	Hexachlorobenzene
2-Methylnaphthalene	*n*-Undecane (*n*-C_{11})	PCB 8	2,4′-Dichlorobiphenyl	γ-HCH[d]
1-Methylnaphthalene	*n*-Dodecane (*n*-C_{12})	PCB 18	2,2′,5-Trichlorobiphenyl	Heptachlor
Biphenyl	*n*-Tridecane (*n*-C_{13})	PCB 28	2,4,4′-Trichlorobiphenyl[b]	Aldrin[d]
2,6-Dimethylnaphthalene	*n*-Tetradecane (*n*-C_{14})	PCB 29	2,4,5-Trichlorobiphenyl[c]	Heptachlor epoxide[d]
Acenaphthylene[a]	*n*-Pentadecane (*n*-C_{15})	PCB 44	2,2′,3,5′-Tetrachlorobiphenyl	*cis*-Chlordane
Acenaphthene[a]	*n*-Hexadecane (*n*-C_{16})	PCB 50	2,2′,4,6-Tetrachlorobiphenyl[c]	*trans*-Nonachlor
1,6,7-Trimethylnaphthalene	*n*-Heptadecane (*n*-C_{17})	PCB 52	2,2′,5,5′-Tetrachlorobiphenyl[b]	Dieldrin
Fluorene[a]	*n*-Octadecane (*n*-C_{18})	PCB 66	2,3′,4,4′-Tetrachlorobiphenyl	Mirex
Phenanthrene[a]	*n*-Nonadecane (*n*-C_{19})	PCB 77	3,3′,4,4′-Tetrachlorobiphenyl[b]	2,4′-DDE
Anthracene[a]	*n*-Eicosane (*n*-C_{20})	PCB 87	2,2′,3,4,5′-Pentachlorobiphenyl[c]	4,4′-DDE[d]
1-Methylphenanthrene	*n*-Docosane (*n*-C_{22})	PCB 101	2,2′,4,5,5′-Pentachlorobiphenyl[b]	2,4′-DDD
Fluoranthene[a]	*n*-Tetracosane (*n*-C_{24})	PCB 104	2,2′,4,6,6′-Pentachlorobiphenyl[c]	4,4′-DDD

Pyrene[a]	*n*-Hexacosane (*n*-C_{26})	PCB 105	2,3,3′,4,4′-Pentachlorobiphenyl	2,4′-DDT
Benz[*a*]anthracene[a]	*n*-Octacosane (*n*-C_{28})	PCB 118	2,3′,4,4′,5-Pentachlorobiphenyl	4,4′-DDT[d]
Chrysene[a]	*n*-Triacontane (*n*-C_{30})	PCB 126	3,3′,4,4′,5-Pentachlorobiphenyl	
Benzo[*b*]fluoranthene[a]	*n*-Dotriacontane (*n*-C_{32})	PCB 128	2,2′,3,3′,4,4′-Hexachlorobiphenyl	
Benzo[*k*]fluoranthene[a]	*n*-Tetratriacontane (*n*-C_{34})	PCB 138	2,2′,3,4,4′,5′-Hexachlorobiphenyl[b]	
Benzo[*e*]pyrene[a]	Pristane	PCB 153	2,2′,4,4′,5,5′-Hexachlorobiphenyl[b]	
Benzo[*a*]pyrene[a]	Phytane	PCB 154	2,2′,4,4′,5,6′-Hexachlorobiphenyl[c]	
Perylene		PCB 170	2,2′,3,3′,4,4′,5-Heptachlorobiphenyl	
Indeno[1,2,3-*cd*]pyrene[a]		PCB 180	2,2′,3,4,4′,5,5′-Heptachlorobiphenyl	
Dibenz[*a,h*]anthracene[a]		PCB 187	2,2′,3,4′,5,5′,6-Heptachlorobiphenyl	
Benzo[*ghi*]perylene[a]		PCB 188	2,2′,3,4′,5,6,6′-Heptachlorobiphenyl[c]	
		PCB 195	2,2′,3,3′,4,4′,5,6-Octachlorobiphenyl	
		PCB 201	2,2′,3,3′,4,5′,6,6′-Octachlorobiphenyl[c]	
		PCB 206	2,2′,3,3′,4,4′,5,5′,6-Nonochlorobiphenyl	
		PCB 209	Decachlorobiphenyl	

[a] Analytes in SRM 1647c (~1–20 µg/mL).
[b] Analytes in SRM 1585 (~1–30 µg/mL). PCB 3 (4-chlorobiphenyl) and PCB 15 (4,4′-dichlorobiphenyl) are also in SRM 1585.
[c] Analytes not in SRM 1493.
[d] Analytes in SRM 1583 (~0.5–1.3 µg/mL). δ-HCH is also in SRM 1583.

of analytes in these recent calibration solution SRMs was expanded from SRMs 1583 and 1585 to meet the needs of the NOAA marine monitoring program; eight additional PCB congeners were added to SRM 2262 at the request of EPA. The PCB congeners and chlorinated pesticide SRMs are issued at two concentration levels (approximately 0.2 and 2 μg/mL) for use in preparing multipoint calibration curves.

In addition to the calibration solution SRMs related to aliphatic hydrocarbons, PAHs, PCBs, and chlorinated pesticides, seven additional solution SRMs have been issued for different classes of organic contaminants. SRM 1644, Generator Columns for PAHs, is a unique SRM for preparing calibration solutions of PAHs in water. SRM 1644 consists of three individual coiled stainless steel tubes containing sand that has been coated with approximately 0.5% by weight of anthracene, benz[*a*]anthracene, or benzo[*a*]pyrene. By passing high-purity water slowly through the column, a saturated aqueous solution is obtained. Since the aqueous solubility of a compound is a well-defined thermodynamic quantity, this saturated solution has a known concentration at a specific temperature.[17,18]

The determination of nitrated PAHs in environmental samples has become important in recent years because of the high direct-acting mutagenicity of some of these compounds. Two calibration SRMs containing nitrated PAHs have been issued by NIST: SRM 1587, Nitrated Polycyclic Aromatic Hydrocarbons in Methanol, and SRM 1596, Dinitropyrene Isomers and 1-Nitropyrene in Methylene Chloride. SRM 1587 contains seven nitro-PAHs (2-nitrofluorene, 9-nitroanthracene, 3-nitrofluoranthene, 1-nitropyrene, 7-nitrobenz[*a*]anthracene, 6-nitrochrysene, and 6-nitrobenzo[*a*]pyrene) at concentrations of approximately 4 to 7 μg/mL. SRM 1596 contains a mixture of three dinitropyrene isomers (1,3-, 1,6-, and 1,8-dinitropyrene) and 1-nitropyrene at concentrations of approximately 3 to 10 μg/mL.

SRM 1584, Priority Pollutant Phenols in Methanol, contains 10 of the 11 phenols on the EPA's list of priority pollutants at concentrations ranging from 15 to 65 μg/mL. SRM 1586, Isotopically Labeled and Unlabeled Priority Pollutants in Methanol, was developed for use by laboratories using EPA Analytical Methods 1624 and 1625 (as well as 624-625 and 524-525). These methods specifically require the use of GC-MS and the use of isotopically labeled internal standards. SRM 1586 is composed of two separate solutions each containing the following compounds: carbon tetrachloride, benzene, chlorobenzene, phenol, nitrobenzene, 2-nitrophenol, 2,4-dichlorophenol, naphthalene, bis[2-ethylhexyl]phthalate, and benzo[*a*]pyrene. One solution contains the ten compounds with carbon-13 (carbon tetrachloride) or deuterium (all other compounds) labeled compounds, whereas the other solution contains unlabeled compounds. The concentrations of the analytes range from approximately 40 to 130 μg/g. Another SRM containing labeled and unlabeled pollutants is SRM 1614, Dioxin (2,3,7,8-TCDD) in *Iso*-octane. This SRM consists of two solutions, one containing ^{13}C-labeled 2,3,7,8-tetrachlorodibenzo-*p*-dioxin (TCDD) and the second containing unlabeled 2,3,7,8-TCDD, both at concentrations of approximately 65 ng/mL. Non-certified values are also reported for a labeled and unlabeled trichlorodibenzo-*p*-dioxin congener at 1 to 3 ng/mL levels. SRM 1639, Halocarbons (in Methanol) for Water Analysis, is a methanol solution containing seven halocarbons: chloroform, chlorodibromomethane, bromodichloromethane, bromoform, carbon tetrachloride, trichloroethylene, and tetrachloroethylene at concentrations ranging from 40 to 6200 ng/μL.

Natural Matrix SRMs

Validation of the complete analytical procedure (including solvent extraction, cleanup of the extracts, isolation of the analytes of interest, and chromatographic analysis) requires the use of SRMs with matrices similar to those typically encountered in the analysis of environmental samples. Thus, the natural matrix SRMs are the most suitable materials for this purpose. A number of natural matrix SRMs have been issued by NIST with certified and noncertified concentrations of PAHs, aliphatic hydrocarbons, phenols, PCBs, and chlorinated pesticides (see Table 2). Two of these materials, SRM 1941 (Organics in Marine Sediment) and SRM 1974 (Organics in Mussel Tissue), are no longer available and have been replaced with similar materials. The natural matrix materials are not recommended for use in the calibration of the analytical system.

Natural Matrix SRMs for PAH Determinations

Since 1980, seven natural matrix SRMs have been issued with certified concentrations of PAHs and other polycyclic aromatic compounds (PACs) (i.e., SRMs 1580, 1582, 1597, 1649, 1650, 1941/1941a, and 1974). Certification is in progress for four additional materials that will be issued in 1996 (i.e., the renewal SRM 1974a, Organics in Mussel Tissue (a fresh frozen tissue material); SRM 2974, a freeze-dried version of SRM 1974a; RM 8045 GESREM II Mussel Tissue (see description below); and SRM 1975, Diesel Particulate Extract). The natural matrix SRMs issued by NIST are listed in Table 2, including the date issued and the type and number of certified and noncertified constituents.

The SRMs listed in Table 2 represent several matrix types, relative PAH concentrations, and sources of the PAHs (i.e., petrogenic or pyrolytic). SRMs 1580 and 1582 are representative of oil matrices with petrogenic PAHs (i.e., formed from low-temperature processes), which have high levels of alkyl-substituted PAHs relative to the unsubstituted PAHs. These two materials have been described in more detail elsewhere.[5–8,20–22] SRMs 1648 and 1649 are two air particulate samples that were collected in the mid-1970s in St. Louis, MO and Washington, D.C., respectively. SRM 1648, Urban Particulate Matter, was issued in 1978 and is certified for only inorganic constituents; however, data have been reported in the literature for PAH concentrations determined at NIST.[19] An extensive characterization and comparison of the PAH content of these two air particulate SRMs have been reported in the literature.[19] The diesel particulate sample (SRM 1650) is representative of heavy-duty diesel emissions. The concentrations of 13 common PAHs in the two fossil fuel and three particulate SRMs are summarized and compared in Table 4. In addition to the certified and noncertified concentrations of PAHs, Table 4 also contains results from additional analyses that have been performed at NIST on these materials since they were issued as SRMs, but are not reported on the Certificate of Analysis. (References to these additional measurements are provided in the footnotes in Table 4.)

Table 5 summarizes the concentrations of PAHs in the coal tar, sediment, and mussel tissue SRMs. SRM 1597, Complex Mixture of PAHs from Coal Tar, is a natural pyrolytic PAH mixture, which is intermediate in complexity when compared to the calibration SRMs and the other natural matrix materials. The concentrations

Table 4 Concentrations (mg/kg) of the PAHs in Natural Matrix SRMs (Fossil Fuels and Particulate Materials)[a]

Compounds	Shale oil SRM 1580	Petroleum crude oil SRM 1582	Air particulate matter SRM 1648	Urban dust organics SRM 1649	Diesel particulate matter SRM 1650
Phenanthrene	232.0 (12.0)[b]	**101.0 ± 5.0**	3.8 (0.2)[i]	4.5 (0.3)[j]	79.0 (1.0)[k]
	255.0 (8.0)[c]		4.6 (0.3)[j]		63.0 (2.0)[l]
Anthracene	88.4 (2.6)[b]				
	85.9 (4.8)[c]		0.36 (0.1)[j]	0.41 (0.04)[j]	0.88 (0.05)[o]
Fluoranthene	**54.0 ± 10.0**	**2.5 ± 0.3**	9.1 (0.8)[k]	**7.1 ± 0.5**	**51.0 ± 4.0**
	9.2 (0.5)[i]				
	8.7 (0.4)[j]				
Pyrene	**104.0 ± 18.0**	8.0 (0.4)[g]	7.4 (0.6)[k]	7.2 (0.2)	**48.0 ± 4.0**
		7.2 (0.2)[g]	7.3 (0.4)[i]	6.3 (0.4)	
		6.8 (0.2)[j]			
Benz[*a*]anthracene	35.2 (2.5)[b]	**3.0 ± 0.3**	3.2 (0.2)[k]	**2.6 ± 0.3**	**6.5 ± 1.1**
	36.1 (3.2)[c]		3.2 (0.2)[i]		
	3.2 (0.1)[j]				
Chrysene	43.9 (3.4)[b,m]		6.6 (0.2)[j]	3.5 (0.2)	22.0 (1.0)[l]
	41.0 (4.5)[c]				
Benzo[*b*]fluoranthene	12.0 (2.0)[d]		8.0 (0.5)[n]	6.2 (0.3)	
Benzo[*k*]fluoranthene	5.0 (1.0)[d]		3.4 (0.5)[j]	2.0 (0.1)	2.0 (0.06)[n]
Benzo[*a*]pyrene	**21.0 ± 6.0**	**1.1 ± 0.3**	3.1 (0.2)[k]	**2.9 ± 0.5**	**1.2 ± 0.3**
			3.1 (0.2)[i]		
			3.4 (0.4)[j]		
Benzo[*e*]pyrene	**18.8 ± 8.0**	3.5 (0.5)[b]	5.2 (0.3)[k]	3.3 (0.2)[i]	9.6 (0.3)[k]
	18.8 (0.8)[b]		5.2 (0.3)[i]		

Perylene	**3.4 ± 2.2**[c]	**31.0 ± 3.0**	0.76 (0.07)[i]	0.84 (0.09)[i]	0.13 (0.02)[l]
	2.5 (0.2)[f]		0.69 (0.02)[j]	0.65 (0.2)[j]	
Benzo[*ghi*]perylene	7.1 (0.8)[b]	1.7 (0.04)[b]	5.3 (0.4)[k]	**4.5 ± 1.1**	**2.4 ± 0.6**
			4.7 (0.1)[i]		
Indeno[1,2,3-*cd*]pyrene	4.6 (0.6)[b]	0.17 (0.04)[b]	4.8 (0.2)[k]	**3.3 ± 0.5**	1.8 (0.1)[k]
			4.4 (0.15)[i]		2.1 (0.1)[k]
			4.7 (0.2)[j]		3.2 (0.3)[l]

[a] Certified concentrations in bold (see Certificate of Analysis). The certified values were obtained from the combined results of at least two analytical results and the associated uncertainties are generally expressed as ± 2 standard deviations of the mean values of the two techniques or the 95% confidence intervals of each of the techniques. Nonbold values are either noncertified values from the Certificate of Analysis or NIST measurements from the literature (see references indicated); values in parentheses are uncertainties expressed as ± 1 standard deviation of the mean value of replicate analyses which recognize only the within-method variability.

[b] GC-MS electron impact.[23]

[c] Determined by LC-FL[24] using methodology described in Reference 8.

[d] LC quenchofluorometric detection.[25]

[e] LC value used in certification (3.9 ± 0.6 μg/g) contained a contribution from benzo[*k*]fluoranthene, which coeluted with perylene on the column used. GC-MS value was 2.8 ± 0.6 μg/g.

[f] LC-FL.[8]

[g] GC-MS methane positive chemical ionization.

[h] GC-MS methane negative ion chemical ionization.[7]

[i] Determined by GC.[9]

[j] LC-FL.[9]

[k] GC-MS negative ion chemical ionization.[26]

[l] LC-FL.

[m] GC-MS value contains contribution from triphenylene.

[n] LC-FL, data from Reference 27.

Table 5 Concentrations (μg/kg) of PAHs in Natural Matrix SRMs (Coal Tar, Sediments, and Mussel Tissue)[a]

Compound	Coal tar (mg/kg) SRM 1597[b]	River sediment SRM 1939[c]	Marine sediment SRM 1941a[d]	Mussel tissue SRM 1974[e]
Naphthalene	1160.0 ± 50.0		**1010.0 ± 140.0**	
Biphenyl	27.4 (0.3)[f]		**175.0 ± 18.0**	
Fluorene	136.0 (1.0)[f]		**97.3 ± 8.6**	12.0 (2.0)
Phenanthrene	**462.0 ± 3.0**	130 (10)	**489.0 ± 23.0**	**45.0 ± 11.0**
Anthracene	**101.0 ± 2.0**		**184.0 ± 14.0**	**6.1 ± 1.7**
1-Methylphenanthrene			**101.0 ± 27.0**	19.0 (5.0)
Fluoranthene	**322.0 ± 4.0**	190 (10)	**981.0 ± 78.0**	**272.0 ± 47.0**
Pyrene	**235.0 ± 2.0**	170 (20)	**811.0 ± 24.0**	**276.0 ± 30.0**
Chrysene	**71.7 ± 1.0**	51[g] (1)	**380.0 ± 24.0**	124.0[g] (11.0)
Triphenylene	**12.1 ± 0.4**		**197.0 ± 11.0**	
Benz[*a*]anthracene	**98.6 ± 3.6**	46 (1)	**427.0 ± 25.0**	37.0 (3.0)
Benzo[*a*]fluoranthene			**118.0 ± 11.0**	4.1 (1.2)
Benzo[*b*]fluoranthene	70.1 (4.2)[h]		**740.0 ± 110.0**	**52.3 ± 9.4**
	61.4 (1.7)[i]			
Benzo[*k*]fluoranthene	47.4 (4.1)[h]		**361.0 ± 18.0**	24.0 (1.0)
	38.8 (1.3)[i]			
Benzo[*e*]pyrene	57.1 (0.5)[f]		**553.0 ± 59.0**	81.0 (6.0)
Benzo[*a*]pyrene	**95.8 ± 5.8**		**628.0 ± 52.0**	**18.6 ± 3.8**

Perylene	**26.1 ± 1.0**		**452.0 ± 58.0**	**8.5 ± 2.4**
Indeno[1,2,3-*cd*]pyrene	**60.2 ± 4.4**		**501.0 ± 72.0**	**14.6 ± 2.7**
Benzo[*ghi*]perylene	**53.7 ± 7.6**		**522.0 ± 67.0**	**20.0 ± 2.3**
Dibenz[*a,j*]anthracene	6.82 (0.27)[j]		**74.3 ± 6.8**	
Dibenz[*a,c*]anthracene	4.54 (0.13)[j]		**43.1 ± 3.7**	
Dibenz[*a,h*]anthracene	6.76 (0.51)[j]		**73.9 ± 9.7**	
Benzo[*b*]chrysene	10.3 (0.43)[j]		**99.0 ± 20.0**	
Picene	6.75 (0.30)[j]		**80.0 ± 9.0**	

[a] Certified concentrations in bold; see Certificate of Analysis for each SRM. Certified concentrations were obtained from the combined results of at least two independent analytical methods, and the uncertainties are generally a 95% prediction interval with an allowance for systematic error among the methods used. Noncertified values are generally results from one analytical technique, and the uncertainties (values in parentheses) are 1 standard deviation of a single measurement assuming all measurements are statistically independent and identically distributed.

[b] See References 28 and 29.

[c] Concentration reported on material as received which contains approximately 2.7% moisture.[15]

[d] Concentration reported on a dry weight basis; material as received contains approximately 2.2% moisture.[12]

[e] Concentration reported on a dry weight basis; material as received contains approximately 88% water; see Reference 13.

[f] GC analysis.[28]

[g] Concentration includes both chrysene and triphenylene.

[h] GC analysis on liquid crystalline column.[8]

[i] LC-FL analysis.[28]

[j] Multidimensional LC-FL analysis.[29]

of over 30 PAHs in SRM 1597 have been reported,[28–30] and the results for 22 of these PAHs are provided in Table 5.

In 1989 and 1990 two marine matrix SRMs were issued with certified and noncertified concentrations for PAHs: SRM 1941 (Organics in Marine Sediment) and SRM 1974 (Organics in Mussel Tissue). (Note: supplies of both of these SRMs are now depleted and renewal materials have been issued or will be issued; see discussion below.) Sediment material for SRM 1941 was collected in the Baltimore Harbor (Maryland);[10] this sediment SRM was issued with 11 certified and 24 noncertified concentrations for PAHs at levels of 40 to 1300 μg/kg.

The development of SRM 1974 represented a unique approach to the production of a biological tissue reference material. The majority of biological tissue reference materials available from NIST and other producers of reference materials (primarily for trace element determinations) are distributed as lyophilized (freeze-dried) matrices. In recent years some limitations of these traditional biological and environmental reference materials have been recognized. The requirements for a reference material regarding physical and/or chemical stability, stable dry weight, and sterility have restricted the matrices that could be used as reference materials or have necessitated sample preparation that often significantly alters the physical or chemical properties of the material. Sample preparation procedures for previous biological marine reference materials have included freeze-drying, fat extraction, radiation sterilization, or more extreme measures such as cooking. The physical characteristics of these materials are, in many instances, significantly different from the sample matrix actually analyzed (this is particularly true for samples analyzed for trace organic constituents); thus, the value of these materials as quality control samples is often limited. To address these concerns, SRM 1974, Organics in Mussel Tissue (*Mytilus edulis*) was issued as a frozen tissue homogenate.[13] Approximately 2500 mussels were collected in Boston Harbor (Massachusetts) and the shells removed to provide 30 kg of tissue. This material was then cryogenically homogenized[31] and the SRM provided as a frozen powder-like homogenate, thereby providing a matrix similar to the sample matrices typically encountered in marine tissue analyses. SRM 1974 was issued with certified concentrations for nine PAHs and noncertified concentrations for 19 additional PAHs. Even though this material was developed primarily for organic constituents, noncertified concentrations of 36 trace elements were reported, thereby providing a reference material that satisfies the matrix requirements of inorganic analysts who routinely analyze fresh tissue samples.

Because of the widespread use of SRM 1941 and 1974, the supplies of these two materials were depleted in 1993, and replacement materials have been issued as SRM 1941a and SRM 1974a (currently in preparation). For the reissue of SRMs 1941 and SRM 1974, larger quantities of sediment and tissue were collected to provide sufficient material to extend the availability of these SRMs to 7 to 8 years. In 1991 a new sediment material was collected at the same Baltimore harbor location as the material used for SRM 1941. The sediment material was freeze-dried and then sieved to obtain the 150 to 250 μm particle size fraction for use as SRM 1941a. The certification of SRM 1941a has been described in detail elsewhere[11,12] and will be discussed below to illustrate the analytical approach currently used to certify PAHs in natural matrix SRMs. In 1992 approximately 6000 mussels were collected at the same Boston Harbor location as the mussels used for SRM 1974. Approximately 80 kg of tissue were obtained from the mussels, and the tissue was cryogenically homogenized to prepare SRM 1974a. The quantities of sediment and mussel tissue collected for these replacement SRMs represented nearly three times the amount of material collected for the original sediment and mussel tissue SRMs. The concentrations of PAHs in SRM 1941a and SRM 1974 are summarized in Table 5. (Final results for SRM 1974a are not yet available, but the PAH concentrations are similar to those in SRM 1974).

Another sediment matrix material, SRM 1939, Polychlorinated Biphenyls (Congeners) in River Sediment, was issued in 1990 primarily for PCB content (see discussion below). SRM 1939 is a river sediment collected from the Hudson River (NY) and is representative of sediment with high levels of PCB congeners and chlorinated pesticides (100 to 7000 μg/kg and 60 to 550 μg/kg, respectively), but low levels of PAHs (50 to 200 μg/kg).[15] The concentrations of PAHs on SRM 1939 are provided only as noncertified values (see Table 5).

In addition to the PAHs listed in Table 5, other PAHs and nitro-PAHs have been determined in the fossil fuel, air and diesel particulate materials, coal tar, and sediment SRMs; these results have been reported in previous reviews.[1,3,29,30] In addition, because of the interest in the mutagenic activity of complex mixtures of PAHs and other organic analytes extracted from environmental samples, reference values for Ames bioassay mutagenicity have been determined for three NIST SRMs: SRM 1597, SRM 1649, and SRM 1650. The mutagenicity reference values were determined as part of an international collaborative study, sponsored by the International Programme on Chemical Safety (IPCS) and the EPA, to study the mutagenicity of complex environmental mixtures in the Ames *Salmonella typhimurium* mutation assay.[32] Samples of SRM 1597, SRM 1649, and SRM 1650 were assayed by 20 laboratories and the results compiled to determine the reference mutagenicity values. A detailed discussion of the treatment of the data to establish the reference values and the use of these reference values has been reported by Claxton et al.[33] These reference values for mutagenicity are reported in the Certificate of Analysis for each SRM and have been summarized in recent reviews.[1,4]

Natural Matrix SRMs for PCB and Chlorinated Pesticide Determinations

The first natural matrix SRMs for PCB measurements were oil and human serum matrices that were fortified with known quantities of Aroclor mixtures. SRM 1581, Polychlorinated Biphenyls in Oil, issued in 1982, was prepared by adding 100 μg/g of Aroclor 1242 and Aroclor 1260 to motor oil and transformer oil matrices. SRM 1589, Polychlorinated Biphenyls (as Aroclor 1260) in Human Serum, was prepared in a similar manner with the addition of 106 ng/g of Aroclor 1260 and was issued in 1985. The serum matrix was also fortified with 0.153 ng/g of 1,2,3,4-tetrachlorodibenzo-*p*-dioxin and 0.081 ng/g of 2,3,7,8-tetrachlorodibenzo-*p*-dioxin. Since 1988, five natural matrix SRMs have been issued with certified and/or noncertified concentrations of natural levels (i.e., not added) of individual PCB congeners and chlorinated pesticides: SRM 1588 (Organics in Cod Liver Oil), SRM 1939 (Chlorinated Biphenyls in River Sediment), SRM 1941/SRM 1941a (Organics in Marine Sediment), SRM 1974 (Organics in Mussel Tissue), and SRM 1945 (Organics in Whale Blubber). The concentrations of selected PCB congeners and chlorinated pesticides in these materials are summarized in Tables 6 and 7, respectively.

Table 6 Concentrations (μg/kg) of PCB Congeners in Natural Matrix SRMs for the Marine Environment[a]

Chlorobiphenyl number[b]	SRM name	Cod liver oil SRM 1588[c]	River sediment SRM 1939[d]	Marine sediment SRM 1941a[e]	Mussel tissue SRM 1974[f]
PCB 18	(2,2′,5-Trichlorobiphenyl)			1.15 ± 0.16	24.0 ± 9 .0
15	(4,4′-Dichlorobiphenyl)				
PCB 26	(2,3′,5-Trichlorobiphenyl)		**4200 ± 290**		
PCB 28	(2,4,4′-Trichlorobiphenyl)	28.3 ± 0.6	**2210 ± 100**	9.8 ± 3.7	62.0 ± 3.0
PCB 31	(2,4′,5-Trichlorobiphenyl)	8.3 ± 0.03	6860 (140)	6.2 ± 2.4	53.0 (4.0)
PCB 44	(2,2′,3,5′-Tetrachlorobiphenyl)	35.1 ± 1.4	**1070 ± 120**	**4.80 ± 0.62**	65.0 ± 8.0
PCB 49	(2,2′,4,5′-Tetrachlorobiphenyl)	29.8 ± 1.0		**9.5 ± 2.1**	116.0 (1.0)
PCB 52	(2,2′,5,5′-Tetrachlorobiphenyl)	83.3 ± 1.3	4480 (60)	**6.89 ± 0.56**	98.0 ± 23.0
PCB 66	(2,3′,4,4′-Tetrachlorobiphenyl)	54.8 ± 2.5	930 (10)[g]	**6.8 ± 1.4**	110.0 ± 5.0
PCB 95	(2,2′,3,5′,6-Pentachlorobiphenyl)	36.5 ± 0.8		**7.5 ± 1.1**	17.0 (1.0)
PCB 99	(2,2′,4,4′,5-Pentachlorobiphenyl)	213.0 ± 2.0		**4.17 ± 0.51**	45.0 (1.0)
PCB 101	(2,2′,4,5,5′-Pentachlorobiphenyl)	**129.0 ± 5.0**	820 (10)	**11.0 ± 1.6[h]**	105.0 ± 11.0[h]
90	(2,2′,3,4′,5-Pentachlorobiphenyl)				
PCB 105	(2,3,3′,4,4′-Pentachlorobiphenyl)	60.2 ± 2.3		**3.65 ± 0.27**	45.0 ± 3.0
PCB 110	(2,3,3′,4′,6-Pentachlorobiphenyl)	75.8 ± 1.9		**9.47 ± 0.85**	94.0 (1.0)
PCB 118	(2,3′,4,4′,5-Pentachlorobiphenyl)	176.0 ± 2.0	510 (10)	**10.0 ± 1.1**	
PCB 128	(2,2′,3,3′,4,4′-Hexachlorobiphenyl)	47.1 ± 2.6	100 (10)	**1.87 ± 0.32**	15.0 ± 2.0
PCB 138	(2,2′,3,4,4′,5′-Hexachlorobiphenyl)	**261.0 ± 29.0**	570 (10)	**13.38 ± 0.97**	110.0 ± 11.0
163	(2,3,3′,4′,5,6-Hexachlorobiphenyl)				
164	(2,3,3′,4′,5′,6-Hexachlorobiphenyl)				
PCB 149	(2,2′,3,4′,5′,6-Hexachlorobiphenyl)	105.0 ± 2.0		**9.2 ± 1.1**	85.0 (1.0)

PCB 151	(2,2′,3,5,5′,6-Hexachlorobiphenyl)	55.2 ± 2.2		2.62 ± 0.22	29.0 (1.0)
PCB 153	(2,2′,4,4′,5,5′-Hexachlorobiphenyl)	**276.0 ± 40.0**		**17.6 ± 1.9**	145.0 ± 8.0
PCB 156	(2,3,3′,4,4′,5-Hexachlorobiphenyl)	**0.93 ± 0.14**			
PCB 170	(2,2′,3,3′,4,4′,5-Heptachlorobiphenyl)	**45.0 ± 5.0**	110 (10)	**3.00 ± 0.46**	4.3 (0.1)
190	(2,3,3′,4,4′,5,6-Heptachlorobiphenyl)				
PCB 180	(2,2′,3,4,4′,5,5′-Heptachlorobiphenyl)	**107.0 ± 4.0**	160 (10)	**5.83 ± 0.58**	13.0 ± 1.0
PCB 187	(2,2′,3,4′,5,5′,6-Heptachlorobiphenyl)	35.3 ± 0.9	180 (10)	7.0 ± 2.6	30.0 ± 1.0
182	(2,2′,3,4,4′,5,6′-Heptachlorobiphenyl)				
PCB 194	(2,2′,3,3′,4,4′,5,5′-Octachlorobiphenyl)			**1.78 ± 0.23**	
PCB 206	(2,2′,3,3′,4,4′,5,5′,6-Nonachlorobiphenyl)			**3.67 ± 0.87**	
PCB 209	(Decachlorobiphenyl)			**8.34 ± 0.49**	

[a] Certified concentrations in bold; see Certificate of Analysis for each SRM. Certified concentrations were obtained from the combined results of at least two independent analytical methods, and the uncertainties are generally a 95% prediction interval with an allowance for systematic error among the methods used. Noncertified values are generally results from one analytical technique; the uncertainties, when expressed as ± values, are a 95% prediction interval, and the uncertainties, when indicated in parentheses, are 1 standard deviation of a single measurement, assuming all measurements are statistically independent and identically distributed.

[b] PCB congeners are numbered according to the scheme proposed by Ballschmiter and Zell[34] and later revised by Schulte and Malisch[35] to conform with IUPAC rules. When two or more congeners are known to coelute, the PCB congener listed first is the major component and the additional congeners may be present as minor components. The quantitative results are based on the response of the congener listed first.

[c] See Reference 33.

[d] Concentration reported on material as received, which contains approximately 1.4% moisture.[15]

[e] Concentration reported on a dry weight basis; material as received contains approximately 2.2% moisture.[12]

[f] Concentration reported on a dry weight basis; material as received contains approximately 88% water.[13]

[g] PCB 95 and PCB 66 were not separated in these analyses.

[h] PCB 101 and PCB 90 were separated in these analyses.

Table 7 Concentrations (μg/kg) of Chlorinated Pesticides in Natural Matrix SRMs[a]

Compound	Cod liver oil SRM 1588[b]	River sediment SRM 1939[c]	Marine sediment SRM 1941a[d]	Mussel tissue SRM 1974[e]
Hexachlorobenzene	**148.0 ± 21.0**		**70.00 ± 25.00**	2.0 (0.2)
α-HCH	**86.0 ± 19.0**			5.8 (0.2)
γ-HCH	25.5 ± 1.7			1.3 (0.1)
trans-Chlordane	**50.0 ± 13.0**			
Heptachlor epoxide	33.0 ± 2.0	510 (20)[f]		26.0 (3.0)
cis-Chlordane	**158.0 ± 8.0**	220 (10)	**2.33 ± 0.56**	26.0 ± 1.0
trans-Nonachlor	**209.0 ± 11.0**	270 (10)	**1.26 ± 0.13**	21.0 ± 5.0
cis-Nonachlor	94.4 ± 1.4			
Oxychlordane			2.59 ± 0.19	
Dieldrin	**150.0 ± 12.0**		1.26 ± 0.37	8.0 ± 4.0
2,4′-DDE	22.0 ± 1.1		**0.73 ± 0.11**	5.8 ± 0.6
4,4′-DDE	**641.0 ± 62.0**	540 (10)	**6.59 ± 0.56**	48.0 ± 2.0
2,4′-DDD	36.3 ± 1.4			**20.0 ± 7.0**
4,4′-DDD	**277.0 ± 15.0**	60 (10)	**5.06 ± 0.58**	
2,4′-DDT	**156.0 ± 5.0**			4.0 ± 1.0
4,4′-DDT	**529.0 ± 45.0**		1.25 ± 0.10	3.0 ± 2.0

[a] Certified concentrations in bold; see Certificate of Analysis for each SRM. Certified concentrations were obtained from the combined results of at least two independent analytical methods, and the uncertainties are generally a 95% prediction interval with an allowance for systematic error among the methods used. Noncertified values are generally results from one analytical technique; the uncertainties, when expressed as ± values, are a 95% prediction interval, and the uncertainties, when indicated in parentheses, are 1 standard deviation of a single measurement, assuming all measurements are statistically independent and identically distributed.

[b] See Reference 14.

[c] Concentration reported on material as received, which contains approximately 1.4% moisture.[15]

[d] Concentration reported on a dry weight basis; material as received contains approximately 2.2% moisture.[12]

[e] Concentration reported on a dry weight basis; material as received contains approximately 88% water.[13]

[f] Concentration includes both heptachlor epoxide and oxychlordane.

The first of these materials, SRM 1588, Organics in Cod Liver Oil, serves as an excellent surrogate for a tissue extract with a high lipid content. This SRM has certified concentrations for five PCB congeners and ten chlorinated pesticides. Measurements for 20 additional PCB congeners and four additional chlorinated pesticides have been performed since the material was first issued in 1989.[36,37] The cod liver oil was also fortified with seven polychlorodibenzo-*p*-dioxins (PCDDs) and polychlorodibenzofurans (PCDFs) at the 0.3 to 1 ng/g level. SRM 1588 was followed by the sediment and mussel tissue materials described above for PAH measurements, i.e., SRM 1941/1941a, Organics in Marine Sediment, and SRM 1974/1974a, Organics in Mussel Tissue (*Mytilus edulis*), and two materials specifically for PCB measurements, SRM 1939, Polychlorinated Biphenyls (Congeners) in River Sediment, and SRM 1945, Organics in Whale Blubber. SRMs 1941 and 1974 were issued with certified concentrations for PAHs and only noncertified concentrations for PCB congeners and chlorinated pesticides.[10,13] However, when the two materials were renewed as SRMs 1941a and 1974a, the certified concentrations included 25-35 PCBs and pesticides as well as PAHs. SRM 1939 was prepared specifically for PCB congener measurements and is representative of a highly contaminated sediment. SRM 1939 has certified and noncertified concentrations of 17 PCB congeners and 5 pesticides.[15] SRM 1945, Organics in Whale Blubber, was issued in 1994 with certified concentrations for 27 PCB congeners and 15 chlorinated pesticides. This SRM was prepared from blubber collected from a stranded pilot whale in 1991. The blubber was cryogenically pulverized and homogenized as described above for the mussel tissue SRMs (SRM 1974/SRM 1974a). The certification of SRM 1945 has been described in detail previously[16] and will be described below briefly to illustrate the current analytical approach used at NIST for certification of PCB congeners and chlorinated pesticides. SRM 1945 is currently the most extensively characterized natural matrix material available with respect to PCB congeners and chlorinated pesticides.

SRM Activities in Progress

SRM 2974, Organics in Freeze-Dried Mussel Tissue (*Mytilus edulis*). Approximately 20 kg of the mussel tissue collected for SRM 1974a was freeze dried and will be issued as SRM 2974. This freeze-dried mussel tissue material was prepared in order to make it more readily available outside the U.S., since shipment of the frozen tissue material (SRM 1974a) is more difficult. SRM 2974 will be issued in 1995 with certified concentrations for PAHs, PCB congeners, and chlorinated pesticides.

GESREM II Mussel Tissue. As part of a joint international effort, NIST is working with the Group of Experts on Standards and Reference Materials (GESREM) to produce a mussel tissue reference material primarily for distribution to developing countries. This reference material, known as Reference Material (RM) 8045, GESREM II Mussel Tissue, is a freeze-dried tissue prepared from mussels collected in Raritan Bay, NJ. GESREM II will be analyzed at NIST and the International Atomic Energy Agency laboratory in Monaco to provide results for selected PAHs, PCBs, and chlorinated pesticides.

Recertification of SRMs 1649 and 1650. SRM 1649, Urban Dust/Organics, issued in 1982, was one of the first natural matrix environmental SRMs developed at NIST for the measurement of organic constituents. This SRM has certified concentrations for only five PAHs and noncertified values for nine additional PAHs. SRM 1650 was issued in 1985 and has certified and noncertified concentrations for a similar number of PAHs and nitro-PAHs. Since these two materials were issued, we have implemented improved and/or additional methods for the measurement of a large number of PAHs (see discussion below). Using current methodologies, the reanalysis of these two SRMs is currently in progress to provide certified concentrations for 15 to 20 PAHs and noncertified values for an additional 15 to 20 PAHs. Certified and noncertified concentrations of selected PCBs and chlorinated pesticides will also be determined for SRM 1649. By significantly increasing the number of certified and noncertified concentrations for PAHs in SRM 1649 and SRM 1650, these materials will be more useful to the organic analytical community.

Diesel Particulate Extract SRM. To evaluate the variability associated with extraction of particulate samples, the analytical environmental chemistry and the bioassay communities have expressed interest in the development of an SRM that is an extract of diesel particulate matter or air particulate matter to complement the existing particulate matrix SRMs that require extraction. Bulk quantities of a new diesel particulate material with relatively high mutagenic activity were obtained for use in the preparation of this SRM. This bulk material was extracted with methylene chloride, and the extract was ampuled as SRM 1975, Diesel Particulate Extract. SRM 1975 will be characterized for selected PAHs and nitro-PAHs as well as for mutagenic activity.

ANALYTICAL APPROACH FOR THE CERTIFICATION OF NATURAL MATRIX SRMs

At NIST the approach to certification of environmental matrix SRMs is based on the use of two or more "chemically independent" analytical techniques. The results of the analyses using multiple techniques, if in agreement, are used to determine the certified concentrations of the measured analytes. The requirement for using two or more different techniques is based on the assumption that the agreement of results from the independent techniques implies that the results are unbiased. If results are obtained from only one analytical technique, the concentrations are generally reported as noncertified or information values. The required independent analytical procedures include different extraction and clean-up/isolation procedures as well as separation and detection techniques. Since the first natural matrix SRM for organic contaminants was issued in 1980 (i.e., SRM 1580), the analytical methods including the cleanup/isolation, the final separation and detection, and the approach to quantitation have improved. The development and evolution of the analytical procedures used during the past 15 years for the certification of PAHs, PCB congeners, and chlorinated pesticides have been described in detail in several publications.[1,3] In this chapter, the certification of two recently issued SRMs will be will be described as examples of the current analytical approach used at NIST for the certification of PAHs, PCBs, and chlorinated pesticides.

Determination of PAHs in SRM 1941a

For the determination of PAHs in the natural matrix SRMs listed in Table 2, various combinations of reversed-phase LC-FL, GC-FID, and GC-MS have been used to provide the necessary two or more independent analytical techniques. SRM 1941a, Organics in Marine Sediment, is the most recent natural matrix SRM issued with certified concentrations for PAHs. Prior to the certification of SRM 1941a, NIST typically provided certified concentrations for only 5 to 12 PAHs; an additional 5 to 25 PAHs were often listed as noncertified concentrations primarily because of the lack of measurements by the required second independent analytical technique.[3,9,10,13,28] The goal in the certification of SRM 1941a was to significantly increase the number of certified compounds and to reduce the uncertainties associated with these certified concentrations. To achieve this goal, four different analytical techniques were implemented to provide measurements for a larger number of PAHs: (1) reversed-phase LC-FL analysis of the total PAH fraction, (2) reversed-phase LC-FL analysis of isomeric PAH fractions isolated by normal-phase LC (i.e., multidimensional LC), (3) GC-MS analysis of the PAH fraction on a 5% phenyl-substituted methylpolysiloxane stationary phase (denoted as 5% phenyl phase), which is the typical GC stationary phase used for PAH analyses, and (4) GC-MS of the PAH fraction on a smectic liquid crystalline stationary phase, which provides excellent shape selectivity for the separation of PAH isomers. These techniques have been reported previously for the measurement of PAHs in environmental SRMs;[3,9,10,13,28,38] however, for the analysis of SRM 1941a, all four techniques were used together to achieve certified values. The analytical scheme for the certification measurements for PAHs in SRM 1941a is illustrated in Figure 1. Two different solvent systems (i.e., methylene chloride and n-hexane/acetone) were used for the Soxhlet extraction, and two different procedures were used for the clean-up and isolation of the PAH fraction from the extracts. These extraction and clean-up procedures have been the typical approach used in the certification of most of the previous natural matrix SRMs. The assumption is made that each technique has different sources of bias; therefore, if the final results are in agreement, it suggests that neither technique was biased.

Two difficulties existed for increasing the number of PAHs with certified concentrations in natural matrix SRMs. For the certification of PAHs in previous natural matrix environmental SRMs, the two basic analytical techniques were GC-MS and LC-FL. GC-MS on a 5% phenyl phase is the typical procedure for the measurement of PAHs, and it generally provides results for 20 to 30 PAHs. However, reversed-phase LC-FL typically provides results for only 10 to 12 major PAHs in complex natural environmental PAH mixtures, thereby limiting the number of PAHs measured by two independent techniques as required to provide certified values. This problem was overcome by using a multidimensional LC procedure to isolate and enrich specific isomeric PAHs that are not easily measured in the total PAH fraction because

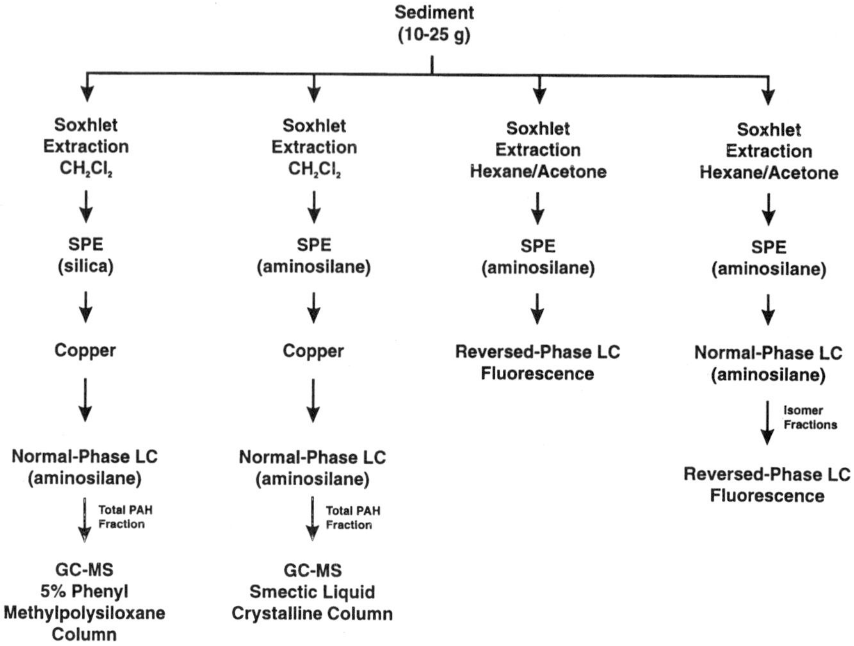

Figure 1 Analytical scheme for the determination of PAHs in SRM 1941a using different analytical techniques.

of low concentrations or low sensitivity and selectivity of the fluorescence detection. The multidimensional LC procedure, which has been described previously,[9,28,39,40] consists of a normal-phase LC separation of the environmental extract on an aminopropylsilane stationary phase, followed by a reversed-phase LC separation on a C_{18} stationary phase. The normal-phase LC step provides a separation of PAHs based on the number of aromatic carbon atoms in the PAH, thereby providing fractions containing only isomeric PAHs and alkyl-substituted isomers.[39,40] These isomeric PAH fractions were then analyzed by reversed-phase LC-FL to separate and quantify the various isomers.

A second problem associated with increasing the number of certified PAH concentrations is the lack of separation of several important isomers on the 5% phenyl phase commonly used in the GC-MS determination of PAHs. For example, the following isomeric PAHs are not completely separated on this stationary phase: chrysene and triphenylene; benzo[*b*]fluoranthene and benzo[*j*]fluoranthene; and dibenz[*a,c*]anthracene and dibenz[*a,h*]anthracene. Reversed-phase LC on a polymeric C_{18} phase provides excellent separation of isomeric PAHs,[40] but a second independent technique is necessary to provide certified results for several of the individual isomers. GC on a smectic liquid crystalline stationary phase exhibits high shape selectivity for PAH separations,[41,42] thereby providing a separation of PAH isomers similar to that achieved in LC with the polymeric C_{18} phase.[43] Thus, the GC-MS analyses on the two columns with different selectivity for PAH separations complement the two LC procedures (i.e., total fraction analysis and the multidimensional approach) to provide results from two or more techniques for over 20 PAHs.

The results obtained for 25 PAHs from the analysis of SRM 1941a using the four analytical techniques are summarized in Table 8. These results illustrate the necessity of using all four techniques to obtain the required information for certification of a large number of PAHs. For the GC-MS analysis on the 5% phenyl phase, results were obtained for over 40 PAHs; however, only those for which data were obtained from the other techniques are provided in Table 8.

The LC-FL analysis of the total PAH fraction provided results for 13 PAHs. However, results for chrysene, benz[*a*]anthracene, benzo[*ghi*]perylene, and indeno[1,2,3-*cd*]pyrene from the LC-FL analysis of the total PAH fraction were not used to determine the certified values (see discussion below). A chromatogram from the LC-FL analysis of the total PAH fraction from the sediment extract is shown in Figure 2. The LC-FL procedure utilizes fluorescence wavelength programming to optimize the selectivity and/or sensitivity for the individual PAHs.[9,40]

The multidimensional LC procedure was used to obtain results for three groups of isomeric PAHs: (1) four-ring isomers of molecular weight 228 (triphenylene, chrysene, and benz[*a*]anthracene); (2) six-ring isomers of molecular weight 276 (benzo[*ghi*]perylene, indeno[1,2,3-*cd*]pyrene, and anthanthrene); and (3) five-ring isomers of molecular weight 278 (dibenz[*a,c*]anthracene, dibenz[*a,j*]anthracene, dibenz[*a,h*]anthracene, pentaphene, benzo[*b*]chrysene, and picene). The reversed-phase LC-FL analyses of these three isomeric PAH fractions are illustrated in Figure 3. The 228 molecular weight fraction was isolated specifically to obtain results for triphenylene, which could not be measured accurately in the total PAH fraction due to low fluorescence selectivity and sensitivity. In addition, the results obtained for chrysene and benz[*a*]anthracene in this fraction were considered to be more reliable than those from the total PAH fraction because of the removal of minor coeluting compounds (compare Figure 2 and Figure 3A for chrysene). The 276 molecular weight fraction was isolated since results for benzo[*ghi*]perylene and indeno[1,2,3-*cd*]pyrene obtained from the analysis of the total PAH fraction were questionable because of low detector sensitivity and possible coeluting compounds (see Figure 2 and Figure 3B). The multidimensional procedure was used previously for the certification of SRM 1597 and SRM 1941 for the 228 and 276 molecular weight PAHs; however, a major step toward increasing the number of PAHs certified in SRM 1941a was the development of the multidimensional procedure for the measurement of the 278 molecular weight isomers.[29] This procedure allowed for the measurement of six additional PAH isomers that had not been determined previously as part of the certification process (see Figure 3C).

The GC MS analysis of SRM 1941a using the liquid crystalline stationary phase provided results for over 30 PAHs including several critical isomers that were not separated using the 5% phenyl methylpolysiloxane phase: chrysene and triphenylene; benzo[*b*]fluoranthene and benzo[*j*]fluoranthene; and dibenz[*a,c*]anthracene and dibenz[*a,h*]anthracene. The comparison of the GC-MS single ion chromatograms for these three isomer groups on the smectic liquid crystalline stationary phase and the 5% phenyl methylpolysiloxane phase are shown in Figure 4 for the analysis of SRM 1941a. The results for these three groups of isomers were compared with the

Table 8 Summary of Analytical Results (μg/kg dry weight) and Certified Concentrations for PAHs in SRM 1941a[a,b]

Compound	GC-MS (5% phenyl)		GC-MS (liquid crystal)		LC-FL (total)		LC-FL (fractions)		Certified values[c]
Naphthalene	1095.0	(62.0)	1031.0	(34.0)	893.0	(66.0)			**1010.0 ± 140.0**
Fluorene	94.3	(7.3)	101.5	(9.5)					**97.3 ± 8.6**
Phenanthrene	492.0	(32.0)	476.0	(31.0)	491.0	(11.0)			**489.0 ± 23.0**
Anthracene	193.0	(12.0)	184.0	(11.0)	182.9	(6.1)			**184.0 ± 14.0**
Fluoranthene	1012.0	(57.0)	1009.0	(58.0)	929.0	(44.0)			**981.0 ± 78.0**
Pyrene	814.0	(33.0)	812.0	(51.0)	808.0	(38.0)			**811.0 ± 24.0**
Benz[*a*]anthracene	423.0	(19.0)	444.0	(26.0)			423.0	(12.0)	**427.0 ± 25.0**
Chrysene	[592.0][d]	(35.0)	396.0	(35.0)			379.3	(7.7)	**380.0 ± 24.0**
Triphenylene			191.0	(10.0)			202.3	(3.0)	**197.0 ± 11.0**
Benzo[*b*]fluoranthene	[979.0][e]	(82.0)	739.0	(50.0)	800.0	(150.0)			**740.0 ± 110.0**
Benzo[*j*]fluoranthene			341.0	(22.0)					
Benzo[*k*]fluoranthene	416.0	(45.0)	362.0	(26.0)	361.0	(21.0)			**361.0 ± 18.0**
Benzo[*a*]fluoranthene	114.0	(13.0)	119.0	(7.8)					**118.0 ± 11.0**
Benzo[*e*]pyrene	522.0	(46.0)	587.0	(45.0)					**553.0 ± 59.0**
Benzo[*a*]pyrene	611.0	(44.0)	656.0	(93.0)	639.0	(34.0)			**628.0 ± 52.0**

Perylene	495.0	(41.0)	434.0	(25.0)	434.0	(22.0)			**452.0 ± 58.0**
Indeno[1,2,3-*cd*]pyrene	446.0	(23.0)	537.0	(38.0)			520.0	(25.0)	**501.0 ± 72.0**
Benzo[*ghi*]perylene	480.0	(34.0)	576.0	(43.0)			522.0	(11.0)	**525.0 ± 67.0**
Dibenz[*a,j*]anthracene	76.2	(4.8)	70.0	(7.7)			76.1	(4.7)	**74.3 ± 6.8**
Dibenz[*a,c*]anthracene	[123.0][f]	(10.0)	41.3	(4.3)			44.8	(3.0)	**43.1 ± 3.7**
Dibenz[*a,h*]anthracene			76.0	(10.0)			73.6	(4.9)	**73.9 ± 9.7**
Pentaphene	47.8	(2.7)	35.4	(2.6)			46.7	(8.4)	**42.0 ± 12.0**
Benzo[*b*]chrysene	94.3	(5.8)	106.0	(13.0)			112.0	(17.0)	**99.0 ± 20.0**
Picene	80.4	(8.2)	75.9	(9.4)			80.8	(6.6)	**80.0 ± 9.0**

[a] Concentrations reported on dry weight basis; material as received contains approximately 2.2% moisture.

[b] Uncertainties (in parentheses) are reported as the standard deviation of a single measurement; subsamples from six bottles were extracted and analyzed in duplicate for each of the four techniques.

[c] The certified values (bold) are weighted means of results from two or more analytical techniques as described by Schiller and Eberhardt.[45] The uncertainty is based on a 95% confidence interval for the true concentration, and includes an allowance for differences among the analytical methods used.

[d] Concentration is the sum of the chrysene and triphenylene.

[e] Concentration is the sum of benzo[*b*]fluoranthene and benzo[*j*]fluoranthene.

[f] Concentration is the sum of dibenz[*a,c*]anthracene and dibenz[*a,h*]anthracene.

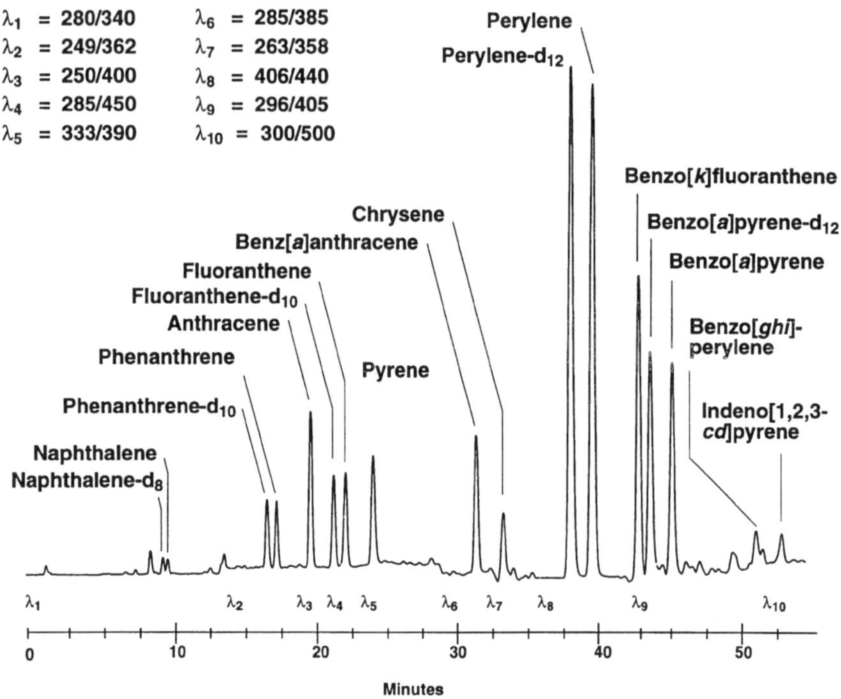

Figure 2 Reversed-phase LC with fluorescence detection analysis of total PAH fraction from SRM 1941a. (From S. A. Wise et al., *Anal. Chem.*, 67 (1995) 1171–1178.)

results from the two LC procedures to provide certified values for several PAHs not previously certified in any natural matrix SRM by NIST. The liquid crystalline stationary phase was first used as part of the certification of selected PAHs in SRM 1597, Complex Mixture of PAHs from Coal Tar.[28] However, results were obtained only for triphenylene and the three benzofluoranthene isomers, and only triphenylene was certified. The liquid crystalline phase provides excellent selectivity for the separation of isomers based on the shape of the PAH and provides selectivity similar to that of the polymeric C_{18} column used for the LC analyses.[43] However, the liquid crystalline phase has some limitations regarding reproducibility of retention times, different selectivity among columns, and loss of stationary phase with extended use.[44] Thus, these phases may not be considered suitable for routine PAH analyses, but can provide valuable information on selected isomers that cannot be obtained otherwise by GC-MS.

As shown in Table 8, the results from the various methods were generally in good agreement with differences among the methods typically in the range of 6 to 10%, with some as low as 1 to 5% (pyrene, phenanthrene, anthracene, benz[*a*]anthracene, chrysene, and triphenylene) and a few as high as 10 to 20% (benzo[*ghi*]perylene, indeno[*1,2,3-cd*]pyrene, and benzo[*b*]chrysene). Pentaphene was the only compound that had differences between methods of greater than 20%. For the critical isomers that are not completely separated on the 5% phenyl phase,

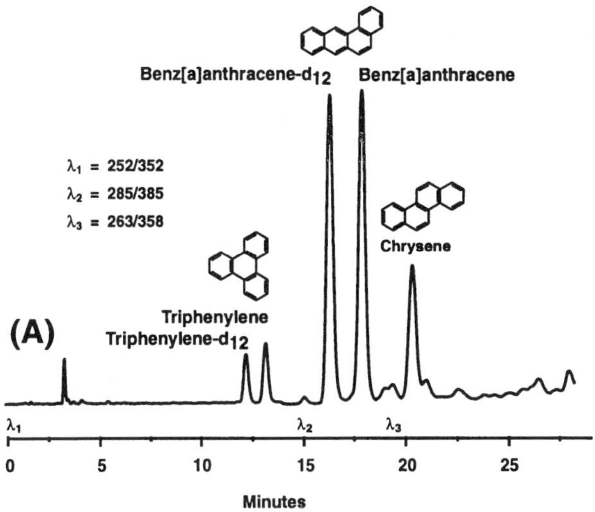

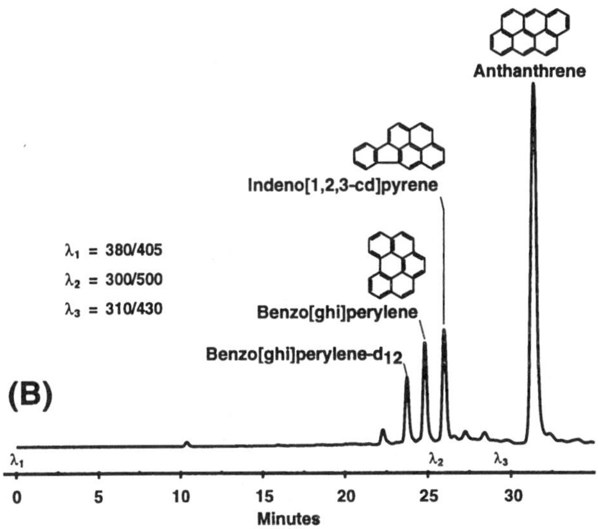

Figure 3 Reversed-phase LC with fluorescence detection analyses of isomeric PAH fractions isolated by normal-phase LC from extracts of SRM 1941a. (A) four-ring isomers of molecular weight 228; (B) six-ring isomers of molecular weight 276; and (C) five-ring isomers of molecular weight 278. (From S. A. Wise et al., *Anal. Chem.*, 67 (1995) 1171–1178.)

the agreement is excellent between the combined results and the sum of the individual isomers determined by the multidimensional LC and the GC-MS (liquid crystalline phase); that is, the sum of the individual values for chrysene (396 and 379 ng/g) and triphenylene (191 and 202 ng/g) agrees with the combined value of 592 ng/g, and the sum of the individual concentrations for dibenz[*a,c*]anthracene

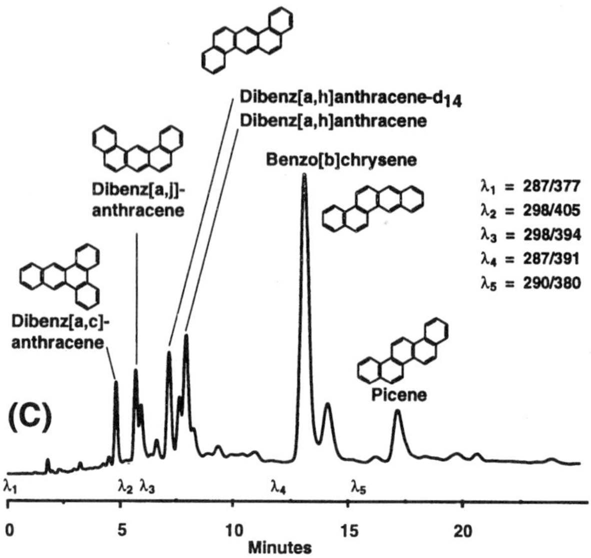

Figure 3 (continued).

(41 and 45 ng/g) and dibenz[*a,h*]anthracene (76 and 74 ng/g) determined by LC-FL (Fraction), and GC-MS liquid crystalline phase agrees well with the combined value of 123 ng/g determined by GC-MS (5% phenyl).

The certified concentrations, shown in Table 8, are weighted means of the results from two or more analytical techniques as described by Schiller and Eberhardt.[45] The weights depend on the relative precision and agreement among the methods. If the methods are in excellent agreement, the weights are inversely proportional to the variances of the method means. However, if the methods do not agree well, the weights are approximately equal. Two examples of results from the weighting scheme are illustrated in Figure 5. After the weights were determined, the variances of the method means and their degrees of freedom were combined to determine the variance of the weighted mean and its degrees of freedom. Because this variance does not include information on between-method differences, an allowance was also included for the observed between-method difference. A 95% confidence interval for the weighted mean, with an allowance for differences between the methods, is the uncertainty associated with the certified value. As shown in Figure 5, the weights for fluoranthene are nearly equal because of disagreement of the results among the three methods; however, for pyrene, where the results from all three methods are in good agreement, the GC-MS (SB-Smectic) results are weighted less because of their lower precision. Although the weighting has little effect on the certified value (since the method means agree well), the unequal weighting of the results reduces the total uncertainty.

One of the goals for the renewal issue of the sediment SRM was to reduce the uncertainties associated with the certified values (i.e., reduce the differences among the results from the different analytical methods). Eleven PAHs were certified in the

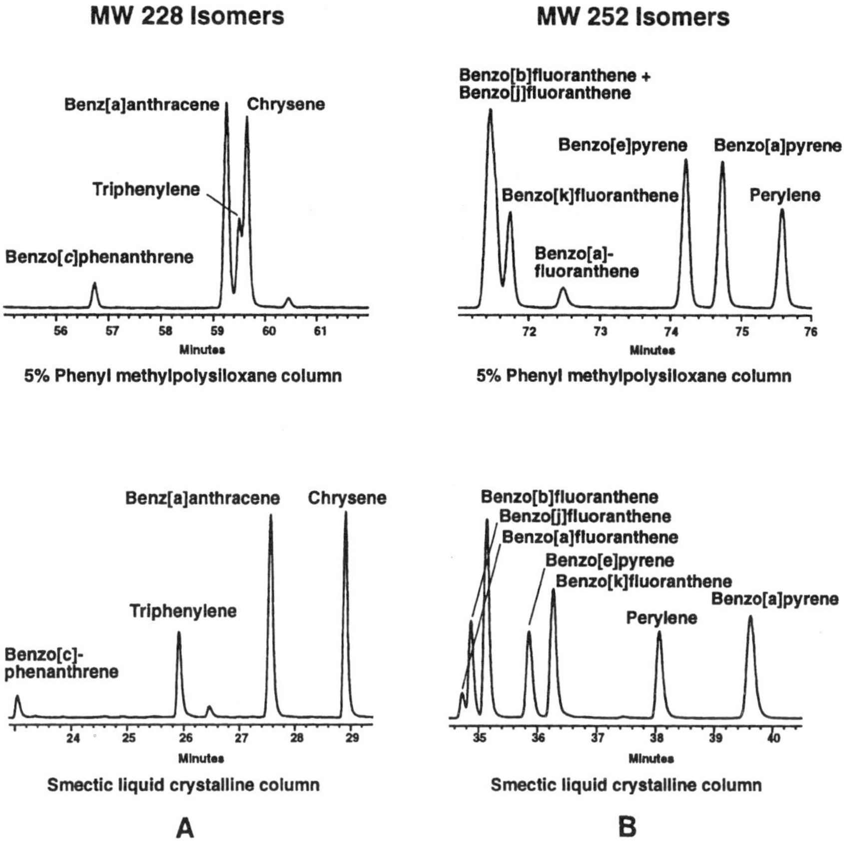

Figure 4 GC-MS analysis (single ion chromatogram) of total PAH fraction from SRM 1941a using 5% phenyl-substituted methylpolysiloxane and liquid crystalline stationary phases. (A) four-ring isomers of molecular weight 228; (B) six-ring isomers of molecular weight 252; and (C) five-ring isomers of molecular weight 278. (From S. A. Wise et al., *Anal. Chem.*, 67 (1995) 1171–1178.)

original SRM 1941 with uncertainties ranging from 7 to 24%. A comparison of the certified values and the associated uncertainties for these 11 PAHs in SRM 1941 and 1941a is provided in Table 9. The uncertainties associated with the certified values for SRM 1941a for the 11 PAHs previously certified in SRM 1941 ranged from 3 to 15% and were reduced compared to SRM 1941 for all but two of the PAHs certified in both materials (perylene and indeno[1,2,3-*cd*]pyrene). The decrease in the uncertainties for the certified values for the natural matrix SRMs as the analytical methods have improved since the first SRM was issued in 1980 until 1988 has been discussed in detail previously.[1,3]

Noncertified concentrations were determined for 14 additional PAHs and are listed in Table 10. Many of these compounds were determined by GC-MS analysis on both stationary phases; however, the results are listed as noncertified values generally because of the magnitude of the disagreement between the results from

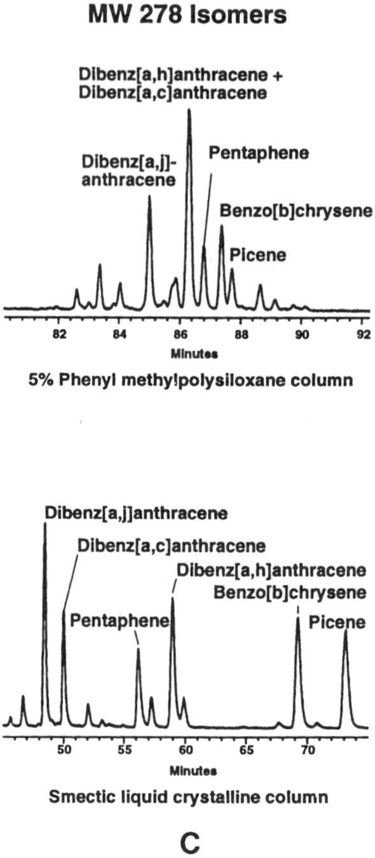

Figure 4 (continued).

the two methods or limited experience in measuring these particular PAHs on the liquid crystalline phase.

Determination of PCBs and Pesticides in SRM 1945

The certification of PCB congeners and chlorinated pesticides in natural matrix SRMs requires the same general approach, using at least two independent analytical techniques, as described above for PAHs. However, in contrast to using both LC and GC for the PAHs measurements, GC is the only suitable chromatographic technique at present for the measurement of PCBs and pesticides because of the low levels in environmental matrices and the selectivity and sensitivity of electron capture and mass spectrometric detection.

Of the SRMs listed in Table 2, only four have certified concentrations for PCB congeners and chlorinated pesticides: SRM 1588, SRM 1939, SRM 1941a, and SRM 1945. When SRM 1588 (Organics in Cod Liver Oil) was certified in 1989,[14] GC-ECD and GC-MS analyses were performed using similar columns, both containing

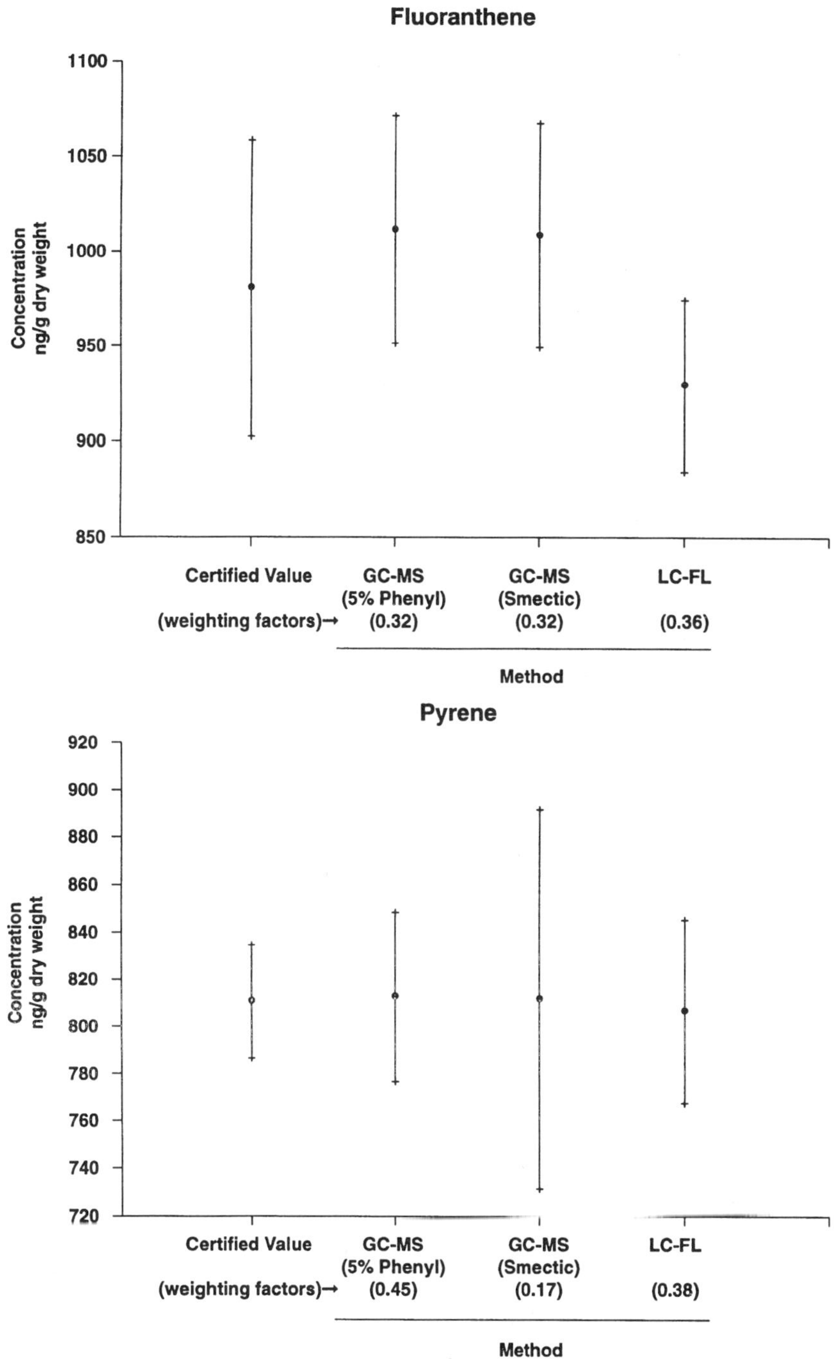

Figure 5 Graphical representations of results for fluoranthene (upper) and pyrene (lower) from three different analytical techniques and the weighting factors used in combining the results to obtain the certified values.

Table 9 Comparison of Certified Concentrations (μg/kg Dry Weight) and Uncertainties for SRM 1941 and SRM 1941a[a]

Compound	SRM 1941	SRM 1941a
Phenanthrene	577 ± 59 (10%)[b]	489 ± 23 (5%)
Anthracene	202 ± 42 (20%)	184 ± 14 (8%)
Fluoranthene	1220 ± 240 (20%)	981 ± 78 (8%)
Pyrene	1080 ± 200 (18%)	811 ± 24 (3%)
Benz[*a*]anthracene	550 ± 79 (14%)	428 ± 25 (6%)
Benzo[*b*]fluoranthene	780 ± 190 (24%)	741 ± 110 (15%)
Benzo[*k*]fluoranthene	444 ± 49 (11%)	361 ± 18 (5%)
Benzo[*a*]pyrene	670 ± 130 (19%)	628 ± 51 (8%)
Perylene	422 ± 33 (8%)	452 ± 58 (13%)
Benzo[*ghi*]perylene	516 ± 83 (16%)	525 ± 67 (13%)
Indeno[*1,2,3-cd*]pyrene	569 ± 40 (7%)	501 ± 72 (14%)

[a] The certified values are weighted means of results from two or more analytical techniques as described by Schiller and Eberhardt.[45] The uncertainty for each certified value is based on a 95% confidence interval for the true concentration and includes an allowance for differences among the analytical methods used.

[b] Values in parentheses are relative percent uncertainties.

Table 10 Noncertified Concentrations of PAHs in SRM 1941a

Compound	Concentration (μg/kg dry weight)[a,b]
Biphenyl[c,d]	175.0 ± 18.0
Acenaphthene[c,d]	41.0 ± 10.0
Acenaphthylene[c,d]	37.0 ± 14.0
Dibenzothiophene[c,d]	70.0 ± 9.4
3-Methylphenanthrene[c,d]	97.0 ± 32.0
2-Methylphenanthrene[c,d]	158.0 ± 32.0
1-Methylphenanthrene[c,d]	101.0 ± 27.0
4H-Cyclopenta[*def*]phenanthrene[c,d]	92.0 ± 15.0
Acephenanthrylene[c,d]	48.1 ± 1.2
Benzo[*ghi*]fluoranthene[d]	97.9 ± 3.1
Benzo[*c*]phenanthrene[c,d]	80.0 ± 39.0
Benzo[*j*]fluoranthene[d]	341.0 ± 22.0
Indeno[1,2,3-*cd*]fluoranthene[c]	20.0 ± 2.3
Anthanthrene[c,e]	129.0 ± 10.0

[a] Concentrations reported on dry weight basis; material as received contains approximately 2.2% moisture.

[b] Concentrations are the mean determined by the technique indicated or the weighted mean from the results of two techniques. The uncertainty for each concentration is based on a 95% confidence interval for the mean plus an allowance for differences between analytical methods when two methods were used; single subsamples from six bottles were extracted, and the extracts were analyzed in duplicate using the technique(s) indicated.

[c] Concentration determined by GC-MS on 5% phenyl phase.

[d] Concentration determined by GC-MS on liquid crystalline phase.

[e] Concentration determined by LC-FL.

a 5% phenyl phase. Due to the similarity of the columns used and the possibility of coelution problems, only five PCB congeners were certified in SRM 1588. In 1990 SRM 1939 (Polychlorinated Biphenyls [Congeners] in River Sediment) was certified using GC-ECD on the 5% phenyl phase and a dual column GC-MS method on a 5% phenyl phase with heart cutting of specific portions of the chromatogram onto a liquid crystalline (SB-Smectic) phase.[15,46] These two analytical techniques were considered to be suitably independent; however, only three PCB congeners, two trichlorobiphenyl congeners and one tetrachlorobiphenyl congener, were certified in SRM 1939, due to the thermal instability of the liquid crystalline phase.[44]

To implement more independent analytical techniques and to expand the number of PCB congeners and chlorinated pesticides with certified concentrations in future SRMs, different analytical techniques were evaluated for the measurement of PCB congeners and chlorinated pesticides in several existing NIST SRMs (SRM 1588, SRM 1941, and SRM 1974) and certified reference materials (CRMs) from the European Commission, Measurement and Testing Programme, formerly the Community Bureau of Reference (BCR), in Belgium (CRM 349, Chlorobiphenyls in Cod Liver Oil and CRM 350, Chlorobiphenyls in Mackerel Oil)).[37] These five materials were analyzed using three GC columns with different selectivity and two different GC detectors, ECD and MS. Using these techniques, concentrations for additional PCB congeners and chlorinated pesticides were reported for these reference materials, and the basis for two independent techniques for the certification of future SRMs was established. The first SRM certified using this approach was SRM 1945, Organics in Whale Blubber. The certification of SRM 1945 is described below to illustrate the current analytical approach for the certification of PCB congeners and chlorinated pesticides in natural matrices.

The analytical scheme for the certification of SRM 1945 is shown in Figure 6. Two different solvents (methylene chloride and *n*-hexane/acetone) were used for the Soxhlet extraction, and two different procedures were used for the clean-up and isolation of the PCB congeners and chlorinated pesticides. One set of samples was analyzed by GC-ECD on the 5% phenyl phase, and then the same sample extracts were analyzed on a dimethylpolysiloxane phase with 50% methyl C-18 (denoted as 50% methyl C-18 phase). A second set of samples was analyzed by GC-MS on a 5% phenyl phase. The 50% methyl C-18 phase provides a useful complement to the PCB congener separations obtained with the 5% phenyl phase with the resolution of PCB 15 from PCB 18, PCB 95 from PCB 66, PCB 90 from PCB 101, PCB 77 from PCB 110, and PCB 190 from PCB 170; all of these pairs normally coelute on the 5% phenyl phase. The 50% methyl C-18 phase has been characterized in detail by Ballschmiter et al.[47] Mass spectrometric (MS) detection discriminates between coeluting peaks with differences of an odd number of chlorines (i.e., 1, 3, 5, and so on). Therefore, of the congener pairs listed above, MS detection discriminates between PCB 15/18, PCB 66/95, and PCB 77/110. The 50% methyl C-18 and the 5% phenyl phases have different selectivity for the PCB separations. In addition to the separation of critical congener pairs mentioned above, other minor congeners demonstrate different selectivity. Based on the differences in sample preparation, separation selectivity, and detection selectivity, the assumption is made that each method has different sources of bias. The results

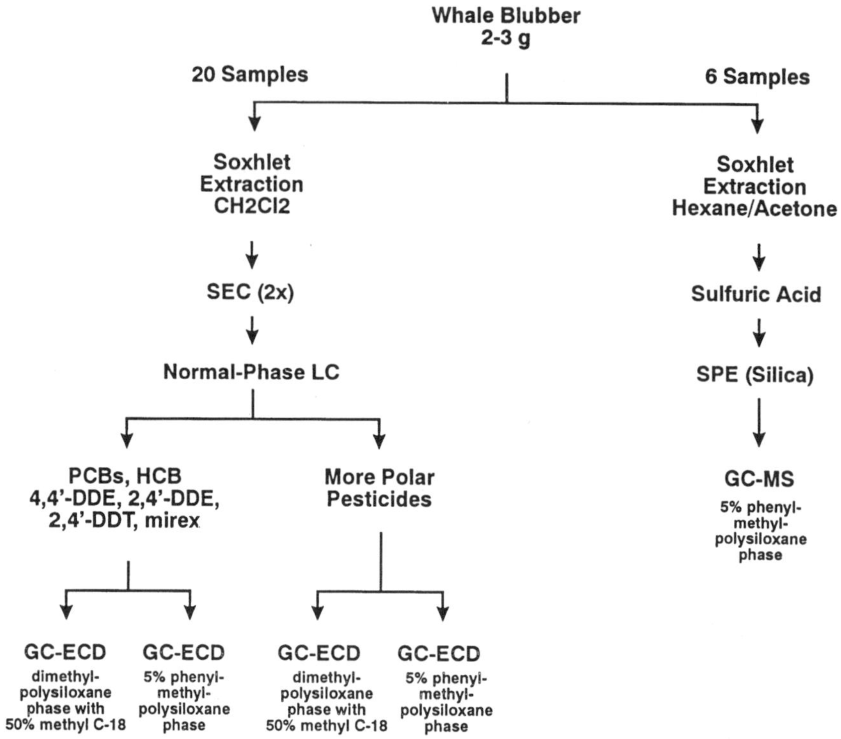

Figure 6 Analytical scheme for the determination of PCB congeners and chlorinated pesticides in SRM 1945 using different analytical techniques.

obtained for 29 PCB congeners and 17 chlorinated pesticides from the analysis of SRM 1945 using the three analytical techniques are summarized in Tables 11 and 12, respectively. Representative chromatograms for the PCB and lower polarity pesticide fraction from the GC-ECD analyses on the 50% methyl C-18 phase and on the 5% phenyl phase are shown in Figures 7 and 8, respectively. The GC-MS single ion monitoring traces for the tetrachloro-, pentachloro-, hexachloro-, and heptachlorobiphenyl congeners are shown in Figure 9. A representative chromatogram for the more polar pesticide fraction from the GC-ECD analyses on the 5% phenyl phase is shown in Figure 10.

Of the critical congener pairs listed above, the most significant impact of the use of the 50% methyl C-18 column and GC-MS analysis is for PCB 66 and 95 since they are present at similar levels. In SRM 1945, however, the concentrations of PCB 15, PCB 90, and PCB 190 were not significant, based on the GC-ECD analysis on the 50% methyl C-18 phase (see Table 11 and discussion below). The critical pairs PCB 101/90 and 170/190 are separated on the 50% methyl C-18 phase, but coelute on the 5% phenyl phase and are not distinguishable by MS detection. The concentration of PCB 90 is less than 10% of PCB 101 in this sample, as indicated by the concentration of 61.6 μg/kg wet weight for PCB 101 determined using the 50% methyl C-18 phase, compared to the concentrations of 66.3 and 68.3 μg/kg

Table 11 Summary of Analytical Results (μg/kg wet weight) for the Determination of PCB Congeners in SRM 1945

	Polychlorinated biphenyls[a]	GC-ECD[b] (50% methyl C-18)		GC-ECD[b] (5% phenyl)		GC-MS[c] (5% phenyl)		Certified value
PCB 18	(2,2′,5-Trichlorobiphenyl)	4.91	(0.32)	3.78	(0.30)	4.76	(0.33)	**4.48 ± 0.88**
PCB 28	(2,4,4′-Trichlorobiphenyl)	12.16	(0.91)	14.14	(0.47)	[16.52][d]	(0.38)	14.1 ± 1.4
PCB 31	(2,4′,5-Trichlorobiphenyl)	2.55	(0.23)	3.70	(0.24)	[16.52][d]	(0.38)	3.12 ± 0.69
PCB 44	(2,2′,3,5′-Tetrachlorobiphenyl)	12.25	(0.64)	13.17	(0.74)	11.25	(0.94)	**12.2 ± 1.4**
PCB 49	(2,2′,4,5′-Tetrachlorobiphenyl)	18.54	(0.84)	22.64	(0.74)	21.28	(0.95)	**20.8 ± 2.8**
PCB 52	(2,2′,5,5′-Tetrachlorobiphenyl)	45.1	(2.5)	42.8	(2.9)	42.4	(2.1)	**43.6 ± 2.5**
PCB 66	(2,3′,4,4′-Tetrachlorobiphenyl)	23.2	(1.8)	[59.7][e]	(2.8)	23.7	(1.9)	**23.6 ± 1.6**
PCB 87	(2,2′,3,4,5′-Pentachlorobiphenyl)	16.25	(0.68)	17.71	(0.91)	16.66	(0.47)	**16.7 ± 1.4**
PCB 95	(2,2′,3,5′,6-Pentachlorobiphenyl)	34.1	(1.7)	[59.7][e]	(2.8)	33.6	(1.6)	**33.8 ± 1.7**
PCB 99	(2,2′,4,4′,5-Pentachlorobiphenyl)	41.5	(2.9)	47.3	(2.8)	47.3	(2.3)	**45.4 ± 5.4**
PCB 101	(2,2′,4,5,5′-Pentachlorobiphenyl)	61.6	(3.3)[f]	66.3	(4.3)	68.3	(2.3)	**65.2 ± 5.6**
90	(2,2′,3,4′,5-Pentachlorobiphenyl)							
PCB 105	(2,3,3′,4,4′-Pentachlorobiphenyl)	31.6	(2.4)	30.0	(2.4)	29.8	(1.3)	**30.1 ± 2.3**
PCB 110	(2,3,3′,4′,6-Pentachlorobiphenyl)	25.9	(2.1)	[36.1][g]	(3.3)	20.7	(1.6)	**23.3 ± 4.0**
PCB 118	(2,3′,4,4′,5-Pentachlorobiphenyl)	74.9	(4.7)	75.6	(4.8)	71.7	(3.7)	**74.6 ± 5.1**
PCB 128	(2,2′,3,3′,4,4′-Hexachlorobiphenyl)	23.2	(2.2)	24.6	(2.1)	22.9	(1.5)	**23.7 ± 1.7**
PCB 138	(2,2′,3,4,4′,5′-Hexachlorobiphenyl)	131.7	(8.0)	127.7	(9.0)	134.2	(5.7)	**131.5 ± 7.4**
163	(2,3,3′,4′,5,6-Hexachlorobiphenyl)							
164	(2,3,3′,4′,5′,6-Hexachlorobiphenyl)							
PCB 149	(2,2′,3,4′,5′,6-Hexachlorobiphenyl)	101.6	(6.6)	106.6	(7.5)	110.6	(4.7)	**106.6 ± 8.4**
PCB 151	(2,2′,3,5,5′,6-Hexachlorobiphenyl)	32.9	(2.3)	26.7	(2.4)	26.7	(1.2)	**28.7 ± 5.2**
PCB 153	(2,2′,4,4′,5,5′-Hexachlorobiphenyl)	214.0	(15.0)	201.0	(16.0)	215.0	(11.0)	**213.0 ± 19.0**
PCB 156	(2,3,3′,4,4′,5-Hexachlorobiphenyl)	9.76	(0.69)	10.93	(0.66)	10.07	(0.72)	**10.3 ± 1.1**
PCB 170	(2,2′,3,3′,4,4′,5-Heptachlorobiphenyl)	39.3	(2.6)[h]	40.2	(2.6)	42.4	(1.9)	**40.6 ± 2.6**
190	(2,3,3′,4,4′,5,6-Heptachlorobiphenyl)							
PCB 180	(2,2′,3,4,4′,5,5′-Heptachlorobiphenyl)	105.5	(6.2)	108.8	(6.7)	109.5	(4.9)	**106.7 ± 5.3**
PCB 183	(2,2′,3,4,4′,5′,6-Heptachlorobiphenyl)	39.5	(2.9)	36.0	(2.8)	34.4	(1.8)	**36.6 ± 4.1**
PCB 187	(2,2′3,4,5,5′,6-Heptachlorobiphenyl)	111.7	(7.0)	103.1	(6.7)	100.9	(5.1)	**105.1 ± 9.1**
PCB 194	(2,2′,3,3′,4,4′,5,5′-Octachlorobiphenyl)	39.2	(2.4)	41.3	(2.6)	39.4	(1.8)	**39.6 ± 2.5**

Table 11 Summary of Analytical Results (μg/kg wet weight) for the Determination of PCB Congeners in SRM 1945 (continued)

Polychlorinated biphenyls[a]		GC-ECD[b] (50% methyl C-18)		GC-ECD[b] (5% phenyl)		GC-MS[c] (5% phenyl)		Certified value
PCB 195	(2,2′,3,3′,4,4′,5,6-Octachlorobiphenyl)	14.0	(1.3)	20.2	(1.2)	19.0	(1.2)	**17.7 ± 4.3**
PCB 201	(2,2′,3,3′,4,5′,6,6′-Octachlorobiphenyl)	16.9	(1.1)	17.2	(1.1)	16.48	(0.82)	**16.96 ± 0.89**
PCB 206	(2,2′,3,3′,4,4′,5,5′,6-Nonachlorobiphenyl)	32.9	(1.9)	30.7	(1.9)	29.7	(1.6)	**31.1 ± 2.7**
PCB 209	(Decachlorobiphenyl)	10.75	(0.87)	11.16	(0.94)	9.95	(0.85)[a]	**10.6 ± 1.1**

[a] PCB congeners are numbered according to the scheme proposed by Ballschmiter and Zell[34] and later revised by Schulte and Malisch[35] to conform with IUPAC rules. When two or more congeners are known to coelute, the PCB congener listed first is the major component, and the additional congeners may be present as minor components. The quantitative results are based on the response of the congener listed first.

[b] Duplicate samples from ten bottles extracted, and each extract analyzed in duplicate. Concentrations are the mean, and numbers in parentheses are 1 standard deviation of a single measurement. The same extracts were analyzed using GC-ECD on both the 50% methyl C-18 phase and the 5% phenyl phase.

[c] Samples from six bottles extracted and analyzed in duplicate. The concentration is the mean, and the number in parentheses is 1 standard deviation of a single measurement.

[d] Numbers in [ ] indicate known coelution of two or more congeners. PCB 28 and 31 coelute under the GC-MS conditions used.

[e] Numbers in [ ] indicate known coelution of two or more congeners. PCB 66 and 95 coelute under the GC-ECD (5% phenyl) conditions used.

[f] PCB 101 and PCB 90 were separated on the 50% methyl C-18 phase.

[g] Numbers in [ ] indicate known coelution of two or more congeners. PCB 77 (3,3′,4,4′-Tetrachlorobiphenyl) and 110 coelute under the GC-ECD (5% phenyl) conditions used.

[h] PCB 170 and PCB 190 were separated on the 50% methyl C-18 column.

Table 12 Summary of Analytical Results (μg/kg wet weight) for the Determination of Chlorinated Pesticides in SRM 1945

Compound	GC-ECD[a] (50% methyl C-18)		GC-ECD[a] (5% phenyl)		GC-MS[b] (5% phenyl)		Certified value
2,4′-DDE	12.46	(0.57)	11.69	(0.70)	12.27	(0.54)	**12.28 ± 0.87**
4,4′-DDE	421.0	(31.0)	453.0	(25.0)	440.0	(34.0)	**445.0 ± 37.0**
2,4′-DDD	20.0	(1.3)	16.2	(1.4)	17.96	(0.93)	**18.1 ± 2.8**
4,4′-DDD	128.0	(10.0)	132.4	(9.1)	138.3	(6.2)	**133.0 ± 10.0**
2,4′-DDT	113.0	(7.0)	98.0	(10.0)	101.6	(6.3)	**106.0 ± 14.0**
4,4′-DDT	238.0	(46.0)			246.0	(12.0)	**245.0 ± 15.0**
Hexachlorobenzene	32.2	(2.1)	33.1	(2.2)	33.8	(1.8)	**32.9 ± 1.7**
γ-HCH	3.90	(0.37)	2.63	(0.37)	3.37	(0.24)	**3.30 ± 0.81**
α-HCH	18.6	(1.2)	14.4	(1.3)	15.3	(1.0)	**16.2 ± 3.4**
β-HCH	ND[c]		ND[c]		8.04	(0.71)	**8.0 ± 1.4**
Heptachlor epoxide	10.74	(0.87)	10.1	(1.2)	11.2	(1.0)	**10.8 ± 1.3**
Oxychlordane	20.6	(1.5)	20.7	(1.0)	18.61	(0.83)	**19.8 ± 1.9**
cis-Chlordane	47.2	(3.2)	46.0	(3.4)	47.2	(2.5)	**46.9 ± 2.8**
cis-Nonachlor	54.2	(3.2)	47.8	(3.1)	44.3	(2.7)	**48.7 ± 7.6**
trans-Nonachlor	229.0	(16.0)	233.0	(16.0)	232.0	(10.0)	**231.0 ± 11.0**
Dieldrin	39.4	(3.0)	35.5	(3.1)	ND[d]		**37.5 ± 3.9**
Mirex	30.1	(2.3)	29.6	(2.3)	26.8	(2.2)	**28.9 ± 2.8**

[a] Duplicate samples from ten bottles extracted and each extract analyzed in duplicate. Concentrations are the mean, and numbers in parentheses are 1 standard deviation of a single measurement. The same extracts were analyzed using GC-ECD on both the 50% methyl C-18 phase and the 5% phenyl phase.
[b] Samples from six bottles extracted and analyzed in duplicate. The concentration is the mean, and the number in parentheses is 1 standard deviation of a single measurement.
[c] ND = not determined since β-HCH is not recovered from the normal-phase LC cleanup step.
[d] ND = not determined since dieldrin is destroyed in the sulfuric acid step.

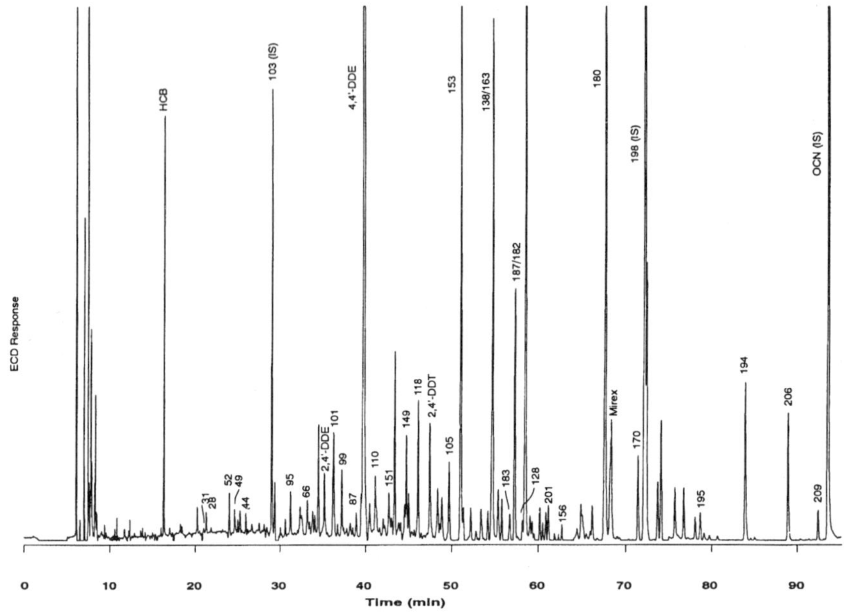

Figure 7 GC-ECD analysis of the PCB and lower polarity pesticide fraction isolated from SRM 1945 (Organics in Whale Blubber) on 50% methyl C-18 dimethylpolysiloxane phase. (From M. M. Schantz et al., *Anal. Chem.*, 67 (1995) 901–910.)

wet weight (see Table 11) for the combined PCB 101/90 as determined on the 5% phenyl phase (both ECD and MS). The concentration of PCB 190 appears to be less than 5% of the concentration of PCB 170 in SRM 1945 based on the results in Table 11.

PCB 110 coelutes with the "planar" congener PCB 77 (3,3′,4,4′-tetrachlorobiphenyl) when using the 5% phenyl phase. The planar non-ortho-substituted PCB congeners are of particular interest due to their enhanced toxicity.[48] However, the planar PCB congeners have not been quantified in any of the NIST SRMs, but work is ongoing to isolate and quantify these compounds. PCB 110 and PCB 77 are separated using the 50% methyl C-18 phase and are distinguishable by MS detection. The concentration of PCB 77 could be as high as 35% of the concentration of PCB 110 in this sample based on the results in Table 11; however, this is unlikely since PCB 77 is only a small fraction of the PCB 110 in the Arochlor 1254 and 1260. The planar non-ortho-substituted congeners should be quantified only after isolation from the nonplanar congeners.

The certified concentrations are shown in Tables 11 and 12 for the PCB congeners and chlorinated pesticides, respectively. These concentrations are weighted means of the results from two or more analytical techniques, determined as described above for the PAH measurements. There are 27 PCB congeners and 15 chlorinated pesticides certified in SRM 1945. This is a substantially larger number of PCB congeners and chlorinated pesticides than has been certified in any previous natural matrix SRM. The relative uncertainties for the certified concentrations range from 5 to 25%,

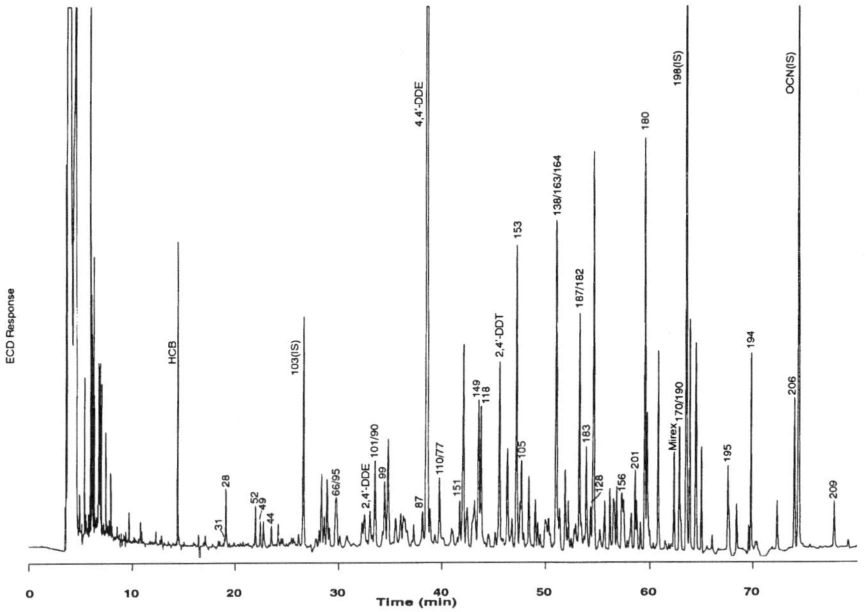

Figure 8 GC-ECD analysis of the PCB and lower polarity pesticide fraction isolated from SRM 1945 (Organics in Whale Blubber) on a 5% phenyl methylpolysiloxane phase. (From M. M. Schantz et al., *Anal. Chem.*, 67 (1995) 901–910.)

with the majority in the 5 to 10% range. Noncertified concentrations were determined for two additional PCB congeners and two additional chlorinated pesticides for which results were obtained from only one analytical technique. The two PCB congeners, PCB 28 and PCB 31, were not separated during the GC-MS analysis. Dieldrin was destroyed during the sulfuric acid cleanup step, and β-HCH was not recovered from the normal-phase liquid chromatographic cleanup step, resulting in only one technique for each of these compounds. The same analytical approach was also recently used to certify 21 PCB congeners and 6 pesticides in the marine sediment reference material (SRM 1941a, Organics in Marine Sediment);[12] this same approach is in progress for the certification of SRM 1974a, SRM 2974, and RM 8045.

CONCLUSIONS

The SRMs described above are intended for use in analytical methods development/validation and to assure long-term measurement comparability. The natural matrix materials are ideally suited for use in developing and validating new methods. These SRMs can be analyzed on a regular basis in conjunction with unknown samples with similar matrices to continually monitor the accuracy and precision of the analytical procedures. Natural matrix SRMs have also been found to be useful for methods development for the measurement of compounds that have not been certified or measured at NIST. In this instance these SRMs represent homogeneous

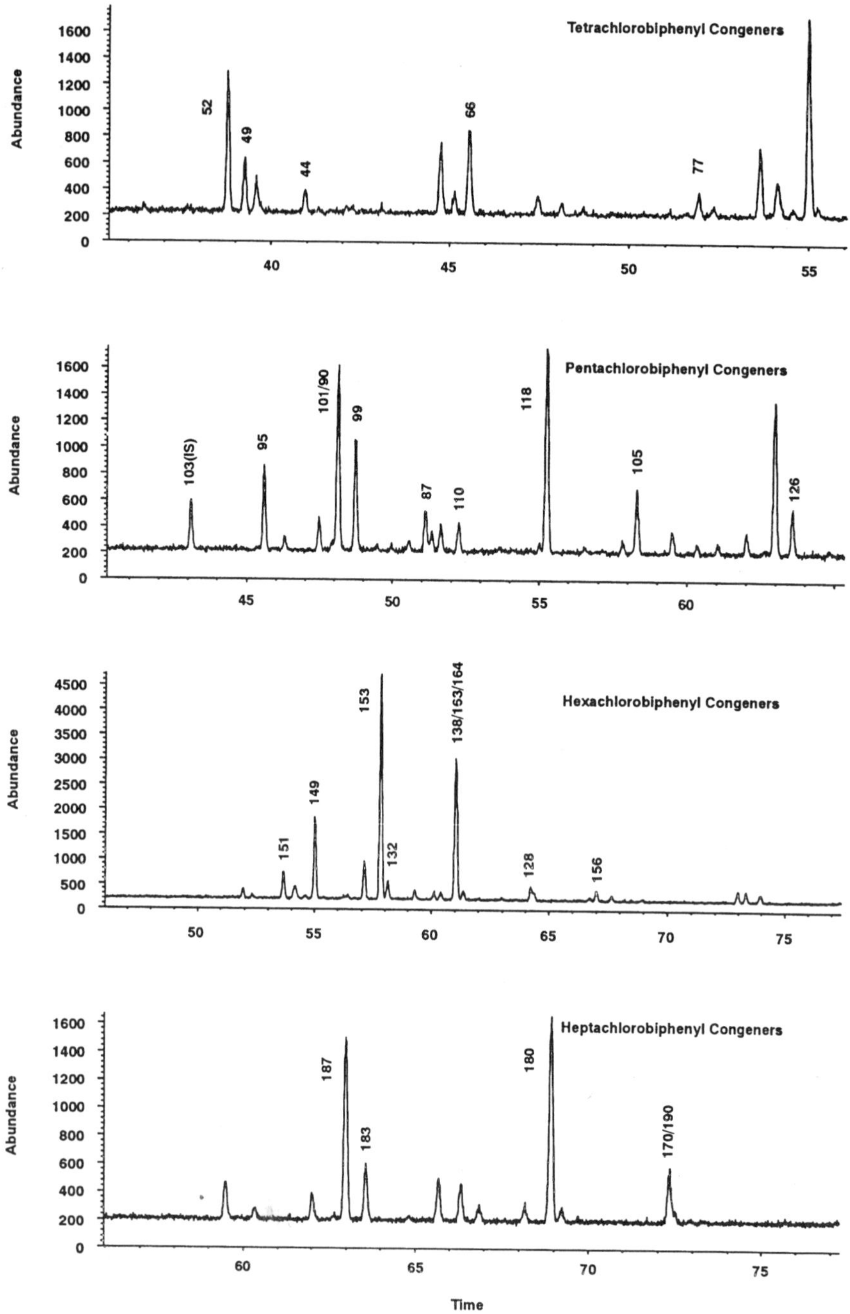

Figure 9 GC-MS analysis of the PCB and lower polarity pesticide fraction isolated from SRM 1945 (Organics in Whale Blubber) on a 5% phenyl methylpolysiloxane phase. Tetrachlorobiphenyl congeners (top) monitored at ion 289.90, pentachlorobiphenyl congeners monitored at ion 323.90, hexachlorobiphenyl congeners monitored at ion 359.85, and heptachlorobiphenyl congeners monitored at ion 393.80. (From M. M. Schantz et al., *Anal. Chem.*, 67 (1995) 901–910.)

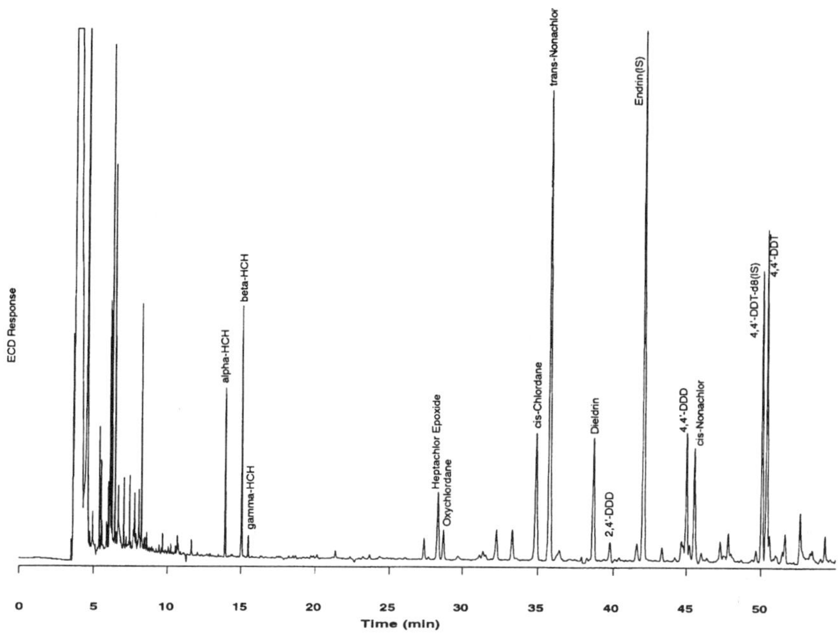

Figure 10 GC-ECD analysis of the more polar pesticide fraction isolated from SRM 1945 (Organics in Whale Blubber) on a 5% phenyl methylpolysiloxane phase. (From M. M. Schantz et al., *Anal. Chem.*, 67 (1995) 901–910.)

natural environmental matrices that are readily available to all laboratories for comparison of analytical results. As laboratories involved in environmental pollution monitoring expand the number of PAHs or PCB congeners measured due to emphasis on toxicity, pollution source identification, etc., published results of analyses of these SRMs by other laboratories will provide a database for comparison within the scientific community. For example, even though these materials have not been analyzed at NIST for PCDDs, PCDFs, or planar PCB congeners, they offer convenient matrices for use in addressing the measurement of these analytes.

ACKNOWLEDGMENT

The following NIST staff members are acknowledged for their participation in the preparation and certification measurements associated with the SRMs mentioned in this paper: B. A. Benner, Jr., J. M. Brown Thomas, G. D. Byrd, S.N. Chesler, F. R. Guenther, M. J. Hays, L. R. Hilpert, W. F. Kline, B.J. Koster, W. E. May, R. M. Parris, R. E. Rebbert, and L. C. Sander. L. M. Oakley and S. B. Schiller are acknowledged for providing statistical evaluation of the certification results for the marine sediment and whale blubber SRMs. The support and assistance of the Standard Reference Materials Program is also gratefully acknowledged.

REFERENCES

1. **S. A. Wise,** Standard reference materials for the determination of trace organic constituents in environmental samples, In: *Environmental Analysis: Techniques, Applications and Quality Assurance*, D. Barcelo, Ed., Elsevier Science Publishers, Amsterdam, The Netherlands, 1993, 403–446.
2. **S. A. Wise, M. M. Schantz, R. M. Parris, R. E. Rebbert, B. A. Benner, Jr., and T. E. Gills,** Standard reference materials for trace organic contaminants in the marine environment, *Analysis*, 20 (6) (1992) M57-M61.
3. **S. A. Wise, L. R. Hilpert, R. E. Rebbert, L. C. Sander, M. M. Schantz, S. N. Chesler, and W. E. May,** Standard reference materials for the determination of polycyclic aromatic hydrocarbons, *Fresenius Z. Anal. Chem.*, 332 (1988) 573–582.
4. **S. A. Wise, M. M. Schantz, B. A. Benner, Jr., R. M. Parris, R. E. Rebbert, L. C. Sander, B. J. Koster, S. N. Chesler, and W. E. May,** Standard reference materials for the determination of polycyclic aromatic hydrocarbons in environmental samples — current activities, *Fresenius J. Anal. Chem.*, 345 (1993) 325–329.
5. **H. S. Hertz, J. M. Brown, S. N. Chesler, F. R. Guenther, L. R. Hilpert, W. E. May, R. M. Parris, and S. A. Wise,** Determination of individual organic compounds in shale oil, *Anal. Chem.*, 52 (1980) 1650–1657.
6. **W. E. May, J. M. Brown-Thomas, L. R. Hilpert, and S. A. Wise,** The certification of selected polynuclear aromatic hydrocarbons in standard reference material 1580, "Organics in Shale Oil", In: *Polynuclear Aromatic Hydrocarbons: Chemical Analysis and Biological Fate*, M. Cooke and A. J. Dennis, Eds., Battelle Press, Columbus, OH, 1981, 1–16.
7. **L. R. Hilpert, G. D. Byrd, and C. R. Vogt,** Selectivity of negative ion chemical ionization mass spectrometry for benzo[a]pyrene, *Anal. Chem.*, 56 (1984) 1842–1846.
8. **W. F. Kline, S. A. Wise, and W. E. May,** The application of perdeuterated polycyclic aromatic hydrocarbons (PAH) as internal standards for the liquid chromatographic determination of PAH in a petroleum crude oil and other complex mixtures, *J. Liquid Chromatogr.*, 8 (1985) 223–237.
9. **W. E. May and S. A. Wise,** Liquid chromatographic determination of polycyclic aromatic hydrocarbons in air particulate extracts, *Anal. Chem.*, 56 (1984) 225–232.
10. **M. M Schantz, B. A. Benner, Jr., S. N. Chesler, B. J. Koster, K. E. Hehn, S. F. Stone, W. R. Kelly, R. Zeisler, and S. A. Wise,** Preparation and analysis of a marine sediment reference material for the determination of trace organic constituents, *Fresenius J. Anal. Chem.*, 338 (1990) 501–514.
11. **S. A. Wise, M. M. Schantz, B. A. Benner, Jr., M. J. Hays, and S. B. Schiller,** Certification of polycyclic aromatic hydrocarbons in a marine sediment standard reference material, *Anal. Chem.*, 67 (1995) 1171–1178.
12. **M. M. Schantz, B. A. Benner, Jr., M. J. Hays, W. R. Kelly, R. D. Vocke, Jr., R. Demiralp, R. G. Greenberg, S. B. Schiller, G. G. Lauenstein, and S. A. Wise,** Certification of standard reference material (SRM) 1941a, organics in marine sediment, *Fresenius J. Anal. Chem.*, 352 (1995) 166–173.
13. **S. A. Wise, B. A. Benner, Jr., R. G. Christensen, B. J. Koster, J. Kurz, M. M. Schantz, and R. Zeisler,** Preparation and analysis of a frozen mussel tissue reference material for the determination of trace organic constituents, *Environ. Sci. Technol.*, 25 (1991) 1695–1704.
14. Certificate of Analysis SRM 1588, Organics in Cod Liver Oil, National Bureau of Standards, 1989, Gaithersburg, MD.

15. **R. E. Rebbert, S. N. Chesler, F. R. Guenther, B. J. Koster, R. M. Parris, M. M. Schantz, and S. A. Wise,** Preparation and analysis of a river sediment standard reference material for the determination of trace organic constituents, *Fresenius J. Anal. Chem.*, 342 (1992) 30–38.
16. **M. M. Schantz, B. J. Koster, L. M. Oakley, S. B. Schiller, and S. A. Wise,** Certification of polychlorinated biphenyl congeners and chlorinated pesticides in a whale blubber standard reference material, *Anal. Chem.*, 67 (1995) 901–910.
17. **W. E. May, S. P. Wasik, and D. H. Freeman,** Determination of the aqueous solubility of polynuclear aromatic hydrocarbons by a coupled column liquid chromatographic technique, *Anal. Chem.*, 50 (1978) 175–179.
18. **W. E. May, S. P. Wasik, and D. H. Freeman,** Determination of the solubility behavior of some polycyclic aromatic hydrocarbons in water, *Anal. Chem.*, 50 (1978) 997–1000.
19. **S. A. Wise, B. A. Benner, Jr., S. N. Chesler, L. R. Hilpert, C. R. Vogt, and W. E. May,** Characterization of polycyclic aromatic hydrocarbons from two standard reference material air particulate samples, *Anal. Chem.*, 58 (1986) 3067–3077.
20. **F. R. Guenther, R. M. Parris, S. N. Chesler, and L. R. Hilpert,** Determination of phenolic compounds in alternate fuel matrices, *J. Chromatogr.*, 207 (1981) 256–261.
21. **F. R. Guenther, S. N. Chesler, and R. M. Parris,** Analysis of nitrogen heterocycles in shale oil by a dual capillary column heart cutting technique, *J. Chromatogr.*, 363 (1986) 199–205.
22. **R. E. Rebbert, S. N. Chesler, F. R. Guenther, and R. M. Parris,** A liquid chromatographic-gas chromatographic procedure to determine the concentration of dibenzothiophene in a crude oil matrix, *J. Chromatogr.*, 284 (1984) 211–217.
23. **L. R. Hilpert,** National Institute of Standards and Technology, unpublished results.
24. **S. A. Wise and K. D. Timmons,** National Institute of Standards and Technology, unpublished results.
25. **P. L. Konash, S. A. Wise, and W. E. May,** Selective quenchofluorometric detection of fluoranthenic polycyclic aromatic hydrocarbons in high-performance liquid chromatography, *J. Liquid Chromatogr.*, 4 (1981) 1339–1349.
26. **L. R. Hilpert,** Determination of polycyclic aromatic hydrocarbons and alkylated polycyclic aromatic hydrocarbons in particulate extracts using negative ion chemical ionization mass spectrometry, *Biomed. Environ. Mass Spectrom.*, 14 (1987) 383–394.
27. **W. E. May, S. A. Wise, L. R. Hilpert, B. A. Benner, Jr., and W. A. MacCrehan,** Report to the Coordinating Research Council (CRC), 1988.
28. **S. A. Wise, B. A. Benner, Jr., G. D. Byrd, S. N. Chesler, R. E. Rebbert, and M. M. Schantz,** Determination of polycyclic aromatic hydrocarbons in a coal tar standard reference material, *Anal. Chem.*, 60 (1988) 887–894.
29. **S. A. Wise, A. Deissler, and L. C. Sander,** Liquid chromatographic determination of polycyclic aromatic hydrocarbon isomers of molecular weight 278 and 302 in environmental standard reference materials, *J. Polycyclic Aromatic Compounds*, 3 (1993) 169–184.
30. **S. A. Wise, B. A. Benner, H. Liu, and G. D. Byrd,** Separation and identification of polycyclic aromatic hydrocarbon isomers of molecular weight 302 in complex mixtures, *Anal. Chem.*, 60 (1988) 630–637.
31. **R. Zeisler, J. K. Langland, and S. H. Harrison,** Cryogenic homogenization of biological tissues, *Anal. Chem.*, 55 (1983) 2431–2434.
32. **J. Lewtas, L. D. Claxton, H. S. Rosenkranz, D. Schuetzle, M. Shelby, H. Matsushita, F. E. Würgler, F. K. Zimmermann, G. Löfroth, W. E. May, D. Krewski, T. Matsushima, Y. Ohnishi, H. N. G. Gopalan, R. Sarin, and G. C. Becking,** Design and implementation of a collaborative study of the mutagenicity of complex mixtures in *Salmonella typhimurium*, *Mutation Res.*, 276 (1992) 3–9.

33. **L. D. Claxton, G. Douglas, D. Krewski, J. Lewtas, H. Matsushita, and H. Rosenkranz,** Overview, conclusions and recommendations of the IPCS collaborative study on complex mixtures, *Mutation Res.*, 276 (1992) 61–80.
34. **K. Ballschmiter and M. Zell,** Analysis of polychlorinated biphenyls (PCB) by glass capillary gas chromatography, *Fresenius Z. Anal. Chem.*, 302 (1980) 20–31.
35. **E. Schulte and R. Malisch,** Calculation of the real PCB content in environmental samples. I. Investigation of the composition of two technical PCB mixtures, *Fresenius Z. Anal. Chem.*, 314 (1983) 545–551.
36. **M. M Schantz, R. M. Parris, S. A. Wise, H. T. Won, and R. Turle,** Polychlorinated biphenyl (PCB) congeners and chlorinated pesticides in a cod liver oil reference material, *Chemosphere*, 24 (1992) 1687–1698.
37. **M. M Schantz, R. M. Parris, J. Kurz, K. Ballschmiter, and S. A. Wise,** Comparison of methods for gas chromatographic determination of PCBs and chlorinated pesticides in marine reference materials, *Fresenius J. Anal. Chem.*, 346 (1993) 766–778.
38. **S. A. Wise, L. R. Hilpert, G. D. Byrd, and May, W. E.,** Comparison of liquid chromatography with fluorescence detection and gas chromatography/mass spectrometry for the determination of polycyclic aromatic hydrocarbons in environmental samples, *J. Polycyclic Aromatic Compounds*, 1 (1990) 81–98.
39. **S. A. Wise, S. N. Chesler, H. S. Hertz, L. R. Hilpert, and W. E. May,** Chemically-bonded aminosilane stationary phase for the high-performance liquid chromatographic separation of polynuclear aromatic hydrocarbons, *Anal. Chem.*, 49 (1977) 2306–2310.
40. **S. A. Wise, L. C. Sander, and W. E. May,** Determination of polycyclic aromatic hydrocarbons by liquid chromatography, *J. Chromatogr.*, 642 (1993) 329–349.
41. **J. S. Bradshaw, C. M. Schregenberger, H.-C. K. Chang, K. E. Markides, and M. L. Lee,** Synthesis and chromatographic properties of polysiloxane stationary phases containing biphenylcarboxylate ester liquid crystalline side groups, *J. Chromatogr.*, 358 (1986) 95–105.
42. **K. E. Markides, H.-C. Chang, C. M. Schregenberger, B. J. Tarbet, J. S. Bradshaw, and M. L. Lee,** Evaluation of smectic biphenylcarboxylate ester liquid-crystalline polysiloxane stationary phase for capillary column chromatography, *J. High Resolut. Chromatogr. Chromatogr. Commun.*, 8 (1985) 516–520.
43. **S. A. Wise, L. C. Sander, H.-C. K. Chang, K. E. Markides, M. L. Lee,** Shape selectivity in liquid and gas chromatography: polymeric octadecylsilane and liquid crystalline stationary phases, *Chromatographia*, (1988) 25, 473–479.
44. **L. C. Sander, M. Schneider, C. Woolley, and S. A. Wise,** Shape selectivity assessment of stationary phases in gas chromatography, *J. Microcolumn Sep.*, 6 (1994) 115–125.
45. **S. B. Schiller and K. R. Eberhardt,** Combining data from independent chemical analysis methods, *Spectrochemica Acta*, 46B (12) (1991) 1607–1613.
46. **F. R. Guenther, S. N. Chesler, and R. E. Rebbert,** The analysis of polychlorinated biphenyls by multidimensional gas chromatography, *J. High Resolut. Chromatogr. Chromatogr. Commun.*, 12 (1989) 821–824.
47. **K. Ballschmiter, A. Mennel, and J. Buyten,** Long chain alkyl-polysiloxanes as non-polar stationary phases in capillary gas chromatography, *Fresenius J. Anal. Chem.*, 346 (1993) 396–402.
48. **K. Ballschmiter, R. Bacher, A. Mennel, R. Fischer, U. Riehle, and M. Swerev,** The determination of chlorinated biphenyls, chlorinated dibenzodioxins and chlorinated dibenzofurans by GC-MS, *J. High Resolut. Chromatogr.*, 15 (1992) 260–270.

CHAPTER 13

The User's Perspective on Environmental Standard Solutions

Gregory E. Pronger and Marilyn A. Melton

Environmental labs use reference materials in many ways: for initial calibration, for calibration checks, for the preparation of matrix spikes, and to analyze total method performance. This chapter will consider the use of reference materials for calibration purposes. An environmental laboratory has three possible sources to solve standard needs: (1) internal preparation from neat materials, (2) use of purchased "stock" solutions, and (3) use of "ready to use" standards. The different standard sources can be evaluated by using the following five factors: (1) ease, (2) quality, (3) safety, (4) cost, and (5) efficiency. This chapter will explain each factor. It will then use these factors to evaluate the three possible standard sources.

INTRODUCTION

Ease

Few people want to do more work than necessary, including those employed within environmental laboratories. Ease of use generally relates directly to decreased time for preparation of the final solutions required, thereby saving labor costs. Points affecting ease of use are whether the catalog is clear and logically organized, whether order placement is easy and whether the concentrations are appropriate for application, particularly compared with one another. For example, GCMS standards are probably most convenient with all compounds at a single concentration. However, for HPLC for instance, this would be inappropriate, due to dramatic differences in sensitivity of the different compounds.

Quality

The laboratory can measure standard quality in many ways. Is the standard accurate? Is the standard consistent lot to lot? Is it stable? Beyond the quality of the standard is the issue of certification. Some federal and state programs require the use of certified standards. These factors influence the total quality of the standard material.

Safety

Most of the analytes measured and solvents used by environmental laboratories are hazardous. The value of a reference material is enhanced by designing it to reduce any potential hazard.

Cost

It is possible to have an item that is prohibitively expensive in spite of long-term benefits to the lab. One example is the use of isotopic dilution, which may offer improved data quality and efficiency but is infrequently utilized due to the cost of the standard.

Efficiency

Because of the competitive nature of environmental laboratories, efficient operation is a necessity. Here we attempt to balance the benefits of ease and quality versus standard cost. The easiest solution to use may be so costly that any savings in labor is negated by the increase in cost.

Using these five points, we can evaluate the three alternatives for standard preparation.

INTERNAL PREPARATION FROM NEAT MATERIALS

Ease

At one point in time, in-house preparation was the only answer to the standards needed by environmental labs. The fact that there is now an industry built around preparing standards for environmental labs indicates the opinion that most labs have of preparing their own standards.

When preparing standards in-house, beyond single analyte solutions, the ease of preparation greatly diminishes. An error when adding one component can ruin the entire solution. Further, the analyst must invest time investigating the compatibility of the compounds to be put into solution and finding an appropriate solvent.

Quality

To ensure the quality of a prepared solution, the lab faces the same challenges as the commercial standard preparation firm. The lab must ensure the quality of reference material. To verify the first solution, the lab must prepare or purchase a second solution. If it prepares the solution for verification purposes, each compound in the verification standard must be from a separate lot. Finally, clear documentation of the process is required.

Safety

Due to the potential exposure to the analytes in their neat form, preparation of reference materials from neat carries the greatest risk factor of the three alternatives.

Cost

Cost savings is the primary reason for environmental laboratories to prepare standards in-house. For example, for most organic compounds, the cost of 0.1 to 1.0 g quantities of neat standard grade material is around \$5 to \$50.[1–7] To calculate the maximum range in cost we take:

$$\$10/g = \$10/g$$

and

$$\$50/0.1\ g = \$500/g$$

This is a cost of \$10 to \$500 per gram. For 10 milliliters of a single analyte solution at 2000 mg/L, the cost would be \$0.2 to \$10.

The labor cost component is highly dependent on the general organization of the lab, from how well the neat standard storage is organized to the level of paperwork necessary to document the standard preparation. A rough estimate would be 15 to 25 min from finding the standard to completing necessary documentation. If the lab's labor cost is \$15/h (salary and benefits), we have an additional labor cost of \$3.75 to \$6.25. The cost is then \$3.95 (\$3.75 labor + \$0.20 materials) to \$16.25 (\$6.25 labor + \$10.00 materials) for 10 milliliters.

Depending on use, the lab may need to perform analytical verification of the prepared standard. If the lab uses their own equipment, we can estimate the increased standard preparation cost in terms of the cost to prepare the verification standard and the labor cost to do the analysis. First, the lab must prepare the verification solution. Since it is only used to verify the stock solution, it will be consumed much more slowly than the stock standards. For purposes of discussion, we can assume the verification standard lasts an entire year. Next, we need to estimate how many stock solutions will be prepared over this period. A fair assumption is that a new stock solution is prepared quarterly, or four times over the life of the verification standard. We can now spread the cost of the verification standard over the four stock standards. The preparation of the verification standard should cost the lab about the same as the original standard. Therefore we have a range of \$0.98/10 mL (\$3.95/4) to \$4.08/10 mL (\$16.25/4). A simple GC or HPLC analysis of the solution should not take more than 15 to 25 min of analyst time. This is an additional cost of \$3.75 to \$6.25. Therefore, verification costs the laboratory \$4.73 (\$0.98 verification standard + \$3.75 analysis) to \$10.33 (\$4.08 verification standard + \$6.25 analysis).

It should be noted that the laboratory could purchase the verification standard. These solutions cost roughly $10 to $30 per milliliter for a 2000 mg/L solution. Used over the same period (a year) and for the same number of in-house prepared stock standards we have $2.50 ($10/4) to $7.50 ($30/4) per verification analysis. Combining the standard costs with analysis costs, we have $6.25 ($2.50 verification standard cost + $3.75 analysis) to $13.75 ($7.50 verification standard cost + $6.25 analysis). This scenario raises the top end of our cost envelope to $13.75.

Total cost then ranges from $3.95 (if the lab does not analytically verify the solution) to $30.00 ($16.25 labor + $13.75 verification cost) for the 10-mL solution or $0.40 to $3.00/mL. As mentioned, commercial single analyte solutions cost from about $10/mL to $30/mL (for the common environmental organic analytes). Generally then, the lab will save money by preparing single analyte solutions in-house.

Beyond a single analyte standard, however, the cost to the lab to prepare the material in-house increases significantly while the cost of a purchased material increases to a lesser extent. Evaluating the cost of in-house preparation, we find that the cost due to the analytes increases proportionally with their number. The lab may realize some economies of scale when preparing a multiple compound material. For example, the time spent to find two standards is probably less than twice the time it takes to find just one. A fair estimate would be 25 to 50% improvement. This means that standard prep time decreases to 7.5 min (15 min $\times$ 0.5) to 18.75 min (25 min $\times$ 0.75) per analyte. Labor cost is then $1.88 to $4.69 per analyte per 10 milliliters.

We can now combine labor costs with the cost of the analytes. This gives us a range of $2.08 ($1.88 labor + $0.20 materials) to $14.69 ($4.69 labor + $10.00 materials) per analyte per 10 milliliters.

Finally, we can factor in the cost of analytical verification. Assuming the analysis cost remains constant (that we can analyze a multianalyte solution for the same cost as a single analyte solution), the analytical cost is still $3.75 to $6.25. On top of the analysis cost is the cost of preparing the verification solution. Again, we can assume the cost of preparing the verification solution is the same as the cost of preparing the original solution. Hence, we have an additional cost of $1.88 to $14.69 per analyte per 10 milliliters. We again assume the verification solution is good for a year and that four stock solutions are prepared over this period. Therefore, we have $0.47 per analyte per 10 milliliters ($1.88/4) to $3.67 per analyte per 10 milliliters ($14.69/4).

Here again the lab could choose to purchase its verification standard. The purchased cost of one milliliter of a multianalyte solution at 2000 mg/L is roughly $20 to $80. Used over the same period and for the same number of stock standard solutions, we have a verification standard cost of $5 to $20. As previously stated, analysis cost is $3.75 to $6.25. Verification using an external source is $8.75 to $26.25. This raises the cost maximum to $26.25.

The main cause of the difference from the "Low" solution cost to the "High" solution cost is the cost of verification.

We have stated the cost of purchasing multicomponent stock standards to be between $20 and $80. Further, we have calculated the additional cost of verification

Table 1

Number of analytes	Solution cost (per milliliter)		
	Low[a]	Average	High[b]
2	$4.16	$29.89	$55.63
3	$6.24	$38.28	$70.32
4	$8.32	$46.66	$85.01
5	$10.40	$55.05	$99.70
10	$20.80	$96.98	$173.15
15	$31.20	$138.90	$246.60
20	$41.60	$180.83	$320.05

[a] Low cost = (number of analytes) × ($2.08), where $2.08 is the low end standard cost per analyte per milliliter.
[b] High cost = (number of analytes) × $14.69 + $20 + $6.25, where $14.69 is the high cost per analyte per milliliter, $20 is the high end cost of the verification solution, and $6.25 is the high end cost of analytical verification analysis.

using purchased solutions as $8.75 to $26.25. The total cost to use commercial stock standards is then $28.75 to $106.25, with an average of $67.50. Based upon this, for solutions of more than about ten compounds, it is clearly less expensive to use commercial solutions.

It should be noted that there are other costs we have not attempted to capture in this analysis. First, there is the cost of disposal of expired (out of shelf life) standard material (either the neat materials themselves or the prepared solutions) and any waste (standards prepared incorrectly). Second, to prepare the standard material in-house, the lab loses an analyst from production while he/she prepares the standard. These two factors are hardly trivial. However, to incorporate their effects are beyond the scope of this chapter.

Efficiency

Generally, in-house preparation of standards would score low here. The environmental lab does not have the economies of scale realized by the commercial standard preparation firm. The commercial standard preparation firm has devoted personnel and equipment that all but the largest environmental labs cannot justify. This allows the standard preparation firm to prepare standards much more efficiently than the environmental firm. Thus, most environmental labs prefer to purchase all but the simplest standards. Many will only prepare standards in-house when faced with a complete lack of commercial availability.

COMMERCIALLY PREPARED STOCK SOLUTIONS

Ease

The use of commercially prepared "stock" standards saves the laboratory the task of preparing the stock solution internally. However, the ease of use of a stock standard is greatly diminished if more than simple serial dilutions are necessary to prepare the daily standard. For some analyses needing multianalyte solutions, it is preferred to have the different analytes at different concentrations (due to differences in instrument sensitivity or differences in regulatory requirements). If the relative concentrations chosen by the vendor are inappropriate, the laboratory probably cannot use the solution. When faced with this problem, the lab may have to prepare the standards in-house.

Efficiency

As previously discussed, significant time is saved by using commercially prepared stock standards rather than preparing the stock standards in-house. If the cost of the purchased standard is not out of line with the cost of in-house preparation, the lab is more efficient with its use. However, for some instrumentation, it may take the combination of several stock standards to prepare the final working standard. This is particularly true for GCMS, where it may take three solutions (common 624[8] type volatiles) to ten solutions (Appendix 9[9] semivolatiles) to prepare the working standards. When evaluating stock standards available commercially, the lab will favor those that reduce the effort needed to prepare the final solutions.

Quality

The commercial laboratory measures the quality of the stock standard in several ways. The most obvious is the accuracy and the lot-to-lot precision of the concentrations of the analytes. Directly affecting the accuracy and precision of the solutions is their stability. If analytes in the commercial solution are interreactive, or even reactive when combined by the lab with other solutions to prepare the final working solution, then the overall quality of the solution is adversely affected in the eyes of the user laboratory.

Further, the labeling must be clear and complete. The laboratory should receive (without need for a special request) sufficient information to document the preparation of the solution, including the solvent, back to the lots of raw materials used by the commercial standard preparation firm.

Finally, any concentration and purity verification information must be clear and complete.

Safety

Compared to standard preparation from raw materials, there is a significant improvement in the safety with the use of commercially prepared stock standards. The analyst must still be careful when using these solutions because they are frequently prepared in solvents with significant health effects (such as methylene chloride, which is a suspected human carcinogen). Also, even at dilute concentrations, some analytes may pose significant health concerns. However, when using the commercially prepared solutions, at no point in the process do laboratory personnel need to handle the analytes in pure form.

Cost

Compared to the cost of the initial investment to purchase neat materials, commercially available stock standards are generally less expensive. A complete set of neat standards for EPA Method 625[8] costs about $900 to $1000.[1–7] Stock solutions (with many analytes combined in the same solution) that cover the entire list cost between $125 and $250.[1–7] Once the investment in neat materials is made, the cost of preparing them in-house depends upon the absolute labor cost and the decreased available labor (as previously covered within the section covering costs for neat materials). Depending upon the laboratory and the standard to be prepared, either alternative could be the correct choice.

READY TO USE SOLUTIONS

Ease

If the solutions are prepared at the concentrations needed by the laboratory, "ready to use" solutions would have to be the simplest answer to the problem of standard preparation. By purchasing solutions that are ready to use, the laboratory reduces those tasks done in-house not directly related to the primary business function. This should have a direct benefit in available hours of laboratory staff.

Efficiency

If the laboratory measures efficiency strictly by the impact on employee time spent preparing standards, "ready to use" standards would necessarily win. However, the laboratory must balance this gain against the increased cost of these solutions. There are four main factors affecting the evaluation of using commercially prepared stock standards vs. commercial ready to use standards:

- Cost of the ready to use standard
- Cost of the commercial stock standard
- Time required to prepare working solutions
- Labor cost

To compare the cost of the "ready to use" solution and the cost of the commercial stock solution, one must calculate the total cost of preparing the final solution from the commercial stock standard. The cost of using the commercial stock standard is its cost per milliliter, including the cost of labor to make the dilutions. Once the final solution is prepared, the subsequent costs (calibration, verification, etc.) are the same. In other words, whatever the laboratory does to ensure their standard is correct must be done, whether using a dilution of a commercial stock standard or a commercial "ready to use" standard.

For example, compare the costs involved in preparing a standard pesticide curve for EPA Method 8080.[10] The analytes (single peak analytes only) are as follows in Table 2.

Table 2 Comparative Costs in Preparing a Standard Pesticide Curve

Analyte[a]	CAS number[b]	MDL[c] (μg/L)
Aldrin	309-00-2	0.004
α-BHC	319-84-6	0.003
β-BHC	319-85-7	0.006
δ-BHC	319-86-8	0.009
γ-BHC	58-89-9	0.004
4,4′-DDD	72-54-8	0.011
4,4′-DDE	72-55-9	0.004
4,4′-DDT	50-29-3	0.012
Dieldrin	60-54-1	0.002
Endosulfan I	959-98-8	0.014
Endosulfan II	33212-65-9	0.004
Endosulfan sulfate	1031-07-8	0.066
Endrin	72-20-8	0.006
Endrin aldehyde	7421-93-4	0.023
Heptachlor	76-44-8	0.003
Heptachlor epoxide	1024-57-3	0.003
Methoxychlor	72-43-5	0.176

[a] Analyte list taken from Reference 10. Only single peak compounds.
[b] Chemical Abstract Services number.
[c] Method detection limit from Reference 10.

Most labs run these compounds in two separate mixes. It is then necessary to prepare two separate five-point calibrations, two separate daily check standards, etc. It takes roughly 0.5 h to prepare a five-point calibration curve. Using our salary costs from before ($15/h), there is an additional $7.50 to prepare the five-point from commercial stock standards. The cost of a commercial standard for the pesticides is roughly $35/mL.[1–7] Solution concentration is usually around 5 mg/L. A common procedure within environmental labs is to prepare their working solutions in 5-mL aliquots. This allows the use of larger volumes when preparing the dilutions that generally are felt to be more accurate. For comparison purposes, one current source of "ready to use" solutions is Radian Corporation.[6–7] They offer pesticide calibration solutions at (nominally) 2, 5, 10, 20, and 50 μg/L concentrations. To prepare an equivalent solution from commercial stock standards at a 5 mg/L concentration, we have the following in Table 3.

The total cost for one set of calibration curves is then $3.04 (materials) and $7.50 (labor), for a total cost for one curve of $8.54. For both curves we have $17.08. The commercial "ready to use" solution costs $25 per kit. For the two curves we then have a cost for the "ready to use" solutions of $50. From this analysis, the lab would be more efficient to use the commercially prepared stock standards. Because labor is a substantial portion of the cost to prepare the working solutions from the commercial stock standards, for a particular laboratory with different salary and benefits, the outcome of this evaluation could change. Further, this analysis does not attempt to assign a value due to lost production to the hour the analyst spent preparing the two five-point curves. If this were done, it would raise the cost of using the commercial stock materials. Again, depending on the lab's individual circumstance, this may or may not affect the outcome of this analysis.

Table 3 Comparative Costs from Commercial Stock Standards

Concentration (μg/L)	μL into 5 mL	Cost
2	2	$0.07
5	5	$0.18
10	10	$0.35
20	20	$0.70
50	50	$1.75
Total		$3.04

Quality

As with any purchased standards, solution stability, accuracy, packaging, labeling, and verification information impact the quality of the "ready to use" solution. However, with commercially prepared stock standards, the laboratory affects the final quality of the standard in the preparation of the final working standards. When the laboratory purchases commercially prepared working standards, the vendor carries this responsibility. Due to economies of scale and greater expertise in standard preparation, the commercial standard company may produce a solution of higher quality than the lab itself. If this is the case, the lab should realize improved quality in their standards by using "ready to use" solutions. Clearly the lab must judge the quality of the purchased solution itself and its own ability to do the necessary dilutions when using a purchased stock solution vs. the quality delivered with a "ready to use" solution.

Safety

Compared to the use of commercially prepared working standards, the use of "ready to use" standards further reduces exposure to the analytes. Since, the solutions are used "as is", there is no need for dilutions. Dropping this step from standard preparation will reduce exposure to the solvent. Overall, we would have an improvement in analyst safety.

Cost

The cost per ampule of purchased "ready to use" standard is generally less than the cost of purchased stock standards (for the same analytes). However, if the cost is calculated over the life of the purchased primary standard, the cost is greater to use "ready to use" solutions.

CONCLUSION

The environmental laboratory has several avenues to pursue to solve their needs for reference materials. This chapter has attempted to outline a thorough process the laboratory could use to find a sound answer to this most basic part of producing good analytical data.

REFERENCES*

1. AccuStandard, Inc., Environmental Standards Catalog (1993).
2. NSI Environmental Solutions, 1993 Catalog.
3. NSI Environmental Solutions, 1993 Product Update.
4. Restek Corporation, Chromatography Products Catalog (1993-94).
5. Supelco, Chromatography Supplies Catalog (1992).
6. Radian Corporation, Analytical Reference Materials Catalog, 1993.
7. Radian Corporation, Snap-N-Shoot Standards Catalog, 1993.
8. *Federal Register*, 49 (49) (Friday, October 26, 1984).
9. *Federal Register*, 52 (131) (Thursday, July 9, 1987).
10. U.S. Environmental Protection Agency, Test Methods for Evaluating Solid Waste, Physical/Chemical Methods, SW-846, 3rd ed., Final Update 1, J18, (January, 1995).

* With referencing the above standard manufacturers (1–7), the author and the author's company are in no way recommending their use over the products of any other manufacturer in this field. Further, in no way do we mean to imply that the product of these firms is better than that of any other manufacturer of these materials. For the purpose of this paper, example costs and concentrations were necessary, and these firms were referenced solely to fulfill that need.

CHAPTER 14

Matrix-Based Standards and Their Uses

Lawrence H. Keith

Matrix-based environmental reference materials (ERMs) are becoming increasingly important for analyses of complex environmental samples such as industrial wastewaters, soils, sediments, fish, and other biological samples. Although we have long recognized that complex environmental samples can exert large biases on analytical results through poor recoveries, interferences, and other matrix effects, the development of matrix-based ERMs has been slow to evolve. These ERMs can be used as performance evaluation (PE) standards and as quality control standards to measure the occurrence of false positives and false negatives. If a revised definition of Method Detection Limit (MDL) is promulgated that uses "representative matrix" in its calculation, then another important use for them will be the estimation of more accurate MDLs for complex environmental matrices.

ENVIRONMENTAL REFERENCE MATERIAL CLASSIFICATIONS

A discussion of "matrix-based" or "matrix" environmental reference materials (ERMs) involves a small but increasingly important segment of the many kinds and classes of standards that are used for quality control (QC) when performing environmental analyses. There are many kinds of QC samples. Some are designed for measurement and control of laboratory sources of bias (systematic sources of variability) or precision (random sources of variability). Others are designed for overall measurement and control of bias or precision (laboratory and sampling combined sources of variability). Matrix ERMs can be used for a variety of QC samples that encompass both laboratory and overall sources of variability.*

* Although sampling variability cannot be measured directly, it can be estimated by subtracting laboratory variance from overall variance (sampling plus laboratory variance). Variance = Standard Deviation squared.

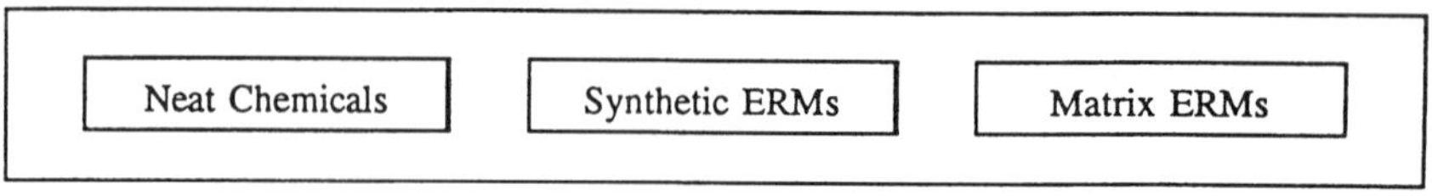

Figure 1 The three major types of environmental reference materials.

THREE MAJOR TYPES OF ERMs

ERMs may be broadly subdivided into three major types of materials (Figure 1). The simplest type of ERM is a neat chemical. In some cases, the neat ERM may be composed of a single component (e.g., benzene, zinc sulfide, or elemental arsenic). In other cases the neat ERM may be composed of multiple chemical components (e.g., chlordane, diesel fuel, or Aroclor 1260).

Synthetic ERMs are mixtures or solutions of one or more neat ERM. They are the most common type of ERM and are frequently used for calibration curves, establishing detector linearity, instrument tuning (e.g., mass spectrometer resolution), chromatographic column resolution, and other types of QC performance evaluations. Examples of synthetic ERMs include solutions of a great variety of organic and inorganic target analytes in all kinds of solvents, including water.

The third and most complex type of ERM is that which is matrix-based. Matrix refers to the environmental material in which the target analytes are contained. The broad division of environmental matrices include air, water, soils, sediments, biota, and solid or liquid wastes. However, each of these major environmental matrix categories contains many important subcategories. For example, there is ambient (outdoor) air, indoor air, and emissions from smoke stacks. Types of water matrices include industrial waste waters, rivers, lakes, precipitation, drinking water, groundwater, etc. Likewise, there are many different kinds of soils, ranging from sandy to loam to clay, each with very different matrix characteristics. Similarly, sediments may vary from small suspended particles in water to layered bottom sediments in rivers, lakes, ponds, etc. Biota matrices of environmental importance include fish, benthic organisms, birds, eggs, grass, leaves, fruits, and vegetables of all types. Solid and liquid wastes are likewise an almost endless variety of matrices found in landfills, dumps, and drums of all types and sizes.

SOME SIMPLIFYING ASSUMPTIONS

Due to the endless complexity of the environmental matrices illustrated above, it is no wonder that the development and use of matrix ERMs has lagged behind that of the other two types of ERMs. Nevertheless, there are certain simplifying assumptions that can help bring the almost endless number of possible matrix ERMs to a manageable few.

The first assumption is that most types of air matrices will not vary significantly enough in their matrix effects to require the development of different kinds of air matrix standards (e.g., an ambient air matrix vs. a stack gas matrix). This may be

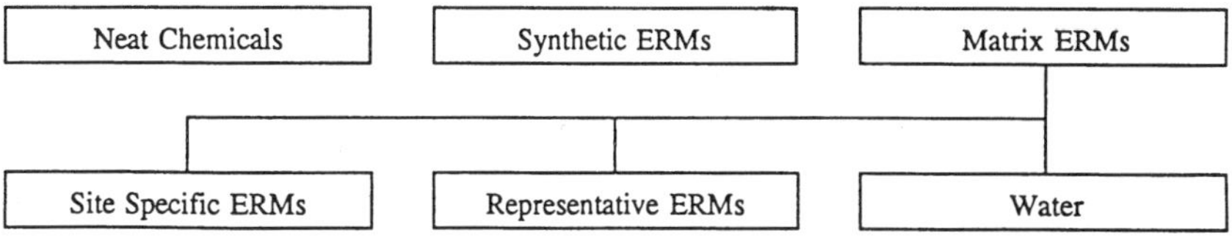

Figure 2 Three categories of matrix-based ERMs.

proven untrue in the future, but for now it appears to be a logical assumption. Therefore, clean "zero" air forms the current basis for most air standards.

The second assumption is that liquid or solid chemical wastes contain few that are found frequently enough in the environment to constitute a need to develop matrix standards for them. A few exceptions are waste transformer oils and waste motor oils.

The third and most critical assumption is that "representative" matrix standards can be developed and used for many of the endless varieties of environmental matrices.

CATEGORIES OF MATRIX-BASED ERMS

There are three basic categories of matrix-based ERMs (Figure 2). These are site-specific ERMs, representative matrix ERMs, and water.

Site-Specific ERMs

Although site-specific ERMs are most frequently associated with landfills, hazardous waste sites, and other local area environmental sampling sites, it is easy to visualize the extension to nonland-based sampling sites. For example, if pollutants from most point sources such as smoke stacks or industrial waste water outfalls are being analyzed, then uncontaminated air or water samples taken upwind or upstream may serve as "clean" background matrix samples. The same applications are appropriate for groundwater, contaminated soils or sediments, and biota samples, including fish, animals, and plants that are specific targets of an investigation.

The greatest advantage of site-specific ERMs is their obvious direct link to matrix effects that may cause false positives or false negatives in the environmental test samples. A *false positive* is the erroneous identification of a specific analyte when, in fact, it is not present at detectable concentrations. Artifacts, contamination during sample preparation or analysis, interferences, and presence of the target analytes at measurable concentrations in background (uncontaminated) samples may all contribute to false positive identifications in samples.

A *false negative* is the erroneous conclusion that a specific analyte is not present in a sample when, in fact, it is present at detectable concentrations. Matrix effects such as poor recovery during sample preparation or interferences may contribute to false negative identifications in samples.

Disadvantages of site-specific ERMs include their expense and availability. Using unpolluted matrix samples collected from a specific site before or during an investigation is more expensive, time consuming, and difficult than using "off-the-shelf" matrix samples. In addition, it may not be practical to obtain site-specific ERMs that are useful for false positive or negative QC samples. If the site-specific matrix sample contains the specific analytes of interest at detectable concentrations, then it cannot be used for determining false positive identifications. Developing site-specific ERMs for false negative QC samples is also difficult because the true value of the target analytes can only be assumed. Matrix effects that contribute to false negatives may be estimated using the method of standard additions but only if the target analytes are near the level of detection of the method being used.

Thus, while site-specific ERMs are the most desirable from a scientific point of view, their practicality is often limited by cost and/or technical considerations.

Representative Matrix ERMs

Representative matrix ERMs are those with no detectable (or acceptably low) amounts of the specific analytes of interest in a closely related matrix to that of the environmental test samples. Thus, one or more soil or sediment or fish types or water types, etc., may be analyzed and then homogenized, sterilized, or treated in appropriate ways to provide aliquots that can be used as representative matrix ERMs.

Representative matrix ERMs may be used as QC blanks or blind PE samples, or may be fortified (spiked) with known concentrations of the analytes of interest. Their advantages are that they are relatively inexpensive to produce (compared to site-specific matrix ERMs) and, when provided with a Certificate of Analysis (COA), are off-the-shelf standards that can be conveniently obtained on short notice and widely used by environmental analytical laboratories. They also are useful for interlaboratory comparisons of precision and bias when QC data from multiple labs are available.

Disadvantages of representative matrix ERMs are that they may not represent specific matrix effects on analytes of interest as accurately as a site-specific ERM will. For example, it is commonly known that clay soils typically give poorer recoveries of many organic pollutants than a sandy soil will. Thus, if a sandy soil standard was fortified with pesticides and the investigation site soil samples were composed of clay, the recoveries calculated from the fortified sandy soil QC samples may be significantly higher than what is being obtained with the clay samples. The sandy soil ERM would thus not be as representative a standard as if a clay soil matrix had been used.

Less variable differences may occur when comparisons of different species of biota are made, but little data exists yet to substantiate this (for example, matrix effects may be small when comparing different species of fish or different fruits, vegetables, and grass).

The most difficult representative matrix ERMs to prepare involve various industrial waste waters. Because of the large and complex variations in components of waste waters from different industries, it is more difficult for one industrial waste water to represent multiple industries. However, useful representative matrix ERMs may be able to be prepared from some major industries such as pulp and paper mills, steel mills, petroleum refineries, etc. Their "degree of representativeness" will need to be established through carefully designed studies. It is unlikely that manufacturing plants that use batch processing (such as dyes, paints, and pharmaceuticals) will be able to use representative matrix ERMs, but too little data is available to make definitive statements yet.

Water

The third category of matrix-based ERMs is water. Although it is not logical to consider reagent grade (distilled) water as a representative matrix for industrial waste water or, worse yet, soils, sediments, biota, and chemical wastes, this practice is, in fact, the prevalent one in use today. Method detection limits (MDLs) are determined from replicate measurements of an analyte in reagent water and then applied, often using arbitrary multiplication factors to guess at matrix effects, to all the other types of matrices.

Category Choices

From the above it is clear that, when sufficient money and time are available, site-specific ERMs are the best category to choose. When time and/or money are limited, the next best choice is a representative matrix-based ERM. The most common criticism of representative matrix ERMs is that the selected material will not adequately represent "my soil samples", or "my fish samples", or "my waste water". We know that, but the alternatives are preparing a site-specific ERM or, much worse, using reagent water. Thus the imperfect but much more accurate representative ERM is the logical practical choice in most cases. Therefore, more emphasis is being placed on the development of good representative matrix ERMs. As more of them become commercially available, the choices between them will be facilitated by closer relationships between environmental test samples and representative matrix ERMs.

CLASSES OF MATRIX-BASED ERMs

Two of the above categories, site-specific and representative matrix ERMs, can be divided into three different classes: matrix blanks, spiked matrix blanks, and naturally contaminated matrices. Reagent water can only be classed as a matrix blank for very clean water and as spiked matrix blanks.

A *matrix blank* is defined as a material composed of the same or similar substances as the test samples, which contains nondetectable or acceptably low amounts of the analytes of interest.

A *spiked matrix blank* is defined as a matrix blank that has been fortified and homogenized with known amounts of the analytes of interest.

A *naturally contaminated matrix* is defined as a material composed of the same or a similar substance as the test samples and which contains the analytes of interest

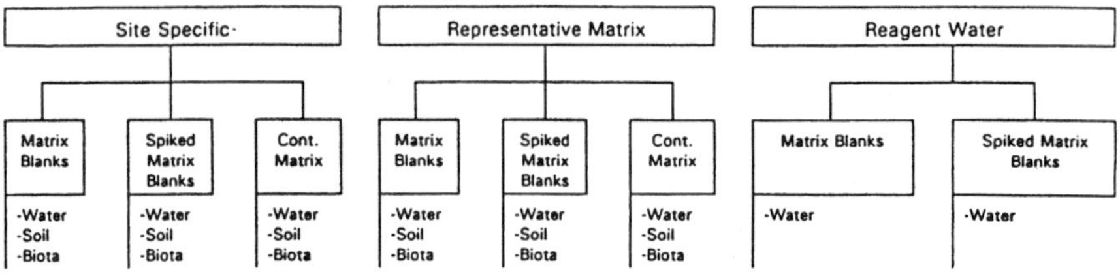

Figure 3 Hierarchy of matrix-based ERMs.

already present within it. When homogenized and submitted to interlaboratory analyses, concentration values of the analytes of interest are assigned.

The true values of the concentrations of the analytes of interest in a naturally contaminated matrix are not known, but consensus values from analysis by multiple laboratories are assigned and become the "accepted" or "true" concentrations. They represent the expected values that an average competent analyst should expect to get within a variance that is calculated based on standard deviation. Naturally contaminated matrix ERMs thus provide excellent performance evaluation standards.

Site-specific and representative matrix ERMs can be composed of a wide variety of matrix types in addition to all three classes of matrix-based ERMs described above. These include water, soil, sediments and other solids, and all kinds of biota, both plant and animal. Reagent water is not only limited to two classes (matrix blanks and spiked matrix blanks) but, of course, to only one type of matrix, "clean" water. This hierarchy is illustrated in Figure 3.

USES OF MATRIX-BASED ERMs

As mentioned above, matrix-based ERMs are excellent sources of performance evaluation standards. These standards, when submitted with environmental test samples to a laboratory (either as known standards or as blind samples) provide very useful measures of a laboratory's performance. Blank matrix ERMs allow a means of checking a method used by the lab for false positives. Spiked blank matrix ERMs allow a means of checking a method used by the lab for false negatives.

Another use of matrix-based ERMs is to calculate Method Detection *Levels* (MDLs). Proposed Method Detection Levels differ from EPA's currently used Method Detection *Limits* (MDLs) in two important ways, which are discussed below. The proposed Method Detection *Level* uses a *representative matrix* in its calculation while Method Detection *Limit* uses only reagent grade water. The proposed Method Detection *Level* allows subtraction of background signals in a blank, while the Method Detection *Limit* compares a sample signal to zero, thus causing bias if background signals are significant.

The U.S. EPA is under Congressional pressure to provide a scientific basis for Practical Quantification Limits (PQLs). These were devised to try to account for matrix effects by using various multiplication factors with EPA's Method Detection *Limit*.[1] The American Chemical Society (ACS) Committee on Environmental Improvement (CEI) has been working with EPA to suggest replacing PQLs with a Reliable Quantification Level (RQL). The proposed RQL is linked to the statistically derived Method Detection *Level* and may be $3^1/_3$ to 4 times the MDL when single measurement are used. Since the recommended MDL is usually 3 standard deviations (SD) above zero, the proposed RQL would eventually be 10 SD or 12 SD. Since 1983 the recommended Limit of Quantitation (LOQ) has been 10 SD above background.[2] Whether the RQL is 10 SD or 12 SD is not significant since both are arbitrary guidelines that establish a concentration level at which quantitative measurements are considered to be essentially valid or reliable numbers.

A glossary of the definitions and acronyms is provided at the end of this chapter. The important factors in the proposed EPA/ACS-CEI definitions are

- The RQL is linked to the Method Detection *Level* and would replace both the EPA Practical Quantitation Limit and the ACS Limit of Quantitation (LOQ).
- The proposed Method Detection *Level* would use a representative matrix in its calculation.
- For matrices other than clean water (e.g., drinking water), representative matrices may be either site-specific ERMs or representative matrix ERMs.

The impending use of representative matrices, where appropriate and where possible, will provide Method Detection *Levels* that are far more accurate than those calculated using only reagent water. They will also provide MDLs used with permits and other regulatory activities that should be more legally defensible and less controversial than those based on calculations using only reagent water. This should result in data that is less biased and more technically accurate. It will also result in a growing need for more and better matrix-based ERMs. These will not replace existing types of ERMs, but rather will supplement them. Matrix-based ERMs will add more complexity and choices to laboratories and environmental chemists, but that is characteristic of all fields of natural science as they evolve. The result is better and more accurate data as we progress from initial simple assumptions to a more sophisticated understanding of the science. We call this progress. We are soon going to make more progress in the science of making and using environmental reference materials.

REFERENCES

1. **U.S. Environmental Protection Agency,** Guidelines establishing test procedures for the analysis of pollutants under the clean water act: final rule and proposed rule, 40 CFR Part 136, *Fed. Reg.*, Part VIII (October 26, 1984).
2. **L. H. Keith, et al.,** Principles of environmental analysis, *Anal. Chem.*, 55 (1983) 2210–2218.

GLOSSARY OF TERMS AND DEFINITIONS

ACS	American Chemical Society
CEI	Committee on Environmental Improvement — This is a standing joint Board-Council Committee of the ACS whose members are appointed by the President of the ACS.
EPA	U.S. Environmental Protection Agency
ERM	Environmental Reference Material — Any substance with known or assigned values that is used as an analytical standard by laboratories.
LOQ	Limit of Quantification — The level above which quantitative results may be obtained with confidence. LOQ is usually recommended to be ten times the standard deviation above a blank.
MDL	Method Detection *Level* — The lowest concentration at which *individual* measurement results for a specific analyte are statistically different from a blank (that may be zero) with specified confidence level for a given method and representative matrix. An *intralaboratory* MDL estimate represents the average detection capability of a single laboratory for a specific analyte, method, and matrix at a given point in time. An *interlaboratory* MDL estimate represents the method detection capability for a specific analyte and specific matrix determined in more than one laboratory. (The recommended MDL is 3 standard deviations, and it is only based on the probability of not having false positive detections of an analyte.)
MDL	Method Detection *Limit* — The minimum concentration of a substance that can be measured and reported with 99% confidence that the analyte concentration is greater than zero. (The MDL is three times the standard deviation, and it is only based on the probability of not having false positive detections of a given analyte in reagent water.)
PE	Performance Evaluation — A standard containing analytes with known or assigned concentration values, which is used to evaluate a laboratory's performance of analysis with a method and specified analytes.
PQL	Practical Quantification Level — The lowest level that can be reliably achieved within specified limits of precision and accuracy during routine laboratory operating conditions. PQL multiplication factors may vary from 10 to 100,000 times the Method Detection *Level*.
RDL	Reliable Detection Level — When sufficient data are available, the RDL is the experimentally determined concentration at which false negative and false positive rates are specified. Otherwise, the RDL is the concentration that is two times the Method Detection *Level* (6 standard deviations). It is the lowest recommended level for qualitative decisions based on *individual* measurements.
RQL	Reliable Quantification Level — The recommended lowest level for quantitative decisions based on *individual* measurements for a given method and representative matrix. (The recommended RQL is two times the RDL or 12 standard deviations.)

CHAPTER 15

In-House Reference Materials as a Means to Quality Assurance: The Canadian Wildlife Service Experience

Bryan Wakeford and Richard Turle

Since 1980, the Canadian Wildlife Service has used in-house reference materials for monitoring the Quality Assurance of its trace organochlorine/PCB determinations, both for in-house operations and for contractors. The reasons for this approach are given. Analytical data received from check sample programs and other exchanges have allowed usable values to be assigned. Statistical analysis of in-house quality control charts have been used to determine causes of variability and to predict the limitations of the analytical method.

INTRODUCTION

Since the late 1960s the Canadian Wildlife Service (CWS) has been aware that toxic chemicals, entering the Great Lakes basin, have had measurable effects on species of birds that nest near and feed in the Lakes. The observations by field biologists led to the establishment of a program to monitor persistent toxic chemicals in the eggs of herring gulls (*Larus argentatus*) and other colonial waterbirds of the Great Lakes. The rationale for this program and the attendant sampling protocols, etc., have been described in detail elsewhere.[1,2]

The main purpose of the long-term monitoring of trace toxic chemicals has been to determine whether concentrations have changed over time. The CWS analytical chemists have had the responsibility of providing accurate, precise, and consistent data for researchers to interpret the long-term trends in concentrations of these chemicals in wildlife populations. Inconsistent or inaccurate data could undermine Environment Canada's reseach efforts and mislead Departmental officers who formulate regulations for chemicals in the environment.

Quality assurance (QA) programs are needed to produce accurate data. Two successful aspects of the CWS laboratory QA program have been the retrospective use of tissues from a specimen bank[18] and the development of reference materials.[3,4] The CWS specimen bank is a repository of portions of virtually all tissues that have been received through the Wildlife Toxicology Program since the early 1970s. Retrospective studies of specimens have been devised by biologists to determine concentrations of chemical compounds using techniques that may not have been available at the time of collection.[15–17] Also, analytical chemists have had the opportunity to determine whether technological changes have resulted in variations in analytical determinations.

The CWS laboratory has been using in-house reference materials as part of its QA program since 1979. They were developed for sample substrates required in the long-term monitoring program. Over this period, their use has increased the CWS's confidence in its analytical data.

REFERENCE MATERIALS MADE AT CWS

Rationale

It has long been recognized that analytical reference materials are an essential part of a good quality control program in chemical analysis.[5] They allow the analyst to check both precision (reproducibility) and accuracy of analytical data. If the substrate type is the same as that of the samples to be analyzed and not easily distinguished from them, the reference material may also be used as a "blind" check on the quality of analysis.

In the late 1970s, the CWS relied heavily on the analysis of organochlorine residues in herring gull eggs from the Great Lakes, as part of the International Joint Commission (IJC) Wildlife Surveillance Program. As the importance of the measurements increased, the CWS program managers required increased confidence in the data, which were largely produced through contracts with private laboratories. An urgent need existed for a reference material for the program. The RM would be submitted blindly to the contractor and serve to assess the accuracy and precision of analysis for each batch of samples. Spiked samples would not serve this purpose, because the patterns of organochlorine residues (particularly for PCBs) in naturally contaminated samples is very different from that of artificially fortified ones.

In 1979, there was only one primary RM available for the analysis of organic substances in environmental samples, a fish oil standard from the NRC Atlantic Regional Laboratory in Halifax, N.S. It was felt that this material could not be used as reference materials for whole tissue substrate. Therefore, the decision was made to prepare a secondary RM and to analyze it rigorously to establish the homogeneity of the material and the concentrations of the compounds of interest.[6]

In 1993, the long-term monitoring is still continuing, and there are still no "ideal" primary reference materials available to be used in the CWS QA program. The National Institute of Science and Technology (NIST) markets one tissue organochlorine and PCB analysis, the SRM 1588 Cod Liver Oil.[12] This RM's lipid has a very different chemical composition from that of avian and mammalian tissue and thus behaves differently in chemical analysis. It has only been certified for a limited number of OC/PCBs. For these reasons, such an RM has only limited use for the CWS program. In the years from 1980 to 1993, there have been a number of secondary RMs prepared for the QA of chemical data of the CWS toxic chemical program.

Production and Characterization of CWS Secondary Reference Materials

This section briefly describes the manner of preparing each of the secondary reference materials that have been and are being used at the CWS. An inventory of remaining material was made in May 1993.

1979 Lake Erie Herring Gull Eggs Reference Material

Eighty-four herring gull eggs obtained fresh from Big Chick Island and Middle Sister Islands in Lake Erie in 1979 were used for the egg pool. Four to five eggs were homogenized (without shells) in a blender. Each homogenate was poured into a 5-gal glass jar. After addition of all the eggs (7.3 kg), the jar was rolled and shaken thoroughly to mix the contents and frozen before the next step.

The egg pool in the glass jar was rehomogenized at room temperature and divided into 34 subpools in numbered glass jars, and frozen prior to further subsampling. Subsequently, each pool was thawed and rehomogenized; 10-g aliquots were poured into 20-ml glass scintillation vials (for organic analysis) with aluminum foil-lined lids and 20-ml plastic scintillation vials (for metals analysis). In total, 571 glass vials and 155 plastic vials were prepared and stored in the CWS walk-in freezer at –40°C.

Characterization: Of the 34 pools, 12 were sampled. One aliquot from each of the pools was analyzed at CWS using a standard methodology. The results for ten organochlorine residues and total PCB were determined.

Uses: This first RM for organochlorine analyses was unfortunately very high in PCBs and thus was not very useful for routine OC/PCB analysis. It has been used in interlaboratory comparisons.

The current inventory is 128 jars weighing a total of 1695 g.

1980 Lake Huron Herring Gull Eggs Reference Material

One hundred eggs were collected from three locations in Lake Huron; East Grant, Ivor Rocks, and Anchor islands. The procedure for blending the material at CWS differed slightly compared with the 1979 lot, in that they were shelled in the field and sent to the CWS in a frozen state in four glass jars each containing 25 eggs.

The total material, which weighed 8.23 kg, was mixed together and dispensed into four jars (acetone and hexane washed). Each jar was then independently homogenized and 5-g aliquots were dispensed into glass and plastic vials. Eight hundred glass vials and 398 plastic vials were prepared and stored in the CWS walk in freezer at –40°C.

Characterization: This material was then analyzed in-house at CWS in a manner similar to the 1979 material.

Uses: The 1980 RM was used in the routine monitoring program for OC/PCB analyses, both to maintain in-house QA and to monitor the performance of private contract laboratories from 1980 to 1989. It was also used for a number of interlab comparison studies.

The current inventory is 145 vials weighing a total of 1450 g.

The 1986 Lake Ontario Herring Gull Eggs Reference Material

One hundred seventeen herring gull eggs were collected in the spring of 1986 at Hamilton Harbour in Lake Ontario. The material was homogenized and mixed into a large batch, weighing 8.9 kg. It was then frozen at –40°C in 15- and 30-g aliquots and in three large portions of about 2 kg each.

Characterization: This material was collected to support the dioxin/furan analyses protocol in the CWS lab. Ten replicate aliquots were analyzed for 2,3,7,8-chlorine substituted polychlorinated dibenzodioxins and furans.

Uses: This RM was used for QA of dioxin/furan research work and for interlab collaborations on dioxin/furan analysis.

The current remaining inventory is three large jars weighing a total of 6420 g.

The 1989 Lake Ontario Herring Gull Eggs Reference Material

This material was prepared in both an undiluted and diluted form.

Undiluted: The same laboratory procedure as was used for the 1986 egg pool was employed, using 138 herring gull eggs collected from Lake Ontario, 50 of which were from Hamilton Harbour. Total weight of the eggs was 10.4 kg.

Diluted: Some egg was diluted 1 + 9 (i.e., 1:10), to give adequate levels of both very low as well as high contaminant concentrations. Domestic chicken eggs were chosen for the dilution because they generally have low PCB levels (ΣPCB = 0.02 and 0.01 mg/kg in the two pooled samples that were analyzed). One hundred seventy-five large chicken eggs (from a local supermarket) were homogenized, 24 at a time, in an homogenizer, aluminum foil under the lid, at low speed for 20 sec. The entire homogenate was poured into a large plastic carboy and homogenized with a stirrer. To the final volume of 9.2 kg, 1.02 kg of the herring gull egg reference pool was added, creating a 10 to 1 dilution of the 1989 Reference Material.

The OC/PCB levels of the diluted material were determined by the analysis of 11 individual vials selected at random from the approximately 1000 prepared. This diluted material has been used for quality control (QC) purposes since 1990. The dioxin values were also determined on the undiluted RM, and this material is used routinely as part of the QC program for dioxin analyses.

The current inventory is 830 jars and vials weighing a total of approximately 10.84 kg.

1990 Breast Muscle Reference Pool

In response to a need for an RM which could be used in a program of analysis of OC/PCBs in waterfowl breast muscle, a new material was developed in 1990. The breast muscle from 94 lesser scaup, from Prince Edward County, Ontario, was removed and placed into ten 1-L glass jars.

Breast muscle, 3.1 kg, was chopped on a Teflon board and mixed manually. Approximately 130.0 g batches were homogenized in a blender and then placed in a steel bowl. Once all the breast muscle was homogenized, the pool was manually mixed with a spatula. The breast muscle was then homogenized in a food processor in two batches. This homogenate was manually mixed and aliquoted. Aliquots were stored at –40°C.

Nine aliquots of the material were analyzed at the CWS lab; the results showed very low levels of the contaminants being analyzed. This material has been used as a means of checking the quality of data from a contract lab for the past 3 years.

The current inventory is 150 vials and jars weighing a total of approximately 5.1 kg.

THE USE OF REFERENCE MATERIALS (RMS) AT THE CWS LABORATORIES

Internal Quality Assurance

Organochlorine/PCBs Analysis

The CWS has used RMs as a means of controlling quality of analytical procedures since the early 1980s. For quality control (QC), one RM sample is analyzed for every project batch or every tenth sample, whichever is less. The results of the RM are then compared to the established acceptance limits which are ± 2 standard deviations (SD) of the long-term mean, which can be plotted in a Shewart chart. If the result obtained is more than ± 2 SD, it is considered out of control. Typically, a RM control sample should not exceed ± 2 SD more than once in 20 analyses or exceed ± 3 SD more than once in 100 analyses. If the RM has more than one parameter which exceeds ± 3 SD then the batch must be reanalyzed.

Reasons for reanalysis are usually traced to poor lipid extraction from the sample or incorrect calibration for the gas chromatograph. The lipid levels are recorded for the RM to ensure a consistency in extraction.

An example of the QC charts that are produced is shown in Figure 1. The mean and upper and lower control limits were established for the diluted 1989 RM in 1990. The data shown are for the results determined in 1993. These data were plotted from a spreadsheet compilation of the all the results for the RM, a portion of which is shown in Table 1. This table is divided into two sections. The upper portion is statistics calculated from the analysis of 11 aliquots of the RM. The overall mean, SD, and CV are supplemented with minimum and maximum limits as expressed by a range of ± 2 times the standard deviation from the mean.

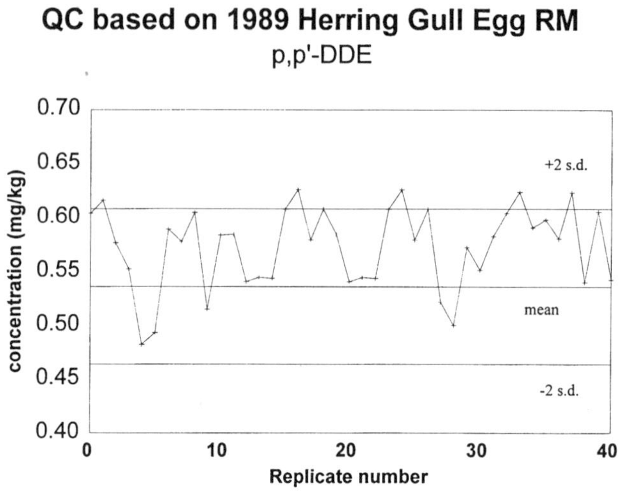

Figure 1 Quality control based on 1989 herring gull egg reference material — p,p′-DDE.

Dioxin/Furan Analysis

The use of reference materials is a portion of the quality assurance program for the dioxin/furan analysis. As with the OC/PCBs analysis, an aliquot of a RM is analyzed with each batch of samples in a project or every tenth sample. The results are compiled on an on-going basis as one means of assessing current methodology. An example of the data set compiled is shown in Table 2. These data were compiled over a 2-year period on the 1986 RM. A more complete description of all the elements of QC in a dioxin/furan analysis is described in an internal report.[7]

Heavy Metal Analysis

A large portion of analytical data generated for the CWS Wildlife Toxicology program involves heavy metal determinations. Generally speaking, the availability of commercial reference materials has been sufficient to meet the QC needs for RMs at CWS. They are available in a variety of organic substrates and are typically characterized for a broad spectrum of metals. The materials that are used routinely are

DORM-1	Dogfish muscle (National Research Council of Canada)
DORT-1	Dogfish liver (National Research Council of Canada)
SRM 1566	Oyster tissue (National Institute of Science and Technology)
SRM 1577a	Bovine liver (National Institute of Science and Technology)

With each project batch or every 20 samples (whichever is less), at least two different reference materials are analyzed in duplicate, and the results are assessed in light of the expected variations (± 2 SD, as in OC/PCBs above). These determinations are a vital portion of the quality assurance (QA) of metals analysis.

Table 1 In-House Quality Assurance Results on 1989 Egg Reference Material (mg/kg)

Analysis	% Fat	% Water	HCB	OCS	*t*-Nonachlor	pp′DDE	Photo-mirex	Mirex	Oxychlordane	*c*-Chlordane	pp′DDD
Mean to date:	10.82	75.83	0.0044	0.0023	0.0094	0.5731	0.0339	0.0851	0.0116	0.0016	0.0016
Standard deviation:	0.41	0.09	0.0004	0.0002	0.0008	0.0387	0.0031	0.0055	0.0010	0.0002	0.0005
Coeff. of variation:	3.7	0.1	8.8	7.7	8.3	6.7	9.1	6.5	8.4	14.2	29.8
n:	27	27	27	27	27	27	27	27	27	27	27
Min. acceptable:	10.01	75.65	0.0036	0.0019	0.0078	0.4958	0.0277	0.0741	0.0096	0.0011	0.0007
Max. acceptable:	11.63	76.02	0.0052	0.0026	0.0109	0.6504	0.0400	0.0961	0.0135	0.0020	0.0026
CV max:	4.9	2.1	141.9	189.6	102.5	17.4	58.9	39.6	93.6	221.4	217.3
3E011	10.6	75.8	0.0045	0.0021	0.0094	0.6044	0.0363	0.0921	0.0094	0.0012	0.0016
3E021	10.7	75.9	0.0046	0.0022	0.0095	0.6164	0.0353	0.0870	0.0119	0.0014	0.0015
3E022	10.5	75.8	0.0044	0.0021	0.0095	0.5768	0.0342	0.0851	0.0120	0.0019	0.0015
3E031	10.2	75.8	0.0043	0.0022	0.0084	0.5528	0.0303	0.0857	0.0145	0.0015	0.0021
3E041	10.7	75.9	0.0038	0.0019	0.0078	0.4827	0.0310	0.0752	0.0102	0.0011	0.0013
3E051	10.7	75.9	0.0039	0.0019	0.0079	0.4938	0.0311	0.0773	0.0094	0.0013	0.0013
3E061	10.5	75.7	0.0044	0.0021	0.0091	0.5895	0.0340	0.0849	0.0112	0.0016	0.0016
3E071	10.8	75.8	0.0044	0.0023	0.0094	0.5779	0.0346	0.0850	0.0121	0.0014	0.0017
3E072	10.8	75.8	0.0048	0.0026	0.0097	0.6052	0.0349	0.0877	0.0102	0.0010	0.0015
3E081	10.4	75.7	0.0041	0.0022	0.0087	0.5153	0.0314	0.0778	0.0117	0.0017	0.0017
3E091	10.5	75.8	0.0039	0.0024	0.0098	0.5840	0.0374	0.0917	0.0116	0.0017	0.0015
3E092	11.4	75.7	0.0051	0.0025	0.0107	0.5847	0.0388	0.0948	0.0122	0.0019	0.0023
3E093	10.[illegible]	75.8	0.0044	0.0021	0.0089	0.5410	0.0327	0.0809	0.0113	0.0015	0.0011
3E094	11.2	75.9	0.0036	0.0021	0.0101	0.5447	0.0317	0.0811	0.0115	0.0017	0.0015
3E101	11.4	76.0	0.0047	0.0024	0.0104	0.5441	0.0370	0.0908	0.0117	0.0017	0.0015
3E102	11.0	75.8	0.0047	0.0023	0.0098	0.6082	0.0350	0.0866	0.0121	0.0017	0.0028

Table 1 In-House Quality Assurance Results on 1989 Egg Reference Material (mg/kg)

Analysis	% Fat	% Water	HCB	OCS	*t*-Nonachlor	pp′DDE	Photo-mirex	Mirex	Oxychlordane	*c*-Chlordane	pp′DDD
3E111	11.3	75.8	0.0046	0.0024	0.0095	0.6262	0.0345	0.0861	0.0115	0.0016	0.0017
3E121	10.9	75.8	0.0044	0.0023	0.0086	0.5796	0.0268	0.0767	0.0120	0.0016	0.0008
3E131	10.6	76.0	0.0045	0.0023	0.0089	0.6086	0.0353	0.0869	0.0118	0.0016	0.0018
3E092	11.4	75.7	0.0051	0.0025	0.0107	0.5847	0.0388	0.0948	0.0122	0.0019	0.0023
3E093	10.1	75.8	0.0044	0.0021	0.0089	0.5410	0.0327	0.0809	0.0113	0.0015	0.0011
3E094	11.2	75.9	0.0036	0.0021	0.0101	0.5447	0.0317	0.0811	0.0115	0.0017	0.0015
3E101	11.4	76.0	0.0047	0.0024	0.0104	0.5441	0.0370	0.0908	0.0117	0.0017	0.0015
3E102	11.0	75.8	0.0047	0.0023	0.0098	0.6082	0.0350	0.0866	0.0121	0.0017	0.0028
3E111	11.3	75.8	0.0046	0.0024	0.0095	0.6262	0.0345	0.0861	0.0115	0.0016	0.0017
3E121	10.9	75.8	0.0044	0.0023	0.0086	0.5796	0.0268	0.0767	0.0120	0.0016	0.0008
3E131	10.6	76.0	0.0045	0.0023	0.0089	0.6086	0.0353	0.0869	0.0118	0.0016	0.0018

Table 2 In-House Quality Assurance Results on 1986 Egg Reference Material

Dioxins and Furans			
Compound determined	Mean (ng/kg wet wt)	Standard deviation	Coefficient variation (%)
2378-TCDD	44	2.3	5
12378-PCDD	8	1.6	21
123678-HxCDD	14	2.8	20
123478-HpCDD	6	1.1	19
OCDD	6	1.1	16
23478-PnCDF	7	1.1	16
123478-HxCDF	3	0.7	22
123678-HxCDF	4	0.9	23
n=27			

Note: TCDD — Tetrachlorodibenzodioxin
PCDD — Pentachlorodibenzodioxin
HxCDD — Hexachlorodibenzodioxin
HpCDD — Heptachlorodibenzodioxin
OCDD — Octachlorodibenzodioxin
TCDF — Tetrachlorodibenzofuran
PCDF — Pentachlorodibenzofuran
HxCDF — Hexachlorodibenzofuran

For contracted work, a laboratory is expected to provide data on appropriate reference materials as part of the QA procedures described in the contract.

Evaluation of Contracted Analytical Work

The CWS RMs have been useful in assessing the capabilities of laboratories to provide accurate data under contract to CWS. The capacity for producing chemical analysis at the CWS lab falls short of the normal demand. This gap is usually closed by letting contracts to outside firms. In order to evaluate the bids from a number of firms, a significant consideration is the capability of the firm to provide high quality data, as well as a competitive price.

In 1992, as part of the evaluation of laboratories bidding on contract to provide OC/PCB data to CWS, portions of the 1989 diluted RM were homogenized and diluted with chicken egg and sent to prospective labs. The homogenates were analyzed at CWS to ensure the results and the homogeneity of the material. Multiple portions were sent to each lab, and the results were assessed statistically. The contract was let to the lab with the best performance measured against predefined criteria, although it was not the lowest cost bid. A similar evaluation of bids was performed in letting the contract for dioxin analysis in the previous year.

Once a laboratory has been awarded a contract for producing data for the CWS, the evaluation of its quality continues; RMs have been extremely valuable for this assessment. The criteria used in assessing the quality of on-going contracts has been described in a CWS Technical Report.[7] This document specifies that RMs will be included as part of CWS's assessment of a laboratory's QC:

> *For each batch of samples a minimum of RMs will be included in the batch and the whole batch blind numbered by CWS or the contractor's quality assurance officer.*

The clause then lists a number of specific criteria, which will be measured from the analysis of the RM.

The use of these criteria is illustrated in Table 3. This table is a portion of the entire spreadsheet that is used to assess OC/PCB results. It is divided into three sections. The upper third lists statistics that have been derived at the CWS lab for the 1989 diluted herring gull RM. The middle lists the individual results that the contract lab has reported on a number of aliquots of this material (blind analysis). The lower third of the table lists statistics generated from the lab data and the criteria that are being used to judge them. Each analyte is assessed by an overall "pass or fail" rule for the set of analyses performed. If a particular analysis is graded a fail, it may point to particular methodological difficulties which must be corrected. If the number of failures for a particular sample is too large, it will require an entire batch of sample tissues be redone. Any number of criteria may be set for judging the adequacy of a contractor's performance from the use of the reference materials. The important point is that these criteria are defensible in light of the experience of the CWS lab with the materials.

DETERMINATIONS OF TRACE COMPONENTS IN REFERENCE MATERIALS

The CWS has used a standard trace chemical analysis technique, which has remained substantially unchanged since 1980.[9,19] The basis for determining the trace levels of organic chemicals in the RMs relies on the hypothesis that multiple analyses of the RMs using the same methodology will yield comparable results over time. The data presented have been compiled using this one standard operating procedure.

The trace component concentrations of organochlorine and PCB compounds, which have been determined for the existing RMs are listed in Table 4. For each of the herring gull RMs from 1980 and 1989 (diluted), a large number of replicates have been included in the calculation of the mean and coefficient of variation (CV). The duck breast muscle data are based on nine replicates, and the undiluted 1989 herring gull RM data are means of triplicated analyses. All the data have been produced at the CWS laboratory in Hull, Quebec.

At present, the 1989 herring gull egg RM (diluted) is the most appropriate to use as an in-house QC material because its OC/PCB levels are close values typically found in wildlife material. There is close agreement, after allowing for dilution, between the diluted egg results and the undiluted material.[10,11] There have been a large number of muscle samples processed in the past few years, and the 1990 duck muscle has been used as a means of ensuring good data from contract laboratories.

Table 3 Quality Assurance — Contract Laboratory (mg/kg)

Determination	pp′DDE	Photomirex	Mirex	Oxychlordane	pp′DDD	*cis*-Nonachlor	pp′DDT	Dieldrin	PCB 118	PCB 146	PCB 153	PCB 105
Reference mean	0.5041	0.0327	0.0803	0.0108	0.0016	0.0062	0.0043	0.0113	0.1043	0.0411	0.2547	0.0227
Standard deviation	0.0452	0.0028	0.0063	0.0013	0.0004	0.0010	0.0015	0.0014	0.0116	0.0033	0.0236	0.0025
Coefficient of variation	9.0	8.6	7.9	11.9	26.6	16.4	34.6	12.7	11.1	8.1	9.2	11.1
Replicates (n)	39	39	39	39	39	39	39	38	39	39	39	39
Minimum acceptable	0.4137	0.0270	0.0676	0.0082	0.0008	0.0042	0.0014	0.0084	0.0811	0.0344	0.2076	0.0177
Maximum acceptable	0.5944	0.0383	0.0930	0.0134	0.0025	0.0083	0.0074	0.0142	0.1275	0.0478	0.3018	0.0278
Lowest	0.3436	0.0239	0.0590	0.0086	0.0011	0.0048	0.0018	0.0086	0.0696	0.0305	0.1934	0.0158
Highest	0.5883	0.0387	0.0996	0.0143	0.0031	0.0089	0.0083	0.0160	0.1263	0.0513	0.3226	0.0274
CV max	78	60	41	96	218	122	142	95	36	54	25	70
1989 EGGRM undil.	0.3683	0.0367	0.0679	0.0095	0.0017	0.0046	0.0027	0.0095	0.0835	0.0335	0.2098	0.0114
1989 EGGRM dil 10:1	0.3674	0.0334	0.0691	0.0097	0.0019	0.0052	0.0021	0.0107	0.0836	0.0329	0.2026	0.0114
1989 EGGRM undil.	0.4372	0.0463	0.0761	0.0116	0.0016	0.0060	0.0567	0.0106	0.0802	0.0354	0.2154	0.0218
1989 EGGRM undil.	0.3476	0.0284	0.0515	0.0116	0.0021	0.0060	0.0362	0.0041	0.0519	0.0268	0.1518	0.0152
1989 EGGRM undil.	0.3578	0.0413	0.0661	0.0101	0.0014	0.0050	0.0515	0.0101	0.0699	0.0317	0.2073	0.0201
1989 EGGRM undil.	0.4033	0.0348	0.0581	0.0107	0.0018	0.0050	0.0490	0.0109	0.0648	0.0291	0.1763	0.0165
1989 EGGRM undil.	0.3076	0.0244	0.0565	0.0118	0.0017	0.0057	0.0399	0.0086	0.0633	0.0290	0.1672	0.0156
1989 EGGRM dil 10:1	0.3334	0.0376	0.0633	0.0112	0.0012	0.0048	0.0398	0.0121	0.0723	0.0343	0.2024	0.0171
Mean	0.3636	0.0354	0.0636	0.0108	0.0017	0.0053	0.0347	0.0096	0.0712	0.0316	0.1916	0.0161
Standard deviation	0.0404	0.0069	0.0079	0.0009	0.0003	0.0005	0.0211	0.0025	0.0111	0.0030	0.0233	0.0037
Min	0.3076	0.0244	0.0515	0.0095	0.0012	0.0046	0.0021	0.0041	0.0519	0.0268	0.1518	0.0114
Min > min acceptable	NO	NO	NO	YES	YES	YES	YES	NO	NO	NO	NO	NO
Max	0.4372	0.0463	0.0761	0.0118	0.0021	0.0060	0.0567	0.0121	0.0836	0.0354	0.2154	0.0218
Max < max acceptable	YES	NO	YES	YES	YES	YES	NO	YES	YES	YES	YES	YES
Criterion 1	FAIL	FAIL	FAIL	PASS	PASS	PASS	FAIL	FAIL	FAIL	FAIL	FAIL	FAIL

Table 3 Quality Assurance — Contract Laboratory (mg/kg)

Determination	pp′DDE	Photomirex	Mirex	Oxychlordane	pp′DDD	*cis*-Nonachlor	pp′DDT	Dieldrin	PCB 118	PCB 146	PCB 153	PCB 105
CV	11.0	19.5	12.5	8.4	16.7	10.0	60.8	25.6	15.6	9.6	12.1	22.9
CV max?	PASS	PASS	PASS	PASS	PASS	PASS	PASS	PASS	PASS	PASS	PASS	PASS
± 2SD?	PASS	PASS	PASS	PASS	PASS	PASS	PASS	PASS	PASS	PASS	PASS	PASS
SE	–27.3	8.3	–20.8	–0.1	4.0	–15.2	692.7	–15.2	–31.7	–23.2	–24.8	–29.0
SE < 50%?	PASS	PASS	PASS	PASS	PASS	PASS	FAIL	PASS	PASS	PASS	PASS	PASS
TE	–11.2	50.6	–1.1	16.7	38.6	1.7	1655.8	28.2	–10.4	–8.6	–6.5	3.5
TE > 50%?	PASS	FAIL	PASS	PASS	PASS	PASS	FAIL	PASS	PASS	PASS	PASS	PASS
Criterion 2	PASS	FAIL	PASS	PASS	PASS	PASS	FAIL	PASS	PASS	PASS	PASS	PASS
OVERALL	PASS	FAIL	PASS	PASS	PASS	PASS	FAIL	PASS	PASS	PASS	PASS	PASS

Table 4 Statistics for OC/PCBs for CWS Reference Materials

Analysis	1980 Herring gull egg			1989 Herring gull egg (diluted 1:10)			1989 H. gull egg (undiluted)	1990 Duck breast muscle		1979 H. gull	
	Mean (mg/kg)	CV	n	Mean (mg/kg)	CV	n	Mean (n=3) (mg/kg)	Mean (n=9) (mg/kg)	CV	Mean (n=12) (mg/kg)	CV
% Lipid:	8.55	7.5	77	9.73	9.6	71	9.08	4.732	5.6	8.320	1.6
% Moisture:	76.71	1.0	77	75.77	0.6	71	75.04	63.056	5.5		
1245-Tetrachlorobenzene				0.0065	97.2	18		0.00187	36.9		
1234-Tetrachlorobenzene				0.0028	105.6	10		0.00074	70.2		
Pentachlorobenzene	0.01146	115.3	53	0.0007	14.6	25	0.0071	0.00019	31.8		
Hexachlorobenzene	0.07850	12.8	135	0.0047	9.8	71	0.0473	0.00132	12.4	0.179	7.3
Octachlorostyrene	0.02695	10.5	75	0.0023	15.1	71	0.0214	0.00018	40.4		
trans-Nonachlor	0.11728	25.0	135	0.0086	9.4	71	0.0763	0.00099	22.9		
pp′ DDE	4.990	9.2	135	0.5131	8.0	71	5.0977	0.01760	16.5	3.990	4.9
pp′ DDT	0.08443	19.6	135	0.0035	46.5	71	0.0266	0.00030	94.3		
Photomirex	0.06191	11.3	125	0.0331	7.9	71	0.3257	0.00120	15.6	0.041	11
Mirex	0.15687	13.3	135	0.0807	6.6	71	0.7983	0.00509	15.0	0.095	5.5
α-BHC	0.06389	109.4	122			1		0.00154	5.7		
β-BHC	0.01840	192.4	134	0.0007	20.5	44	0.0055	0.00020	66.1		
γ-BHC (Lindane)	0.00752	56.6	82					0.00042	33.0		
Oxychlordane	0.20270	12.1	135	0.0106	11.1	71	0.0992	0.00089	32.0	0.152	2.6
trans-Chlordane	0.20354	21.3	66	0.0003	110.9	26		0.00010	0.0		
cis-Chlordane	0.01959	87.1	125	0.0013	16.5	71	0.0116	0.00057	57.2	0.022	6.5
pp′ DDD	0.01274	133.2	135	0.0015	23.6	71	0.0125	0.00080	20.7	0.049	4.1
cis-Nonachlor	0.05643	12.4	128	0.0058	15.9	71	0.0564	0.00028	24.0	0.085	2.6
Heptachlor epoxide	0.13291	10.2	135	0.0043	21.3	70	0.0453	0.00354	17.0	0.119	5.9
Dieldrin	0.27689	11.5	135	0.0108	15.3	70	0.1136	0.01426	14.4	0.254	3.5
Arochlor 1254:60 (1:1)	23.43	9.5	135	3.088	7.7	71	33.6367	0.08303	18.4	94.4	6.6
Arochlor 1260	9.91	12.9	128	1.573	10.1	71	16.7033	0.05480	19.8	37.0	7.8

Table 4 Statistics for OC/PCBs for CWS Reference Materials (continued)

	1980 Herring gull egg			1989 Herring gull egg (diluted 1:10)			1989 H. gull egg (undiluted)	1990 Duck breast muscle		1979 H. gull	
Analysis	Mean (mg/kg)	CV	n	Mean (mg/kg)	CV	n	Mean (n=3) (mg/kg)	Mean (n=9) (mg/kg)	CV	Mean (n=12) (mg/kg)	CV
PCB congeners				1.5033	7.6	71					
31	0.07197	39.8	3								
28	0.05718	21.8	53	0.0045	43.4	71	0.0715	0.00022	70.4		
52	0.08970	18.0	115	0.0143	16.5	71	0.1237	0.00039	20.1		
49	0.02207	63.5	41	0.0016	36.1	30		0.00000	ERR		
47	0.17965	48.2	76				0.0578	0.00642	24.8		
44	0.02008	23.1	48	0.0029	20.4	58	0.0259	0.00006	94.9		
42	0.00770	28.7	3								
64	0.01170	15.8	104	0.0010	10.5	71	0.0095				
				0.0179	122.2	48					
70	0.01921	23.5	47					0.00026	62.2		
66	0.35230	11.4	122	0.0336	8.2	71	0.3507	0.00159	22.1		
60	0.16485	46.3	36	0.0153	48.0	59					
101	0.23129	8.8	122	0.0352	7.4	71	0.3430	0.00067	15.0		
99	0.70921	8.7	126	0.0704	6.6	71	0.7947	0.00227	17.5		
97	0.05028	12.7	120	0.0054	7.7	71	0.0550	0.00038	124.7		
87	0.04091	47.9	72	0.0086	22.1	70	0.0666	0.00018	150.9		
110	0.20288	11.2	126	0.0247	26.1	71	0.2271	0.00120	87.4		
151	0.03046	68.0	27	0.0052	46.8	71	0.0391	0.00006	94.9		
				0.0353	5.9	48					
118	0.90434	10.2	126	0.1050	9.0	71	1.0540	0.00461	20.1		
146	0.30425	11.0	126	0.0404	8.2	65	0.4376	0.00128	14.5		
153	1.96177	9.0	126	0.2451	8.9	71	2.8903	0.00977	18.4		
105	0.19765	23.8	125	0.0229	9.3	71	0.2507	0.00077	16.0		
141	0.04372	20.2	125	0.0094	12.8	71	0.0879	0.00013	106.1		
137	0.07232	12.9	126	0.0076	20.0	71	0.0736	0.00010	50.0		
138	1.72208	9.4	126	0.2261	7.7	71	2.4627	0.00604	18.4		

158	0.07701	11.7	122	0.0090	9.2	71	0.1020	0.00010	0.0
129	0.05297	27.1	126	0.0108	10.3	71	0.1061	0.00044	74.7
182	0.47898	11.9	126	0.0925	11.1	71	0.9941	0.00331	23.2
183	0.29989	20.1	126	0.0442	9.6	71	0.4706	0.00130	21.4
128	0.31453	17.2	120	0.0309	23.9	71	0.2944	0.00068	24.2
185	0.01660	0.0	4	0.0016	74.9	32	0.0085	0.00000	ERR
174	0.04600	23.4	118	0.0098	17.8	71	0.0918	0.00038	84.5
171	0.14294	19.2	120	0.0215	10.0	71	0.2326	0.00066	43.2
200	0.04116	19.9	108	0.0065	11.2	71	0.0694	0.00022	37.5
172	0.07636	19.0	126	0.0107	9.2	71	0.1066	0.00023	30.3
180	-.15727	12.8	126	0.1830	10.1	71	1.9423	0.00633	19.8
170	0.40634	15.6	126	0.0671	11.1	71	0.6896	0.00251	23.3
201	0.22219	14.1	126	0.0399	10.3	71	0.4136	0.00359	20.9
203	0.16744	14.6	126	0.0249	10.8	71	0.2571	0.00218	20.0
195	0.03733	27.4	84	0.0072	15.6	71	0.0704	0.00033	156.6
194	0.13021	20.6	126	0.0215	12.1	71	0.1925	0.00126	23.9
206	0.03018	27.5	78	0.0051	15.9	71	0.0486	0.00341	21.9
Sum of PCB congeners	10.680	9.4	126	1.5033	7.6	71	15.513	0.06488	15.5

RELIABILITY OF THE CWS ANALYTICAL PROCEDURE

Accuracy

All the trace organic data concerning the CWS RMs have been generated using one general procedure. The accuracy of the CWS OC/PCB method could be validated if there existed appropriate primary RMs. The closest material that is currently available is the SRM 1588 cod liver oil from the National Institute of Science and Technology (NIST). This material has been certified for a number of components; however, the CWS has provided data to the NIST on a number of additional congeners. The NIST has published the CWS data along with its own,[12] to expand the number of components for which some data can be reported. The data produced by the CWS on the components that have been certified in SRM 1588 are listed in Table 5. CWS is confident that its analytical procedure produces accurate data. Further feedback on the accuracy of the published CWS results is expected from other labs that may be using this material to determine specific PCB congeners.

Table 5 Concentrations (μg/kg) of Organochlorine Pesticides and Selected PCB Congeners in NIST SRM 1588 Cod Liver Oil

Compound	NIST certified value	2 SD	CWS	2 SD
HCB	148	21	137	15
trans-Nonachlor	209	11	256	33
pp′DDE	641	62	596	35
pp′DDT	529	45	587	104
α-BHC	86	19	55	2
trans-Chlordane	50	13	38	11
cis-Chlordane	158	8	120	15
pp′DDD	277	15	225	59
PCB 101	129	5	147	6
PCB 153	276	40	305	104
PCB 138	261	29	292	84
PCB 180	107	4	128	56
PCB 170	45	5	62	24

Note: CWS mean values based on four determinations.

Interlab Comparisons

There have been a number of interlaboratory comparisons made using CWS RMs, which can serve to measure the precision of data between the laboratories that employ similar methodologies to determine OCs and PCB levels in environmental samples.

Results from a series of exchanges of the 1979 herring gull egg RM have been summarized in Table 6. This RM, whose concentrations of PCBs were considered too elevated for it to be used for ongoing QC, has shown its usefulness as a means of validating the analytical methodology at CWS. The CWS results are shown in comparison with those of the U.S. Fish and Wildlife Service (U.S. FWS) at Patuxent, MD, and with a private contract lab, The Ontario Research Foundation (ORF). Also shown are the means of results from a number of labs that participated in an interlab comparison under the sponsorship of the Federal Interdepartmental Committee on Pesticides (Canada). The mean values of all the reported results agree closely. These studies served to support CWS's confidence in the preparation of the RMs and in the analytical methodology used to determine results in its research program.

In 1990, there was a distribution of the 1989 RM to 13 laboratories in Canada, who participated in an interlab comparison under the sponsorship of the Canadian Association of Pesticide Control Officials (CAPCO).[13] In this study, labs were asked to provide data on seven organochlorine pesticides as well as nine individual PCB congeners. The overall means, standard deviation, and CVs of these 13 labs in comparison with the data from six replicate analyses by the CWS lab are shown in Table 7. The CVs among the labs were higher than those in the CWS lab, as would be expected, but the means were in good agreement. Figures 2 and 3 show the means of the OCs and PCB congeners that were reported.

These studies have shown that the RMs that have been produced in the CWS lab are homogeneous and that the results among a number of labs that are employing similar analytical procedures are in close agreement.

RELATIONSHIPS FOUND OVER TIME

Theoretical Expectations of Variations of Coefficients of Variation

One of the criteria which is used by CWS in the assessment of contract labs results is a limit on the maximum coefficient of variation (CV) allowed. Because CV is known to increase as the concentration of an analyte decreases, CWS has used the concept of CV_{max} to allow for the systematic increase of CV as concentration declines.[14]

The CV_{max} criterion is based on the work of Frehse and Timme, in which statistical arguments were used to develop the following equation

$$CV_{max} = CV_o\ (x_o/x)^{\log f}\ (\%)$$

where CV_{max} = maximum permissible CV, CV_o = CV at the detection limit, x = mean of replicate analysis of blind sample, x_o = detection limit, and f = the factor by which CV_{max} diminishes per order of magnitude.

For example, if f = 2 and CV_o = 100, the equation becomes

$$CV_{max} = (x_o/x)^{0.3} \cdot 100\ (\%)$$

In 1980 the CWS determined that for the concentration levels that were being measured, the form of the equation that the available pesticide residue data could best fit was as follows:

Table 6 Interlaboratory Comparison of 1979 Herring Gull Reference Material

Analysis	CWS Laboratory			ORF			F.I.C.P.			USFWS		
	Mean (mg/kg)	CV	n	Mean (mg/kg)	CV	n	Mean (mg/kg)	CV	n	Mean (mg/kg)	CV	n
Hexachlorobenzene	0.192	10	21	0.153	13	15	—	—	—	—	—	—
pp′ DDE	4.830	16	21	3.903	10	13	4.640	13	5	5.380	7	5
pp′ DDT	0.048	41	10	—	—	—	0.065	9	5	—	—	—
Mirex	0.122	11	21	0.102	25	13	0.139	14	3	—	—	—
Oxychlordane	0.194	4	21	0.192	15	13	0.202	15	2	0.184	6	5
cis-Chlordane	0.029	10	15	0.022	13	17	—	—		—	—	—
pp′ DDD	0.058	15	15	0.035	53	13	0.069	5	4	—	—	—
Heptachlor epoxide	0.137	12	21	0.110	22	13	0.129	25	8	0.123	21	4
Dieldrin	0.278	9	21	0.257	14	13	0.280	36	5	0.258	21	4
Arochlor 1254:60 (1:1)	83.12	21	11	72.94	10	13	88.60	3	4	—	—	—
Arochlor 1260	33.99	13	21	—	—	—	—	—	—	34.40	7	5

Note: All results expressed in mg/kg wet weight; ORF — Ontario Research Foundation; F.I.C.P. — Federal Interdepartmental Committee on Pesticides; USFWS — United States Fish and Wildlife Service.

Table 7 CAPCO Wildlife Check Sample Program — 1990 — Results in ppm (μg/g) Wet Weight Basis

Lab #	% Fat	HCB	Oxy-chlordane	*c*-Nonachlor	p,p′DDE	Mirex	H.E.	Dieldrin	PCB 52	PCB 101	PCB 118	PCB 138	PCB 153	PCB 180	PCB 182	PCB 183	PCB 194	Total PCB
1	10.46	0.006	0.016	0.009	0.394	0.082	0.007	0.014	0.013	0.030	0.106	0.213	0.234	0.155	0.078	0.039	0.019	
2	9.00	0.001	0.011	—	0.425	0.061	0.004	0.009	—	—	—	—	—	—	—	—	—	2.191
3	10.95	0.004	0.024	0.030	0.448	0.153	0.007	0.016	—	—	—	—	—	—	—	—	—	
4	7.00	0.005	—	0.007	0.368	0.036	—	0.010	0.002	0.012	0.044	0.079	0.148	0.109	0.082	0.007	0.014	
5	10.50	0.006	0.006	0.005	0.490	0.051	0.015	0.016	0.014	0.062	0.092	0.215	0.320	0.160	0.085	0.051	—	
6	—	0.006	0.015	0.028	0.305	0.066	0.006	0.012	—	—	—	—	—	—	—	—	—	
7	—	0.004	0.020	0.010	0.325	0.110	0.005	0.004	0.017	0.050	0.097	0.176	0.203	0.126	0.071	0.035	0.014	
8	11.07	0.005	0.008	0.004	0.412	0.073	0.004	0.008	0.012	0.031	0.097	0.202	0.225	0.158	0.081	0.046	0.018	
9	8.91	0.009	0.013	0.017	0.093	0.065	0.006	0.015	0.014	0.033	0.080	0.074	0.085	0.070	0.028	0.078	0.018	
10	9.42	0.004	0.018	0.009	0.449	0.084	0.007	0.012	0.015	0.031	0.079	0.259	0.316	0.165	0.086	0.048	0.023	
11	9.45	0.004	0.010	0.005	0.528	0.062	0.011	0.019	0.011	0.030	0.075	0.168	0.155	0.124	0.050	0.035	0.014	
12	8.90	0.004	—	—	0.515	0.110	0.010	0.020	0.040	0.070	0.140	0.290	0.325	0.220	0.155	0.080	0.030	3.200
13	9.95	0.004	—	—	0.270	0.080	—	—	—	—	—	—	—	—	—	—	—	2.095
Mean:	9.60	0.005	0.014	0.012	0.386	0.079	0.007	0.013	0.015	0.039	0.090	0.186	0.223	0.143	0.079	0.046	0.019	2.495
SD:	1.17	0.00	0.01	0.01	0.12	0.03	0.00	0.00	0.01	0.02	0.03	0.07	0.09	0.04	0.03	0.02	0.01	0.61
CV:	12.2	38.1	40.5	77.6	30.6	38.4	44.8	37.1	67.0	47.2	29.1	39.1	38.2	29.5	43.1	48.6	30.1	24.5
CWS Reference values																		
Mean n = 6	9.79	0.004	0.010	0.006	0.498	0.082	0.004	0.011	0.012	0.032	0.103	0.241	0.271	0.209	0.097	0.047	0.025	3.294
SD	0.232	0.000	0.000	0.000	0.025	0.004	0.001	0.001	0.001	0.001	0.006	0.014	0.014	0.017	0.005	0.003	0.003	0.190
CV	2.4	4.8	3.0	5.5	4.9	5.3	1.7	6.3	4.1	4.3	5.5	5.8	5.1	7.9	5.6	5.6	12.5	5.8

Notes: Total PCB determinations.

Lab #2 calculation based on Aroclors 1254:1260 1:1.

Lab #13 calculation based on a 4:1 mix of Aroclor 1254/1260.

Lab #12 and the CWS Lab quantitation based one congener, PCB 138, using a 1:1 mixture of Aroclor 1254/1260.

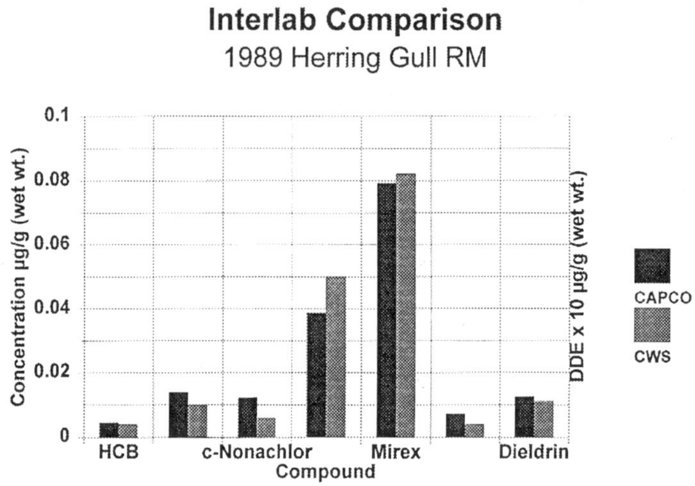

Figure 2 Interlab comparison — 1989 herring gull reference material — HCB, *c*-nonachlor, mirex, dieldrin.

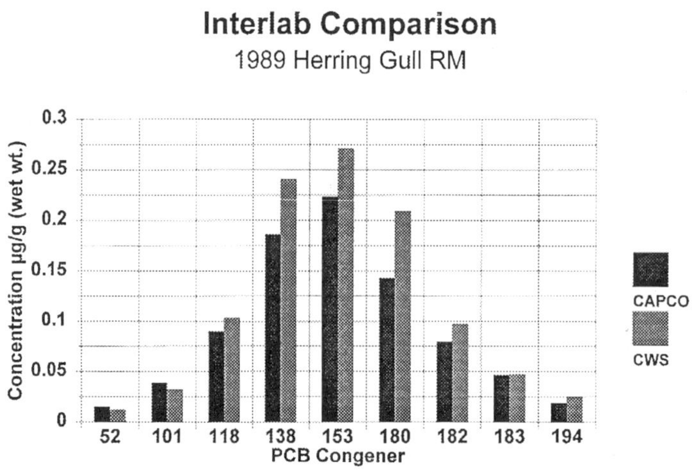

Figure 3 Interlab comparison — 1989 herring gull reference material — PCB congener.

$$CV_{max} = 13.7\ x^{-0.431}\ (\%)$$

based on x_0 = 0.01 mg/kg, CV_0 = 100%, and CV_{max} = 15% at 1 mg/kg. This is equivalent to f = 2.7. For example, for an analyte mean concentration of 0.0895 mg/kg, the expected CV_{max} would be 38.8%. These equations allow CV_{max} to increase as concentrations decline in a systematic manner.

This model was tested by CWS using data from the 1980 RM to determine whether the form of the equation was valid. Data were available for both organochlorines and individual PCB congeners (Table 4). The best fit of the data for orga-

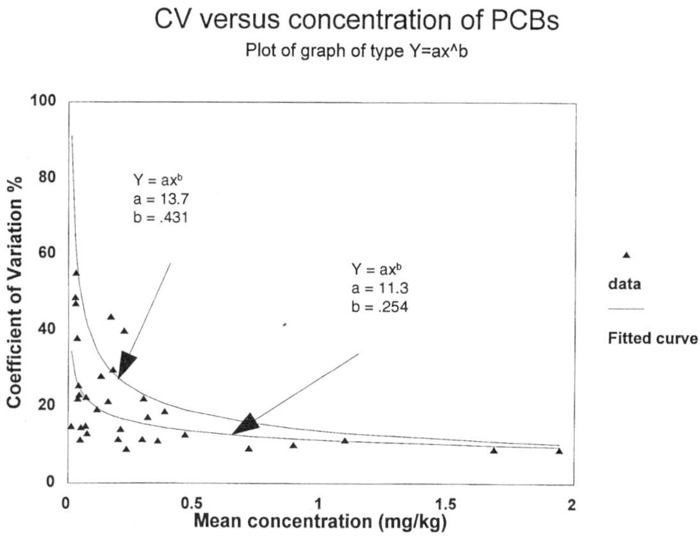

Figure 4 CV vs. concentration of PCBs. Plot of graph of type $Y = ax^b$.

nochlorines was found to be similar in form but with slightly different coefficients. For organochlorine residues the curve was

$$CV = 11.56\ x^{-0.303}$$

For the PCBs, the resulting curve was of the form

$$CV = 11.33\ x^{-0.254}$$

When the two data sets were combined, the equations coefficients were again slightly different:

$$CV = 10.286\ x^{-0.316}$$

Figure 4 is a graph showing the calculated curves and the points for the individual PCB means.

This statistical observation has been useful for the CWS because it enables QC data from contract laboratories to be evaluated in a more objective manner.

Stability of the RM

The 1980 RM was used for 10 years as a QC material. Data are available to ascertain whether the concentrations of trace organic chemicals remain constant in a biological matrix when they are stored at –40°C.

The components of the RM were carefully determined by CWS in 1980. From 1981 to 1985, the material was used to monitor the QC of a laboratory from which

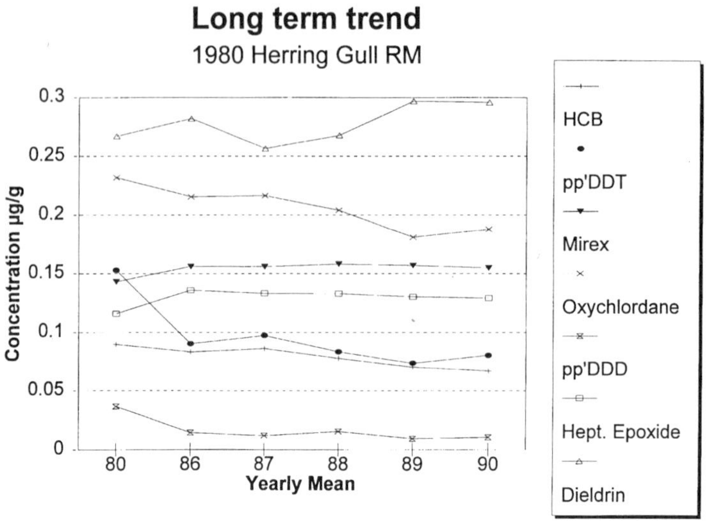

Figure 5 Long-term trend — 1980 herring gull reference material — HCB, pp′DDT, mirex, oxychlordane, pp′DDD, hept. epoxide, dieldrin.

CWS contracted analyses. From 1986 to 1990 the material was used to monitor internal QC of the CWS trace analyses as well as contracted work.

The graphs in Figures 5 and 6 summarize some of the data that have been collected. Only the data from the CWS laboratory have been plotted. Figure 5 shows the values for seven organochlorine (OC) compounds, which were routinely monitored throughout the period and which are present in similar concentration levels. Generally, the overall indication is that most OC compounds are stable in this matrix when held at –40°C over 10 years.

No attempt has been made to determine whether these differences are statistically significant. There was no overall trend toward desiccation noted; the moisture level remained essentially constant. The differences may perhaps be explained by minor adjustments made to the methodology used over the period. The goal of the storage of tissue at –40°C was not that it be unchanging but that changes would be so slow that the material could be used for routine quality control. These data indicate that the overall goal was served. It does, however, point to potential research efforts in the future.

Variations of PCB Data for the 1980 RM

Beginning in 1986, the CWS began determining and reporting PCBs as individual congeners and not as Aroclors. The 1980 RM had not originally been examined for the congeners, but because the pattern that was typically seen in the wildlife tissues was markedly different from Aroclor patterns, this made sense. Indeed the majority (>60%) of the total PCB present in the RM was due to only five congeners — PCBs 99, 118, 138, 153, and 180. All the PCB congeners were plotted on QC charts, as

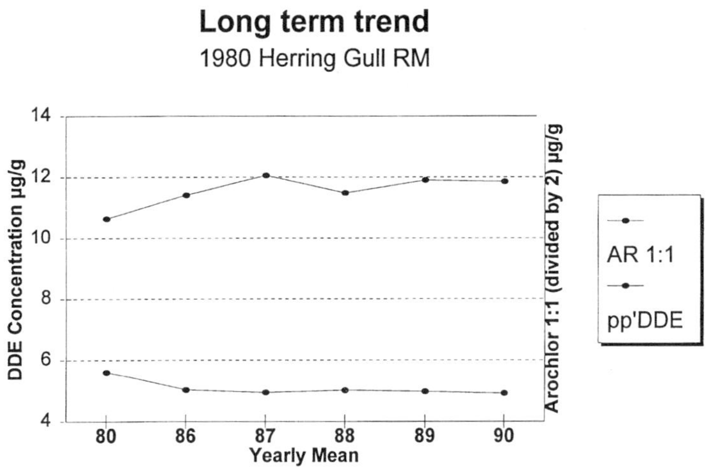

Figure 6 Long-term trend — 1980 herring gull reference material — AR 1:1, pp′DDE.

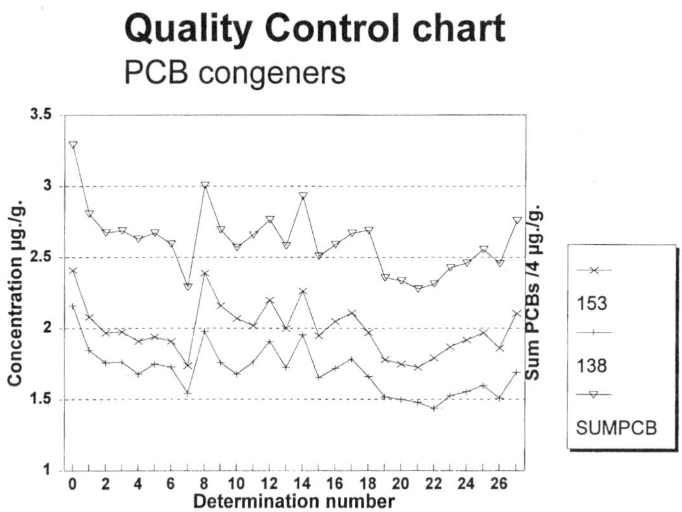

Figure 7 Quality control chart — PCB congeners, 153, 138, ΣPCB.

the organochlorines had been. In examining these charts during the 1986–1989 period, some patterns were noticed.

For example, it was noticed that the QC charts looked similar for congeners that were at high concentrations. Figure 7 is a plot for two major congeners, 153 and 138, and for the sum of all the congeners analyzed during 1987. The patterns are essentially the same. On the other hand, Figure 8 shows the QC charts of four other congeners, 97, 141, 105, and 110, for the same period. These congeners are all present at concentrations of ten times less than the major congeners. The variation is much more random.

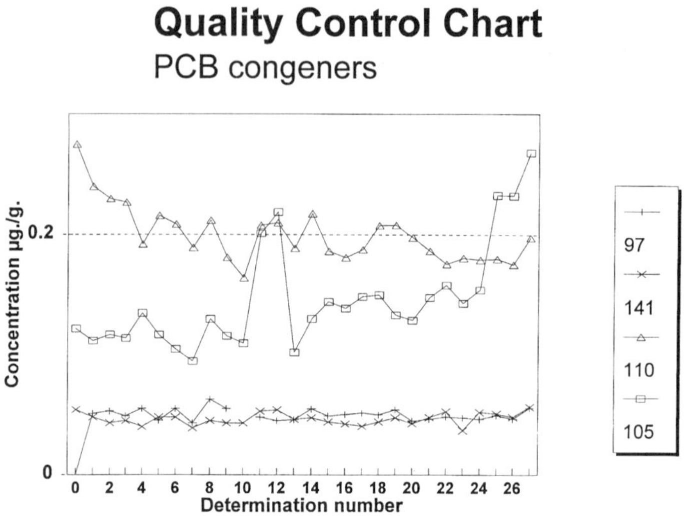

Figure 8 Quality control chart — PCB congeners, 97, 141, 110, 105.

Generally, this can be explained if we presume that two processes are operating. At higher concentrations, variations in the process that affect the sample as a whole predominate. These variations are the steps in the procedure, such as weighing, extraction, evaporation, and dilution. In the determination step, the variation in the calibration or in the volume actually injected into a gas chromatograph (GC) is most likely to affect the sample as a whole. At very low concentrations, those factors causing variation that affect each congener differently predominate. For example, at the determination stage, the concentration and type of interference is more likely to add to or reduce the real concentration. Further, the ability of the GC integrator to determine the exact start or end of a peak declines with concentration. Often, because of poor resolution, part of the area integrated for one congener will be included in an adjacent peak. Instrument noise is more likely to affect chromatographic processing of congeners near the instrument detection limit than congeners present at high concentration. These arguments provide reasonable explanations for the QC chart patterns that were noted.

Total PCB concentrations can be obtained by summing all the concentrations of the individual congeners. From the above arguments, it is no surprise that the CV for ΣPCB (8.2%) value is slightly lower than that obtained by the previous one-peak method based on PCB 138 (8.9%). This is despite the ΣPCB value being obtained by summing 40 results. Because PCBs 99, 118, 138, 153, and 180 constitute 61% of the ΣPCB value and the 16 PCBs with concentrations present at greater than 0.2 mg/kg constitute 88% of the ΣPCB value, we would anticipate those factors that affect the sample as a whole to predominate in determining the CV. Some of the variation in concentration in one congener caused by peak processing is effectively cancelled out by another, thereby reducing the CV lower than that obtained for the congeners in highest concentration. Thus the CV of the ΣPCB reflects the variation inherent in the extraction, separation, and measurement steps of the method as a whole. It is not likely that the CV for a manual method such as this could be reduced except by automation. The fact that the CVs are approximately the same is of no concern when calculating conversion factors to calculate ΣPCB values based on existing Aroclor 1:1 1254/1260 data.[4]

CONCLUSION

The CWS laboratory has been successful in developing and using secondary reference materials since 1980. The reasons for their development grew out of a practical concern for the quality of results being generated for research purposes and the fact that appropriate certified RMs were unavailable.

The majority of the RMs that have been developed have been herring gull egg homogenates. These materials have proven to be relatively easy to produce in reasonable quantities in a laboratory without specialized facilities dedicated to mixing and homogenizing large amounts of materials. They have been shown to be stable for the purposes of analysis of trace organochlorine compounds for a period of over 10 years.

The CWS has used its RMs for internal quality control, for monitoring the quality of data from contract laboratories, for interlaboratory exchanges, for evaluating laboratories submitting bids for contract work from CWS, and for research into methodology and storage conditions for wildlife tissues. In all these uses, the RMs have proven to be valuable. They have not been produced with a view to marketing the products to other lab organizations. A sufficient number of each RM will be maintained in the CWS specimen bank for future research work.

In the future, as needs arise, more RMs will be produced at the CWS lab. For the time being, there is sufficient material from the 1989 RM (diluted and undiluted) to be used to ensure quality control of projects demanding trace organochlorine chemical analysis for several more years The 1990 duck breast muscle RM will continue to be used as well because there is an ever-increasing demand for the analysis of these types of tissue for trace organic chemicals, even though the OC/PCBs are at levels toward the low range of contaminants found in Canadian wildlife.

It is possible for a laboratory to produce its own secondary RMs. However, this should only be done if there is no suitable certified material available and with the knowledge that there may be a considerable lapse of time between preparing the material and having confidence that it is homogeneous and useable.

ACKNOWLEDGMENTS

Most of the analytical results presented in this paper were the work of Mr. Henry Won. We also wish to thank Ross Norstrom for his ideas and Brian Collins for his help with statistics. Finally, we wish to acknowledge the work of analysts of many laboratories who have contributed their results on CWS reference materials through their participation in interlaboratory comparisons.

REFERENCES

1. **P. Mineau, G.A. Fox, R. J. Norstrom, D. V. Wesloh, D.J. Hallett, and J. A. Ellenton,** Using the herring gull to monitor levels of organochlorine contamination in the Canadian Great Lakes, In: *Toxic Contaminants in the Great Lakes*, J. O. Nriagu and M. S. Simmons, Eds., John Wiley & Sons, New York (1984) 426–452.
2. **C. A. Bishop, D. V. Wesloh, N. M. Burgess, J. Struger, R. J. Norstrom, R. Turle, and K. A. Logan,** An Atlas of Contaminants in Eggs of Fish-Eating Colonial Birds of the Great Lakes (1970–1988), Technical Report Series No. 152, The Canadian Wildlife Service, Ontario region (1992).
3. **R. Turle, R. J. Norstrom, and H. T. Won,** Quality assurance as applied to long-term monitoring of chemical residues and specimen banking, In: Progress in Environmental Specimen Banking, Stephen A. Wise, Rolf Zeisler, and George M. Goldstein, Eds., National Bureau of Standards, U.S., Special Publication 740 (Apr. 1988).
4. **R. Turle, R. J. Norstrom, and B. Collins,** Comparison of PCB quantitation methods: reanalysis of archived specimens of herring gull eggs from the Great Lakes, *Chemosphere*, 22 (1–2) (1991) 201–213.
5. **John K. Taylor,** Handbook for SRM Users, National Bureau of Standards, U.S., Special Publication 260 (Sept. 1985).
6. **H. T. Won and R. J. Norstrom,** Analytical Reference Materials: Organochlorine Residues in CWS-79-1, a Herring Gull Egg Pool from Lake Erie, Canadian Wildlife Service Manuscript Report No. 41 (1980).
7. **R. J. Norstrom and M. Simon,** Quality Control Program — Polychlorinated Dibenzo-p-dioxins (PCDDs) and Polychlorinated Dibenzofurans (PCDFs) in Wildlife Tissues, CWS Analytical Report CRD-88-10.
8. **R. Turle and R.J. Norstrom,** The CWS Guideline to Practical Quality Assurance for Contracted Chemical Analysis, Technical Report Series No.21, Canadian Wildlife Service Headquarters (1987).
9. **R. J. Norstrom and H. T. Won,** Long term preservation of egg tissue homogenates for determination of organochlorine compounds: freezing versus freeze-drying, *J. Assoc. Off. Anal. Chem.*, 68 (1) (1985) 129–135.
10. **H. T. Won,** Analysis of Herring Gull Egg Pool as QA Samples for OCs and PCBs, CWS Analytical Report, CWS-89-14 (July 1989).
11. **H. T. Won,** Analysis of Diluted Herring Gull Egg Pool and Chicken Egg Pool for OCs and PCBs, CWS Analytical Report 90-03 (May 1990).
12. **M. M. Schantz, R. M. Parris, S. A. Wise, H. T. Won, and R. Turle,** Polychlorinated biphenyl (PCB) congeners and chlorinated pesticides in a cod liver oil reference material, *Chemosphere*, 24 (12) (1992) 1687–1698.
13. **H. T. Won,** Report of the CAPCO Wildlife Check Sample Program — 1990 (May 1990).
14. **H. Frehse and G. Timme,** *Residue Rev.*, 73 (1980) 27–47.
15. **J. E. Elliott, R. J. Norstrom, and J. A. Keith,** Organochlorines and eggshell thinning in northern gannets (*Sula bassanus*) from Eastern Canada, 1968-1984, *Environ. Pollution*, 52 (1988) 81–102.
16. **J. E. Elliott, D. G. Noble, R. J. Norstrom, and P. E. Whitehead,** Organochlorine contaminants in seabird eggs from the Pacific Coast of Canada, 1971–1986, *Environ. Monitoring Assessment*, 12 (1989) 67–82.
17. **P. A. Pearce, J. E. Elliott, D. B. Peakall, and R. J. Norstrom,** Organochlorine contaminants in eggs of seabirds in the northwest Atlantic, 1968–1984, *Environ. Pollution*, 56 (1989) 217–235.

18. **J. E. Elliott,** Specimen banking in support of monitoring toxic contaminants in Canadian wildlife, In: International Review of Environmental Specimen Banking, NBS Special Publication 706, S. A. Wise and R. Zeiser, Eds., U.S. Dept. of Commerce, Washington, D.C. (1985) 4–12.
19. **D. B. Peakall, R. J. Norstrom, A. D. Rahimtula, and R. D. Butler,** Characterization of mixed function oxydase systems in the nestling herring gull and its implications for bioeffects monitoring, *Environ. Toxicol. Chem.*, 5, (1986) 379–385.

CHAPTER 16

The Use of Reference Materials and Standards in the Analysis of Environmental Samples for Organic Contaminants

Ian H. Rogers and M. Coreen Hamilton

A Reference Material (RM) has been defined as a material or substance, one or more properties of which are sufficiently well established to be used for the calibration of an apparatus, for the assessment of a measurement method, or for assigning values to materials.[1] This chapter discusses the various categories of RMs that are potentially available and what they can be used for in an environmental analytical laboratory. The coverage of the topic is not exhaustive but is based on our own needs and experiences in organic contaminant analysis at the trace and ultratrace levels, especially in regard to biological tissues. The figures illustrate aspects of quality assurance/quality control (QA/QC) that provide practical and meaningful evidence to convince those, who need to be convinced, that the laboratory has put its analytical methods under a state of statistical control. It is just as vital to be able to show this in a government reference laboratory as it is in a contract laboratory within the private sector. If attention is not paid to such fundamental aspects of organic ultratrace analysis, the scientific community will rapidly lose confidence in the operations of the laboratory.

The discussion in this chapter is centered on the ultratrace analysis of tissue samples for polychlorinated dibenzo-*p*-dioxins and dibenzofurans. There are good reasons for this. First, we have had more experience in this type of analysis than in analysis of any other toxic residues. In the second place, really stringent QA/QC standards have been established for this type of work, so this protocol illustrates how every step of the procedure needs to be verified, with nothing left to chance. The requirement for such precautions arises from the extreme toxicity of 2,3,7,8-tetrachlorodibenzo-*p*-dioxin (TCDD) to mammalian systems and its description as "the most toxic chemical known to man". For this reason ultratrace levels in the environment have real significance, and accurate work can be accomplished only through adherence to strict QA/QC requirements. Tolerance criteria have been established for fish marketed for human consumption, and the Canadian standard has been set at 20 pg/g fresh weight of 2,3,7,8-TCDD equivalents. Obviously, the results must be accurate, as poor data could be used by regulators to close apparently contaminated fisheries, and various legal consequences would arise. An even more serious situation could occur if the health of consumers were jeopardized through contaminated fish being permitted for sale because of poor quality analytical data. The analysis of dioxins and furans in tissue samples is the most demanding of all the environmental residue analyses that currently exist.

It is not logical to discuss RMs without also considering the related topic of chemical standards. The certification of a reference material by a body, such as the National Institute of Standards and Technology (NIST), implies traceability to some recognized standard.The availability of suitable standards of an acceptable level of purity at accurately determined concentrations is fundamental both to the ability to produce RMs and to use them. Standards are needed as individual compounds and as cocktails to meet various demands of the analytical protocol. Examples of both these requirements follow. The use of individual standards provides flexibility, as these can be mixed in any proportion desired. The availability of a certified solution of 2,3,7,8-TCDD is extremely valuable in setting up the analytical conditions to measure unknown samples. For other purposes it is convenient to use cocktails, as in the calibration of the mass spectrometer. The commercial availability of cocktails saves the analytical laboratory from having to establish a "high hazard" area, where solid samples of individual congeners could be weighed out. Many laboratories prefer not to have to handle solid samples of dioxins or furans.

Since this chapter examines issues relating to RMs from the viewpoint of the user, it is neccessary to distinguish two types of user. The first is a manager of a single operational unit, who needs to be assured that the analyses for which he has responsibility are accurate and verifiable — in short, that they were produced under statistically controlled conditions. The other user has a much wider perspective in which the operations of several laboratories need to be compared to ensure that the results they severally produced are also verifiable and consistent. The ramifications of these two cases will be discussed in more detail.

The two types of user outlined above are not the only ones that come to mind. For example, there are statisticians or government regulators who have to use the analytical data for their own specific purposes. They are one stage removed from having to handle RMs, or indeed to conduct analyses. They indirectly need RMs and standards because they rely on other scientists to ensure that the data they use are fully reliable. Another case concerns a contractor and a contractee. The contractor will specify what level of reliability the data must have, and it is the concern of the contractee to produce it. In this chapter we regard the primary user as the party with the most compelling need. However, it is also legitimate to keep in mind the needs of the secondary users, because these provide the real driving force behind the whole topic.

THE ROLE OF CHEMICAL STANDARDS IN THE ULTRATRACE ANALYSIS OF DIOXINS AND FURANS

The need for pure standards of whatever one is trying to measure is basic and hardly needs to be emphasized. The requirement to analyze complex mixtures of dioxins and furans has tested the ability of chemists to synthesize and purify all the possible native congeners. Even more of a challenge has been the need to prepare the analogous ^{13}C-surrogates, a process that has been completed for the 2,3,7,8-substituted members.

The most sensitive measurement method for ultratrace analysis of organic residues is the isotope dilution technique. The mass spectrometers are calibrated against cocktails that contain representative native target dioxins and furans at the various levels of chlorination from four to eight. Usually a five-point calibration is done in which five concentrations of the cocktail are analyzed, and the resulting data allow the plotting of standard response curves, bracketing the range of concentrations over which the analyst expects to be operating. Cocktails suitable for calibration of instruments for dioxin and furan analyses are commercially available. Most laboratories have now switched to the use of gas chromatographs integrated to high-resolution mass spectrometers (GC-HRMS) with an advantage of at least an order of magnitude in sensitivity compared to lower resolution quadrupole instruments (GC-LRMS).

Mass spectrometric methods use window mixtures to define the elution range of a family of congeners of a given chlorine content. This is illustrated in Figure 1, which shows the use of window mixtures to define the elution range of tetra- and pentachlorinated dioxins and furans. These mixes include the first and last congeners to elute in each case so that the mass spectroscopist can concentrate on his chosen targets. Peaks outside the window are neither dioxins nor furans. The use of window mixtures in the dioxin analysis is critical, as there is only, at best, an interval of 0.1 min between the elution of the last of the TCDDs and TCDFs and the first of the P_5CDFs.

After the windows have been established, and before the analyst can proceed to measure dioxins and furans in real samples, protocols[2,3] specify the need to demonstrate that the chromatographic separation process is working properly. This is done by the analysis on a daily basis of a mixture of 2,3,7,8-TCDD or 2,3,7,8-TCDF with their nearest eluting neighbours at equal concentrations. Suitable cocktails containing these mixtures at the appropriate concentrations are commercially available.

Proof that the mass spectrometer is ready to measure dioxins accurately is provided through the satisfactory analysis of an aliquot of SRM 1614, available from NIST. This is a solution containing 2,3,7,8-TCDD at a certified concentration and is the only such certified standard currently available for dioxin and furan measurements.

Mention has already been made of the need for ^{13}C-analogs of target species. A cocktail containing representatives of the various 2,3,7,8-substituted dioxins and furans, at each level of chlorination from four to eight, is spiked into every environmental sample prior to solvent extraction. These surrogates are used as internal standards and progress through all the stages of clean-up together with the native

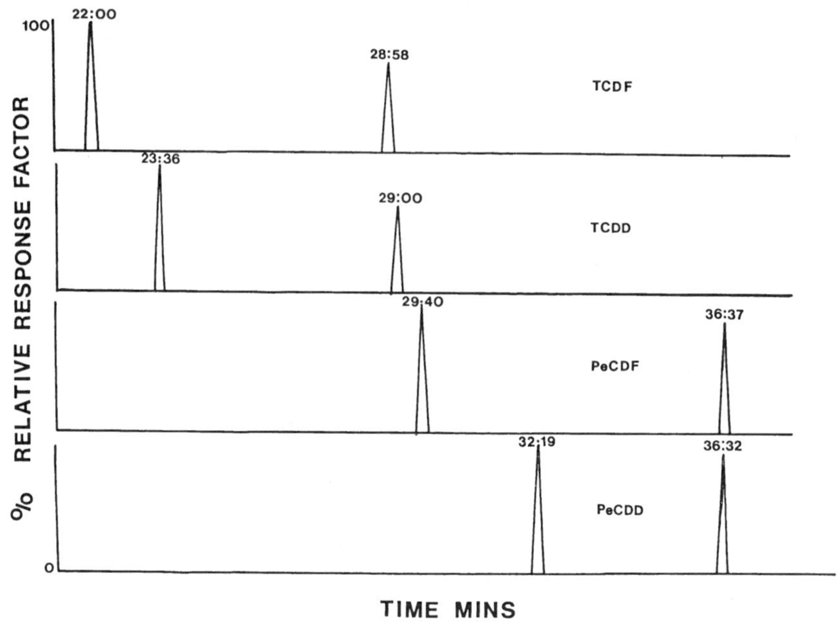

Figure 1 The use of a window mixture to define the range of elution times for the various congeners of the tetrachlorodibenzofurans (TCDFs), the tetrachlorodibenzo-*p*-dioxins (TCDDs), the pentachlorodibenzofurans (PeCDFs) and the pentachlorodibenzo-*p*-dioxins (PeCDDs). Note the narrow time interval between the elution of the last TCDD and the first PeCDF.

species. It is assumed that they behave similarly in all respects. Immediately prior to the final measurement by GC-HRMS, two other surrogate compounds are added as recovery standards. The use of these standards allows measurement of the recovery of the internal standards (^{13}C-labeled surrogates). The recovery standards routinely used in the measurement of dioxins and furans are ^{13}C-1,2,3,4-tetrachloro- and ^{13}C-1,2,3,7,8,9-hexachlorodibenzo-*p*-dioxin. The protocol for the Environment Canada "River Road Method" dictates that the recoveries of the internal standards (^{13}C-labeled surrogates) must be within the range of 40 to 130% for TCDD and TCDF and 30 to 130% for the higher chlorinated dioxins and furans.

Standards are needed for the preparation of spiked samples useful as a component of the QA/QC program. These may be spiked blanks or spiked matrix samples, and the matrix may be clean or may already contain a known level of the compounds of interest. Such tests establish the precision of a method or keep it under a state of control.

A problem has arisen with the move to the use of GC-HRMS. When properly tuned and maintained, these instruments can measure low femtogram amounts of 2,3,7,8-TCDD. It is exceedingly difficult to produce ^{13}C-surrogates of a level of purity such that any native impurity does not show under these conditions. This was not a problem when GC-LRMS was used in the dioxin analysis. With the current trend to the development of mass spectrometers of even higher sensitivity, it is predictable that this problem will become more acute unless methods are changed to call for lower spiking levels.

THE AVAILABILITY AND USE OF STANDARDS TO MEASURE OTHER ENVIRONMENTAL RESIDUES

The discussion has so far centered on the measurement of dioxin and furan residues via GC-HRMS. The needs for native standards for spiking, method development, and instrument calibration are common to other frequently used approaches in organic residue determinations such as gas or liquid chromatography and also to GC-LRMS. In the latter case, surrogate standards are also required. Gas chromatography offers a choice between the use of internal and external standards. In general it is accepted that the internal standard route gives results with better precision and is the method of choice.

The native congeners of all the PCBs have been prepared and are marketed. The current situation is less comprehensive for individual ^{13}C-PCB surrogates. Three nonortho-substituted PCBs are available and two mono-ortho-substituted analogs have been introduced, as well as several di-ortho-substituted congeners. Further ^{13}C-labeled PCB components are expected to become available in a few months.

In 1985 the National Research Council of Canada (NRCC) marketed a certified set of four cocktails (CLB-1), each of which contained 14 or 15 native PCB congeners, 51 in all. These were selected to ensure separability by electron capture gas chromatography (GC-EC). The identity of other congeners, not present in the cocktails, could be inferred from their RRTs on capillary GC columns. However, it was stated by NRCC that not all of the 207 PCB congeners could be unambiguously resolved by GC on any one specific column.

Earlier work on the quantitation of PCBs was conducted by GC-EC using packed and later capillary columns. The standards in use were a series of Aroclor mixtures prepared commercially by the chlorination of biphenyl to various chlorine contents. The results of this analytical work were not precise and have been described in the literature as "semiquantitative". Problems arose in trying to compare environmental residues, which had been subjected to weathering or biodegradation, to the original composition as manufactured. Residues were quantitated by comparing the total integral to that of the appropriate Aroclor or Aroclor mixture, whose pattern it most closely resembled.

Today there is widespread interest in the analysis of nonortho- and mono-ortho-substituted PCBs, because of their effects on biota. Although slightly less toxic than 2,3,7,8-TCDD, they are more plentiful than the dioxins and furans in the receiving environment and often contribute more to the total toxic burden than do these residues. Nonortho- and mono-ortho-substituted PCBs are quantitated by either GC-LRMS or GC-HRMS. PCB fractions, containing mono- or di-ortho chlorine substituents, are recoverable from extracts of contaminated biota by carbon-fiber column chromatography. It is difficult to separate IUPAC #169 from the dioxins, so it is usually measured with them. The other coplanar nonortho-substituted, including IUPAC # 77, 126, and 127, are measured in a separate fraction.

No window mixtures are currently available for the analysis of coplanar or mono-ortho PCBs. It is neccessary to use an Aroclor pattern to define a window, and this has to be checked frequently when calibrating the instrument.

The use of GC-EC with capillary columns is still popular for the analysis of total PCBs. It is neccessary to standardize the GC resolution against some appropriate mixture of congeners that is difficult to separate. Individual laboratories have their own selected standard mixtures for this purpose. The choice of an internal standard for the analysis is also nonspecific, as ^{13}C-labeled PCB congeners cannot be distinguished from the native species by GC-EC.

The analysis of toxaphene residues in environmental samples is a difficult problem, which was recently addressed in an international workshop. The best method is electron capture negative ion (commonly called chemical ionization) mass spectrometry. Although it is not yet possible to purchase suitable native congeners, some 20 congeners have been synthesized at the University of Kassel in Germany and are currently available for purchase. No labeled standards are available for the analysis of toxaphene.

Most analyses of pesticide residues are conducted by GC-EC or by GC-LRMS. All the authentic native compounds are available. Window mixes are not required, as the instruments are programmed to search for the residues in relation to the chosen internal standard. Each laboratory must demonstrate chromatographic resolution using some suitable test mixture. Only a very few labeled pesticide standards are available, which hampers the use of more accurate isotope dilution internal standard methods for the quantitation of trace levels. The use of labeled surrogates as internal standards opens the door for greater precision in analytical results, as is demonstrated in Figure 2 by the control charts of relative response factors (RRFs) for hexachlorobenzene (HCB), a chlorinated pesticide, as measured by GC-EC and GC-LRMS. There is greater variability in the HCB response monitored relative to an internal standard, tetrachloro-m-xylene, which elutes close to the HCB (Figure 2b) than is observed for the response of HCB relative to ^{13}C- labeled HCB (Figure 2a). When a labeled surrogate is not available, the choice of an internal standard can be critical to the precision of the results. For example, the HCB response relative to another internal standard, PCB #209, shows even greater variability (Figure 2c), presumably due to the longer elapsed time between the elution of HCB and PCB #209 from the GC.

A plentiful supply of ^{13}C-labeled and deuterated surrogate standards is available for the analysis of chlorophenols, chloroguaiacols, and chlorocatechols. These are ideal for the quantitation of such residues by GC-LRMS. The native standards are all available. Again a recovery standard mixture is needed, and some laboratories use the mixture of the three tetrabromophenols as surrogates for this purpose. The analysis of chlorophenolic residues after derivatization is a special problem and is discussed later.

The only other common target for residue determinations is the polynuclear aromatic hydrocarbon class (PAHs). All the native standards and many deuterated surrogates are available. Analyses are usually conducted by GC-LRMS, but high performance liquid chromatography (HPLC) is also used. An example is given in Figure 3, in which deuterated analogs are used to measure five-ring PAHs by GC-LRMS. The

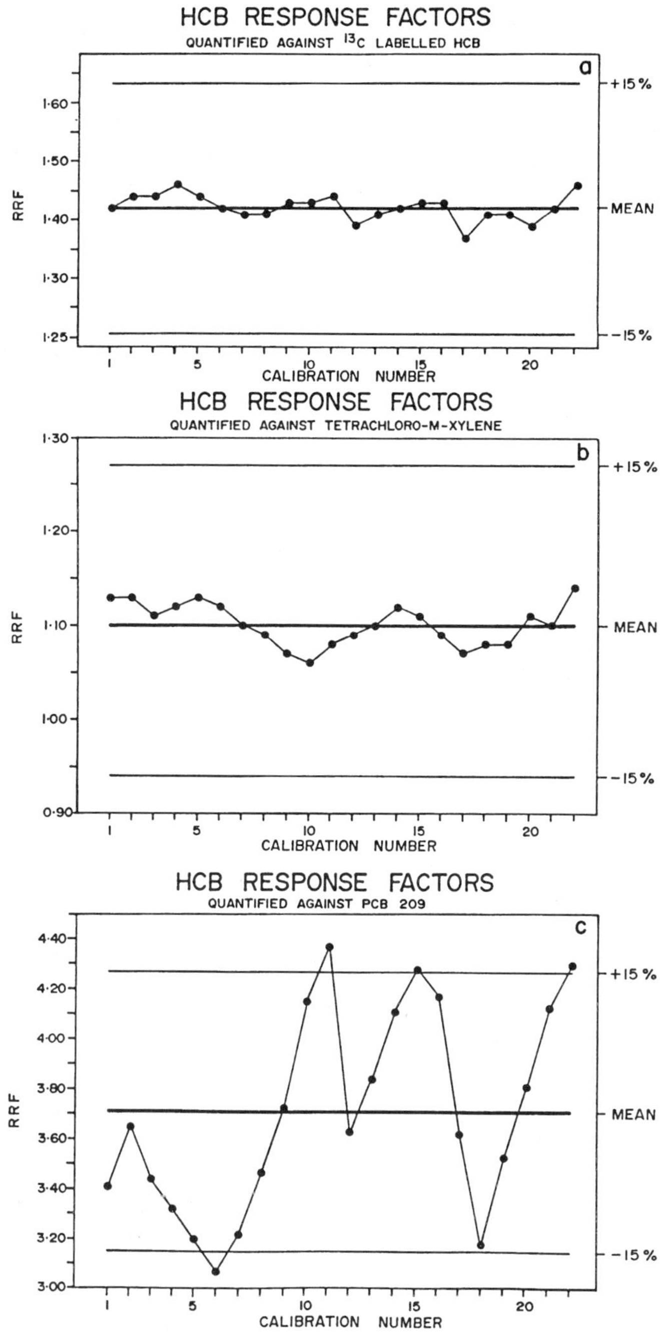

Figure 2 Analysis of hexachlorobenzene (HCB) residues using three different internal standards. (a) ^{13}C-Labeled HCB, (b) Tetrachloro-m-xylene, (c) PCB# 209. In (b) and (c) the method was GC-EC, whereas GC-LRMS was used in (a). Note the tighter analytical precision achieved with the labeled standard.

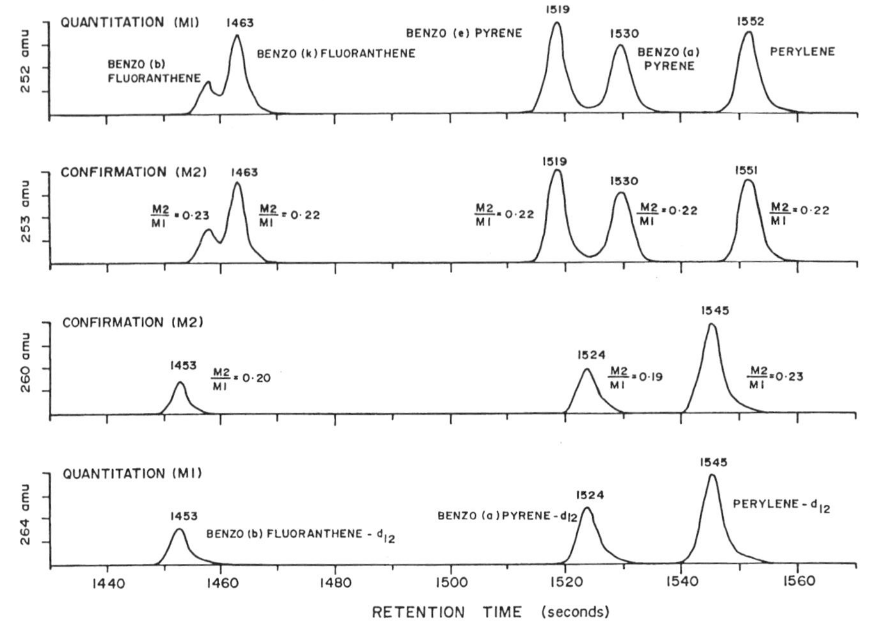

Figure 3 Analysis of polynuclear aromatic hydrocarbons (PAHs) by GC-LRMS using a mixture of deuterated PAHs as internal surrogate standards.

chief problem encountered is the presence of alkyl derivatives of the PAHs in many environmental extracts. Only some native and labeled compounds are available for these; therefore, they are difficult to quantify. No window mixes are required, but each laboratory needs to set up a suitable resolution mixture. Three certified standard solutions are available from NIST.

THE VALIDATION OF MULTICOMPONENT ANALYTICAL STANDARDS

Often the manager of an analytical unit finds there are no cocktail standards listed that are suitable for a specific method. He has no choice but to prepare his own from materials on hand. The first step is to plan what mixture of compounds should be present and at what concentrations. The next is to check the purity of all the individual components that will be part of the cocktail, since impurities could give rise to systematic bias.

The components should be weighed out and the solution prepared, with dilutions being made as neccessary. An internal standard should be added to a diluted sample of the solution and an analysis performed to check that all the components of interest have been added and nothing omitted. The analysis would verify the absence of impurities. The relative concentrations of the components would then be checked to see if everything seemed to be correct. The results would be examined to look for any mathematical errors, and the peak areas and relative response factors would be carefully checked.

To verify the composition of the standard, careful quantification of each component would be conducted to check the accuracy of the concentrations. This could be done in one or more of a number of ways, which could include comparison against reference solutions of the individual components, comparison to an old stock which had been previously validated, interlaboratory exchange of solutions, and the analysis of reference materials or multicomponent reference solutions.

TYPES OF REFERENCE MATERIALS

Not all RMs have the same status. Some are prepared in-house for internal use only. These may be referred to as internal reference materials (IRMs). Others are prepared by an outside reference laboratory or agency and are made available through marketing arrangements. Such products might have the status of certified reference materials (CRMs), in which one or more property values have been measured by a technically valid procedure and are accompanied by or traceable to a certificate or other documentation, issued by a certifying body.[4] Other RMs may have been prepared by NIST. These are referred to as standard reference materials (SRMs) and are considered the ultimate standards for accuracy, based on cross-checking by two or three independent analytical methods in the certification process. The preparation of SRMs has been described in an earlier chapter. They are costly materials, available in limited supply, and should not be used for all purposes since this would too rapidly deplete the stocks. For many purposes other types of RM are suitable, whereas SRMs are most properly reserved for the determination of accuracy and systematic bias in an analytical process.

In some interlaboratory comparison exercises, RMs may be prepared and distributed by a central or reference laboratory. Clearly such material must be homogeneous and ideally the RM to be distributed to the participants should have been analyzed by two or three reference laboratories to arrive at a consensus value with a measured experimental error.

CRITERIA FOR THE SELECTION OF REFERENCE MATERIALS

A number of considerations must be brought to bear in the selection of an RM appropriate for the measurement of some chosen suite of target analytes. In the first place, the RM must contain the compounds that are to be measured, and ideally about three concentrations should be available. These should bracket the concentration range in which the unknowns are expected to lie. It is important that the matrix of the RM matches that of the unknowns so that the recoveries of the target compounds may be expected to match. Analysis of PCBs recovered from a sediment-based RM are not very helpful when the task is to measure PCBs in fish tissue. However, unfortunately, analysts have too often in the past been forced to make such extrapolations, owing to the unavailability of more suitable RMs. This limitation may become less acute, as today there is considerable interest in the scientific community in expanding the list of available RMs to relieve the limitations of matrix type and concentration.

Another factor to be considered is the possible presence of interferences from naturally occurring biochemicals or from other manmade contaminants. Interferences can give rise to systematic bias arising from the presence of false positives.

Standard practice in the measurement of tissue contaminants is to run spiked samples and blanks with each batch of unknown samples. Whenever a suitable naturally contaminated RM is available, it should be analyzed in preference to a spiked sample because it can be expected to more closely resemble the contamination that occurs in real samples. Presumably a molecule of contaminant inside a cell or embedded in a cell membrane would be more difficult to extract into a solvent than would the same material if it were merely physically mixed with disrupted biological tissues.

An RM being considered for use in a particular analysis should match in complexity, and more or less in concentration, the pattern of target compounds present in the unknown samples. Failure to obtain such a match may result in the RM falling outside the linear range for the target analytes in the test samples. Too complex a range of contaminants may, on the other hand, result in false positives.

A final consideration is that the level of the target compounds or other contaminants in the RM must be compatible with the work being done in the analytical laboratory. If the concentrations are too high, there is a risk of contaminating the work area so that ultratrace analyses for those compounds could not be continued for a period of time. This problem would be manifested through an inability to obtain satisfactory blanks.

USES OF RMs FOR INTERNAL CONTROL

It is standard practice for an analytical laboratory in the field of organic residue analysis to develop its own IRMs. Sometimes these are simply residual samples of what has already been analyzed and placed in archival storage. There is nothing wrong with this practice, except that there should be a large supply of material and often there is not. Without the availability of a plentiful supply of an IRM, the manager of the operational unit is hard put to justify the consistency of the data output over long time periods. There will always be a time when a switch has to be made from one IRM to another, but every effort should be made to reduce these potential discontinuities to a minimum.

The most important characteristics of an IRM are that the material should consist of the same matrix as the samples to be analyzed and that it must be homogeneous. A sample of the IRM should be included with each batch of unknown samples to allow the precision of the analytical method to be kept under review. From time to time a sample of a CRM or an SRM should be included as part of a batch of samples to permit the knowledge of the true composition of the IRM to be upgraded. Such upgrading allows information to be generated on the accuracy of the results on an ongoing basis. By adopting this approach to his need for QA/QC, the manager of the laboratory is able to keep his use of scarce CRMs and SRMs to a minimum. Comparison of results measured on the IRM with results from the simultaneous analysis of CRM or SRM samples confers traceability of the data to whatever standard was used to prepare the certified samples.

Training and proficiency of technical staff are matters of vital concern to laboratory management. Besides sending staff on refresher courses or upgrading their knowledge, a manager can assess their proficiency in the workplace by logging in check samples for analysis. These may be "blind" samples, in which the analyst is aware that his work is being assessed, but he does not know the true value. It may be argued by some that this is not a true test of normal performance because an individual will take special care with his work when he is aware that he is being tested. To counter this possibility, the manager can log in "double blind" samples, in which the analyst is unaware that he is being tested. There are practical problems in disguising the nature of these samples so that the analyst is not alerted, and the best way to log in the materials is with the connivance of a third party. The samples themselves may be IRMs or spiked IRMs. They could also be CRMs or SRMs; however, cost factors would probably dictate otherwise.

A person with responsibility for a QA/QC program needs to ascertain that his methods are always under a state of statistical control. This is best done with the aid of control charts. To construct a control chart for a method, the results of the ongoing analysis of IRMs should be plotted against the mean value of the property being measured. The t-test dictates that further results should fall within 2 standard deviations above or below the mean value 19 times out of 20, assuming a normal distribution. This is readily checked by reviewing the results of the past 20 determinations. If more than one measurement in the last 20 fell outside the limits, corrective action may need to be taken. The reason might be that a reagent is changing in concentration with time, or a sample may be deteriorating during storage. Further analyses of unknowns should be delayed until the cause of the problem has been understood and corrected. An illustration of a control chart is taken from the data on the composition of a sample of salmon tissue spiked with a known concentration of dioxins. The results from ten recent analyses for 2,3,7,8-TCDD are plotted in Figure 4. In this case, the method is under control, with the data points distributed evenly about the mean value. Control charts are most useful for keeping records of other parameters in the analytical laboratory, such as, for example, instrument performance and calibration with respect to time. It is also beneficial to plot the values for target compounds in blank samples.

The most important information that can be derived from the analysis of CRMs or SRMs is the assessment of the accuracy of an analytical method. The data obtained from these measurements allow the analyst to measure systematic bias or the variance between the measured value and the true value. There is no other way of obtaining this information, so analyses of these types of RMs constitute an essential step in the validation of an analytical protocol. Another use for CRMs or SRMs is the ability to upgrade the value of IRMs through simultaneous analysis, thereby relating the data to an accepted standard. The analysis of an SRM, in particular, allows the analyst to compare his work directly with the work of the experts in the reference laboratory, in this case NIST. This is a more direct comparison with the work of others than can be obtained from participation in an interlaboratory comparison exercise, in which the comparison can only be made retrospectively.

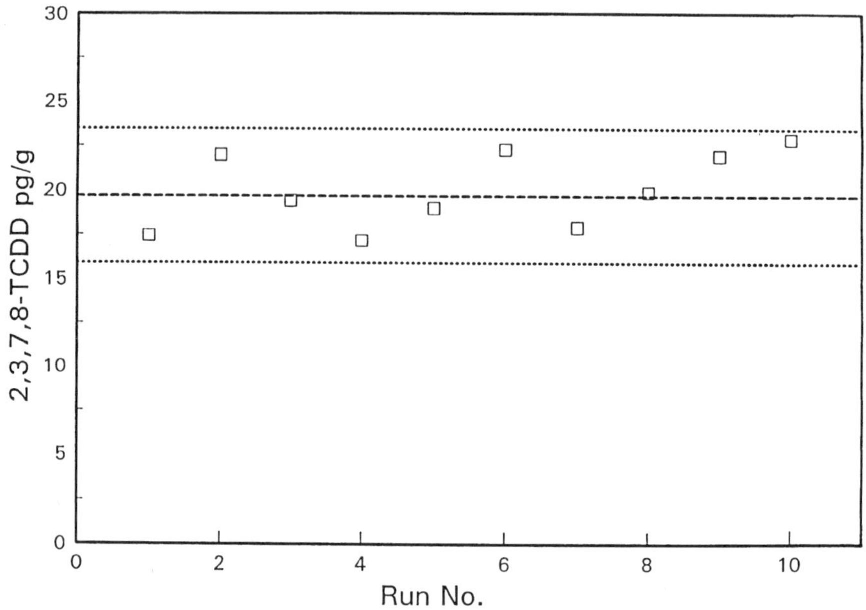

Figure 4 The use of a control chart to assure that an analytical method is under statistical control. In this case a reference material, comprising salmon tissue, was analyzed ten times for 2,3,7,8-TCDD. The mean value is shown as a dashed line. The control boundaries are ±2 SD and are shown as dotted lines.

EXTERNAL USES OF RMs

As discussed under the previous subheading, CRMs or SRMs can be used to assess the accuracy and bias of an analytical method in the validation process. To do so, the RMs should be of three concentrations, covering the range of the target species. This use may be most helpful to organizations like the Association of Official Analytical Chemists, which regularly and continually introduces new or updates existing analytical methods. Government organizations like the Environmental Protection Agency, which introduce many analytical methods, would also use RMs in this way. International scientific organizations, such as the International Council for the Exploration of the Seas (ICES), have conducted sample exchange studies aimed at the improvement of the ability of laboratories around the world to measure PCBs in marine fish or mammal samples. In such studies, consensus value samples are distributed to the participants as well as internal and reference standards.

Comparisons of the proficiencies of environmental testing laboratories need to be made from time to time. The decision maker in such applications might be a government employee with analytical work to be contracted to the private sector. He might, on the other hand, be a representative of a group of companies or even of an industry endeavoring to comply with a government edict on the measurement of environmental contamination resulting from its waste discharges. Such a situation arose in Canada when dioxin contaminated the receiving environment around bleached chemical pulp mills. The need to evaluate can be met through the running of a few test samples, including a CRM and a spiked sample. In such evaluations it is also important for the reference laboratory to send out common standards to help remove any confusion in the data. Accreditation of laboratories doing environmental analyses in the United States is a legal requirement. Because the results must be acceptable to the Environmental Protection Agency, this arm of government accredits the laboratories, which are then retained by industry to conduct the analytical work required of them under the terms of their waste discharge permits.

Another use for RMs stems from the need to keep the results from a group of laboratories under long-term control. This is needed if there is to be an attempt to measure changes in environmental levels of contamination over time. Such changes may be small and are based on mean values of a population of samples. It is therefore critical to be sure that the method is behaving consistently over long timeframes. The group of laboratories in question might all belong to the same company, which would need to know that they all performed the work within an acceptable margin of error. Sample exchange programs would be set up to provide the needed data, and at least some of the samples should be CRMs or high quality IRMs.

Without the availability of RMs, these external needs for high quality analytical data could not be met. Although the analysis of QA/QC samples costs money and time, there is no alternative. However, if the data produced by the group of laboratories being tested were consistently good, the frequency of having to perform such analyses could be reduced.

THE USE OF RMs FOR ANALYTES OTHER THAN THOSE CERTIFIED

The primary user of RMs needs them for two major reasons. He can establish the accuracy of his data by the simultaneous analysis of CRMs or SRMs together with his unknowns. He can maintain control over his data by analyzing the same sample over and over again and plotting the results. Because of the limitations of which CRMs and SRMs are available in suitable matrices, users have been forced to try to solve their needs in some interesting ways. In one of these approaches, chemists have set about the task by working to reach consensus values on CRMs for analytes other than those for which the materials were originally certified. The logic of this approach is several-fold. The CRMs are already known to be homogeneous, they exist in reasonably large quantity, they are stable chemically, and their availability lends itself to the setting up of round-robins. The other major qualifications are that the matrix in which the analytes occur is suitable to the task at hand and that the analytes are present therein at concentrations appropriate to the levels anticipated in real samples.

Reaching consensus values on such RMs may best be achieved through a combination of in-house work and interlaboratory data comparisons. Both approaches should use the same analytical standards. However, the application of different analytical protocols is preferable in that this can increase the credence in the consensus values. Assuming that satisfactory agreement was reached by the participants, the CRM would then be available to measure the accuracy of analytical data.

One SRM that has attracted a good deal of attention for extended certification is a cod liver oil SRM, 1588, available from NIST. This material was certified originally for its content of five PCB congeners and ten pesticide residues. Through collaborative testing by scientists at NIST and the Canadian Wildlife Service, values have now been published for a total of 40 PCB congeners and 20 chlorinated hydrocarbon residues.[5]

Another collaborative study is currently under way in which scientists from several Canadian laboratories are seeking to achieve consensus on the levels of individual components of toxaphene in SRM 1588 cod liver oil.[6] The need to do this is rather compelling, as there are presently no RMs available for toxaphene components in tissue samples.

POST-DERIVATIZATION MEASUREMENT OF ORGANIC RESIDUES

Chlorophenols

A method for the measurement of chlorophenols, chloroguaiacols, chlorocatechols, and related compounds in pulp mill bleachery waste waters was devised by workers at Paprican.[7] The natural chlorophenolic residues were acetylated *in situ* and extracted into hexane. 2,6-Dibromophenol was added as an internal standard prior to the acetylation step. Dilute samples, such as river water, were first concentrated on columns of Amberlite XAD-2 resin prior to acetylation. Relative response factors and relative retention times were derived from experiments in which 2,6-dibromophenol was spiked into solutions of individual chlorophenol standards before acetylation. Extracts were quantified by GC-EC, and the identities of individual components were confirmed by GC-LRMS. For chlorination or caustic extraction stage effluents, the standard deviation varied from 4 to 16% but fell to 12 to 45% using dilute samples of bleached kraft mill effluent (BKME). Recovery values of 80 to 93% were recorded with samples of caustic extraction stage effluent, but difficulty was experienced in recovering tetrachlorocatechol (41%) from biotreated BKME. This method has been widely used in Canada for the past decade for surface water and mill effluent samples. The same technique, with modifications, has been adopted by NCASI.[8] In this version, 2,2′-difluorobiphenyl is added as a recovery standard.

A good selection of the natural chlorophenols, chloroguaiacols, chlorocatechols, chlorovanillins, chlorosyringols, and chlorosyringaldehydes has been available for a few years from one of the chemical suppliers. Quite recently the ^{13}C-labeled analogs were introduced to the market and have made possible the extension of the isotope dilution method to the analysis of these residues. This has now been applied to the measurement of pulp mill effluents, marine sediments, and organic soil.[9] The procedure is applicable to chlorophenolic residues differing widely in volatility and stability and involves the alkaline extraction of naturally contaminated or spiked sediment samples to which ascorbic acid is added as an antioxidant. A cocktail of appropriate surrogates is added to the alkaline solution, followed by filtration and acidification of the filtrate. The chlorophenols are then extracted into dichloromethane, acetylated, and purified by silica column chromatography. The eluate is spiked with a recovery standard and concentrated prior to analysis. The GC-LRMS instrument is calibrated against a standard cocktail containing all of the compounds of interest as well as the surrogate and recovery standards. Three quantification schemes were investigated, based on three different groups of surrogates. The results confirmed that the recoveries of chlorophenols from sediments and soils are very sensitive to the chemical composition of the sample. The method gave acceptable recoveries of 93 ± 33%, 84 ± 33%, and 95 ± 30%, respectively for spiked effluent, marine sediment, and organic soil samples at the 10 ng/g or 200 ng/L levels for 10 g or 1 L aliquots. A comparison of the results with those obtained by GC-EC showed that the latter method tended to give recoveries significantly higher than acceptable values, suggesting the presence of interferences.

CONCLUSION

The past 10 to15 years has seen a rapid development in the variety of residues that are required to be measured at the ultratrace level and in how these measurements can be accurately and repeatedly made. The greatest single stimulus to the field was probably the U.S. EPA Consent Decree, which resulted in a legal obligation for industry to measure and report what residues it discharges to the receiving environment. The need for such information to be acceptable in the courts gave rise to improvements in the sensitivity of instrumentation and in the way in which the analyses are done. Part of this change is the much wider recognition of the usefulness of RMs, and the availability of these in a greater selection of matrices and concentrations is still evolving. At the same time a larger variety of labeled standards and surrogates was introduced to allow isotope dilution methods to be applied to an increasing list of environmental residues. There is still much scope for improvement.

Another driving force behind the changes has been the need to measure dioxin and furan levels in food for human consumption. Because of the unparalleled toxicity of 2,3,7,8-TCDD to experimental animals, the requirement to measure this substance at levels never previously dreamed of has developed expertise in analytical methods which can now be extended to other highly toxic compounds. This may also result in the recognition of other significant residues that will have to be measured. The net outcome of all this scientific effort will be a cleaner and healthier environment, which should benefit all mankind.

ACKNOWLEDGMENT

The authors thank Dr. Michael Ikonomou for helpful discussions during the preparation of this manuscript.

REFERENCES

1. **R. J. Mesley, W. D. Pocklington, and R. F. Walker,** Analytical quality assurance. A review, *Analyst*, 116 (1991) 975–990.
2. ISO Guide 30, Terms and Definitions Used in Connection with Reference Materials, International Organization for Standardization, Geneva, Switzerland (1981).
3. Anon, Method 1613: Tetra- through Octa-Chlorinated Dioxins and Furans by Isotope Dilution HRGC/HRMS, Revision A, United States Environmental Protection Agency, Office of Water Regulations and Standards (1990).
4. Anon, Reference Method for the Determination of Polychlorinated Dibenzo-Para-Dioxins (PCDD's) in Pulp and Paper Mill Effluents, Environment Canada, Conservation and Protection, River Road Environmental Technology Centre (1992).
5. **M. M. Schantz, R. M. Parris, S. A. Wise, H. T. Won, and R.Turle,** Polychlorinated biphenyl (PCB) congeners and chlorinated pesticides in a cod liver oil reference material, *Chemosphere*, 24 (1992) 1687–1698.
6. **P. Andrews,** personal communication (April 1993).
7. **R. H. Voss, T. Wearing, and A. Wong,** Chlorinated phenolics in pulp mill effluents, In: *Advances in the Identification of Organic Pollutants in Water*, Vol. 2, Ann Arbor Science, Ann Arbor, MI (1981) Chap. 53.
8. Anon, NCASI Methods for the Determination of Chlorinated Phenolics in Pulp Industry Wastewaters, NCASI Technical Bulletin No. 498 (1986).
9. **L. E. Sojo, D. Hoover, M. C. Hamilton, and B. R. Fowler,** Analysis of Chlorophenols, Chlorocatechols, Chloroguaiacols and Related Compounds by GC/MS, Proceedings of the TAPPI Environmental Conference (1993) 101–124.

CHAPTER 17

Environmental Certified Reference Materials

Scott N. Willie

INTRODUCTION

A *reference material* is defined by The International Standards Organization as a material or substance of one or more properties, which are sufficiently well established to be used for the calibration of an apparatus, the assessment of a measurement method, or for assigning values to materials.[1] A *certified reference material* (CRM) is defined as a reference material one or more of whose property values are certified by a technically valid procedure, accompanied by or traceable to a certificate or other documentation which is issued by a certifying body.

As stated in Chapter 8, a suitable CRM should have a number of specific requirements.[2] It should

1. be of a suitable matrix, chemically comparable to the samples being analyzed. Ideally, the CRM will be of an identical matrix;
2. be certified with respect to the analytes of interest;
3. have analyte concentrations near those of the samples being analyzed;
4. be homogeneous with respect to the analytes of interest;
5. have a stable shelf life;
6. have accurate and precise analyte concentrations;
7. be traceable to a reliable source.

The use of CRMs must be an integral part of any laboratory quality assurance and quality control program. However, it is obvious that due to the above stringent requirements and the almost limitless possibility of matrices, there will always be a shortage of suitable CRMs for the wide variety of environmental samples.

This chapter of available CRMs is divided into two sections, inorganic and organic materials useful for the analysis of environmental samples. Although, the certified values are not listed in this chapter, the suppliers usually list the certified values and 95% confidence interval in their product catalogue. Some CRM suppliers also report noncertified concentrations, which are values where insufficiently reliable data is available to issue a certified value. In addition, a useful compilation of more than 900 reference materials for marine science is available from the National Oceanic and Atmospheric Administration.[3]

CRMs CERTIFIED FOR INORGANIC COMPONENTS

Plant Materials

This section describes sixteen CRMs that are of botanical origin. CRM 060 and 061 are both aquatic plant reference materials (*Lagarosiphon major* and *Platihypnidium riparioides*, respectively) and CRM 062, olive leaves, are available from the Institute for Reference Materials and Measurements (IRMM).[4] Each of these three materials is certified for seven trace metals (Mn, Cu, Zn, Cd, Hg, Pb), and noncertified values are given for twelve major and minor elements. CRM 129 is a hay powder certified for nine elements, Ca, Cd, I, K, Mg, N, P, S, and Zn. Beech leaves (CRM 100) and spruce needles (CRM 101) are certified for Cl, N, P, S; noncertified values are also given for six trace metals. CRM 279 is a sea lettuce (*Ulva lactuca*) collected from saline water. It is certified for seven elements (Cu, Zn, As, Se, Cd, I, Pb), and four noncertified values are reported. A rye grass (CRM 281) is certified for As, B, Cd, Cu, Hg, Mn, Mo, Ni, Pb, Sb, Se, and Zn. A white clover, CRM 402, is certified for As, Co, Mo, and Se. These IRMM materials are provided in the form of lyophilized powders.[4,5] A lichen (CRM 482) is certified for eight elements (Al, Cd, Cr, Cu, Hg, Ni, Pb, Zn). Another lichen CRM from IAEA (IAEA-336) is certified for 17 elements with noncertified results for 15 other elements.

NIES No. 9 is a Sargasso seaweed (*Sargassum fulvellum*) collected from Shimoda Bay, Shizouka Prefecture, Japan, by the National Institute for Environmental Studies (NIES).[6] This material was freeze-dried, ground, and sieved to pass 80-mesh. It is certified for 16 trace metals (Na, Mg, K, Ca, V, Mn, Fe, Co, Cu, Zn, As, Rb, Sr, Ag, Cd, and Pb), and noncertified values are given for six elements.

The National Institute of Standards and Technology[7,8] (NIST) has produced a suite of five botanical SRMs for trace metals, apple leaves (SRM 1515), peach leaves (SRM 1547), pine needles (SRM 1575) tomato leaves (SRM, 1573c), and spinach leaves (SRM 1570a). Certified values are reported for Al, As, Ca, Cu, Pb, Mn, Hg, N, K, and Sr in all three materials. In addition, noncertified values are reported for eighteen elements in some or all of these materials.

Biological Tissue Materials

Nineteen CRMs of biological tissue origin are summarized in this section and listed in Table 1. CRM 278 is a mussel tissue issued by IRMM. It is certified for 10 elements (Cr, Mn, Fe, Cu, Zn, As, Se, Cd, Hg, and Pb) with 14 noncertified elements reported. CRM 414 is a freeze-dried plankton powder certified for V, Cr, Mn, Ni, Cu, Zn, As, Se, Cd, Hg, and Pb, with noncertified values for six elements.[9] CRM 422, a cod muscle tissue, is certified for As, Cd, Cu, Fe, Hg, I, Mn, Pb, Se,

Table 1 CRMs of Biological Tissue Origin

		Plant Materials	
Aquatic plant	BCR	CRM 060	Trace elements
Aquatic plant	BCR	CRM 061	Trace elements
Olive leaves	BCR	CRM 062	Trace elements
Hay powder	BCR	CRM 129	Trace elements
Beech leaves	BCR	CRM 100	Trace elements
Sea lettuce	BCR	CRM 279	Trace elements
Lichen	BCR	CRM 482	Trace elements
Lichen	IAEA	IAEA-336	Trace elements
Saragasso seaweed	NIES	No. 9	Trace elements
Apple leaves	NIST	SRM 1515	Trace elements
Peach leaves	NIST	SRM 1547	Trace elements
Pine needles	NIST	SRM 1575	Trace elements
Tomato leaves	NIST	SRM 1573c	Trace elements
Spinach leaves	NIST	SRM 1570a	Trace elements
Sea plant	IAEA	IAEA-307	Radionuclides
		Biological Materials	
Mussel tissue	BCR	CRM 278	Trace elements
Plankton powder	BCR	CRM 414	Trace elements
Cod muscle	BCR	CRM 422	Trace elements
Bovine muscle	BCR	CRM 184	Trace elements
Bovine liver	BCR	CRM 185	Trace elements
Tuna fish	BCR	CRM 463, 464	Total mercury and methylmercury
Copepoda, dried	IAEA	MA-A-1/TM	Trace elements
Fish flesh	IAEA	MA-A-2/TM	Trace elements
Garpike fish	IAEA	MA-B-3/TM	Trace elements
Tuna homogenate	IAEA	IAEA-350	Trace elements
Cockle flesh	IAEA	IAEA-134	Radionuclides
Mussel tissue	NIES	No. 6	Trace elements
Fish tissue	NIES	No. 11	Total and organo Sn
Oyster tissue	NIST	SRM 1566a	Trace elements
Bovine liver	NIST	SRM 1577b	Trace elements
Dogfish muscle	NRC	DORM-2	Trace elements, methylmecury
Dogfish liver	NRC	DOLT-2	Trace elements, methylmercury
Lobster hepatopancreas			
(defatted)	NRC	TORT-2	Trace elements, methylmercury
(nondefatted)	NRC	LUTS-1	Trace elements, methylmercury
Tuna homogenate	IAEA	IAEA-352	Radionuclides
Fish	IAEA	MA-B-3/RN	Radionuclides
Human lung	NIST	SRM 4351	Radionuclides
Human liver	NIST	SRM 4352	Radionuclides
Cod liver oil	BCR	CRM 349	PCBs
Mackeral oil	BCR	CRM 350	PCBs
Garpike fish	IAEA	MA-B-3/OC	PCBs, chlorinated pesticides
Shrimp homogenate	IAEA	MA-3/OC	PCBs, chlorinated pesticides
Copepoda, dried	IAEA	MA-A-1/OC	PCBs, chlorinated pesticides
Tuna homogenate	IAEA	IAEA-351	PCBs
Cod liver oil	NIST	SRM 1588	PCBs, chlorinated pesticides
Mussel tissue	NIST	SRM 1974	PAHs, (PCBs, chlorinated pesticides)
Whale blubber	NIST	SRM-1945	PCBs, chlorinated pesticides
Mussel tissue	NRC	MUS-1	Domoic acid

Table 1 (continued)

		Biological Materials	
Mussel tissue	NRC	MUS-2	Okadaic acid/dinophysistoxin-1
Fish homogenate	NRC	CARP-1	Dioxins, furans, PCBs
		Food and Agriculture Products	
Wholemeal flour	BCR	CRM 189	Trace elements
Brown bread	BCR	CRM 191	Trace elements
Skim milk powder			
(natural)	BCR	CRM 063R	Trace elements
(spiked)	BCR	CRM 150	Trace elements
(spiked)	BCR	CRM 151	Trace elements
Milk powder	IAEA	A-11	Trace elements
Milk powder	IAEA	IAEA-152	Trace elements
Rye flour	IAEA	V-8	Trace elements
Whey powder	IAEA	IAEA-155	Trace elements
Rice flour	NIST	SRM 1568a	Trace elements
Wheat flour	NIST	SRM 1567a	Trace elements
Total diet	NIST	SRM 1548	Trace elements
Milk powder (nonfat)	NIST	SRM 1549	Trace elements
Milk powder			
(natural)	BCR	CRM 187	PCBs
(spiked)	BCR	CRM 188	PCBs
		Waters	
Water (low spike)	BCR	CRM 398	Trace elements
Water (high spike)	BCR	CRM 399	Trace elements
Seawater	BCR	CRM 403	Trace elements
Rainwater	BCR	CRM 408	Majors and nutrients
Rainwater	BCR	CRM 409	Majors and nutrients
Freshwater	BCR	CRM 479	Nitrate (low)
Freshwater	BCR	CRM 480	Nitrate (high)
Water	NIST	SRM 1643c	Trace elements
Rainwater			
(high pH)	NIST	SRM 2694-I	Trace elements
(low pH)	NIST	SRM 2694-II	Trace elements
Seawater, open ocean	NRC	NASS-4	Trace elements
Seawater, coastal	NRC	CASS-2	Trace elements
Estuarine water	NRC	SLEW-1	Trace elements
Riverine	NRC	SLRS-2	Trace elements
Pacific Ocean	IAEA	IAEA-298	Radionuclides
		Sediments	
Estuarine	BCR	CRM 277	Trace elements
Lake	BCR	CRM 280	Trace elements
River	BCR	CRM 320	Trace elements
Sediment	BCR	CRM 462	Organotin
Lake	IAEA	SL-3	Trace elements
Marine	IAEA	IAEA-356	Trace elements
Pond	NIES	No. 2	Trace elements
Marine	IAEA	SD-M-2/TM	Trace elements
Lake	IAEA	SL-1	Trace elements
Estuarine	NIST	SRM 1646	Trace elements

Table 1 (continued)

River	NIST	SRM 2704	Trace elements
Marine	NRC	MESS-2	Trace elements
Marine	NRC	BCSS-1	Trace elements
Harbor	NRC	PACS-1	Trace elements, organotin
Lake	IAEA	SL-2	Radionuclides
Marine	IAEA	SD-N-2	Radionuclides
Stream	IAEA	IAEA-313	Radionuclides
Stream	IAEA	IAEA-314	Radionuclides
Fresh water	NIST	SRM 4354	Radionuclides
Columbia River	NIST	SRM 4350B	Radionuclides
Coastal	IAEA	IAEA-357	PCBs, PAHs, chlorinated pesticides
River	NIST	SRM 1939	PCBs, chlorinated pesticides
Marine	NIST	SRM 1941	PCBs, PAHs, chlorinated pesticides
Coastal	NRC	CS-1	PCBs
Marine	NRC	HS-1 and 2	PCBs
Marine	NRC	HS-3, 4, 5, and 6	PAHs
Marine	NRC	SES-1	PAHs
		Soils	
Calcareous loam soil	BCR	CRM 141R	Trace elements
Sandy soil	BCR	CRM 142R	Trace elements
Soil (sewage sludge)	BCR	CRM 143R	Trace elements
Soil	IAEA	SOIL-7	Trace elements
Agricultural soil	NIST	SRM 2709	Trace elements
Agricultural soil	NIST	SRM 2710	Trace elements
Agricultural soil	NIST	SRM 2711	Trace elements
Soil	IAEA	SOIL-6	Radionuclides
Soil	IAEA	IAEA-312	Radionuclides
Peruvian soil	NIST	SRM 4355	Radionuclides
Rocky Flats	NIST	SRM 4353	Radionuclides
		Sewage Sludge	
Domestic	BCR	CRM 144	Trace elements
Industrial	BCR	CRM 146	Trace elements
Mixed	BCR	CRM 145	Trace elements
Sludge	BCR	CRM 088	PAHs
		Miscellaneous	
Coal ash	BCR	CRM 038	Trace elements
Incineration	BCR	CRM 176	Trace elements
Fly ash	BCR	CRM 490	Dioxins and furans
Vehicle exhaust	NIES	No. 8	Trace elements
Fly ach	NIST	SRM 1633b	Trace elements
Urban particulate	NIST	SRM 1648	Trace elements
Coal ash	NIST	SRM 2689	Trace elements
Coal ash	NIST	SRM 2690	Trace elements
Coal ash	NIST	SRM 2670	Trace elements
Waste oil	BCR	CRM 420	PCBs
Coal tar	NIST	SRM 1597	PAHs

and Zn. CRMs 184 and 185 are bovine muscle and liver, respectively. They are certified for Cd, Cu, Fe, Hg, Mn, Pb, Se, and Zn. Two tuna fish tissue CRMs (CRM 463, CRM 464) are certified for total and methylmercury.

A suite of biological reference materials has been prepared by the International Atomic Energy Agency (IAEA).[10,11] MA-A-1/TM is homogenized copepods (*Calanus cristatus*) collected in the Norwegian Sea. This sample was deep frozen and freeze dried under vacuum.[12] The lyophilized copepods were ground in a porcelain ball mill and sieved through a 250-µm stainless steel sieve before rehomogenization and bottling in 35-g aliquots. MA-A-2/TM is a fish flesh, *Mugil cephalus* (*Mugilidae* family), homogenate collected from the French Atlantic coast. This sample was prepared by grinding and sieving through a 500-µm mesh nylon sieve before homogenization. Both of these are certified for 14 trace elements (Cr, Mn, Fe, Co, Ni, Cu, Zn, As, Se, Ag, Cd, Sb, Hg, and Pb). IAEA MA-B-3/TM is prepared from fillets from a garpike fish (*Belona belona*) collected from the western Baltic Sea. Certified values for 17 elements (Na, Mg, P, Cl, K, Ca, Mn, Fe, Cu, Zn, As, Se, Br, Rb, Sr, Hg, and Pb) are given for this material. IAEA-350 is a CRM prepared from tuna collected from the northern Mediterranean Sea.[13] The fish fillets were lyophilized, then ground in a porcelain ball mill before passing through a 0.5-µm sieve. After further homogenization, the material was packaged in 35-g quantities. Certified values are given for 11 trace elements (Cr, Mn, Fe, Ni, Cu, Zn, Se, Br, Cd, Sb, and Hg).

NIES No. 6 is a CRM generated from the soft tissue of common mussels (*Mytilus edulis*) collected in Matoya Bay, Mie Prefecture, Japan. The material was cryogenically ground, freeze-dried, sieved, and blended before bottling. It is certified for 14 trace metals (Na, Mg, K, Ca, Cr, Mn, Fe, Ni, Cu, Zn, As, Ag, Cd, and Pb). A comprehensive description of the analysis procedures has been reported by Okamoto and Fuwa.[14] NIES has produced a fish tissue reference material prepared from sea bass fillets[15] for use in the determination of problem tin species (NIES No. 11). The material was homogenized and freeze-dried before being ball-milled and blended. Twenty gram samples were vacuum packed into polyethylene bags. Certified values of 1.3 ± 0.1 mg/kg and 2.4 ± 0.1 mg/kg were determined for tributyltin and total tin, respectively. A reference value of 6.3 mg/kg for triphenyltin is also reported.

NIST has reissued its oyster tissue CRM as SRM 1566a. This material was cryogenically ground, freeze dried and certified for 25 elements (Na, Mg, Al, P, S, Cl, K, Ca, V, Cr, Mn, Fe, Co, Ni, Cu, Zn, As, Se, Sr, Ag, Cd, I, Hg, Pb, and U), the largest list offered for a single biological CRM. Noncertified concentrations are given for another 15 elements. Bovine liver (SRM 1577b) has also been reissued, and certified concentrations have been determined for Na, Mg, P, S, Cl, K, Ca, V, Mn, Fe, Co, Cu, Zn, As, Se, Sr, Ag, Cd, I, Hg, Pb, and U.

The National Research Council of Canada (NRC) has issued four biological CRMs for trace metals.[16] DORM-2 and DOLT-2 are dogfish (*Squalus acanthias*) muscle and liver reference materials, respectively. They were collected from the Canadian Atlantic coast, homogenized, and acetone extracted to produce partially defatted powders (5 and 24% fat, respectively). These materials are certified for 16 trace elements (Na, Mg, Cl, K, Cr, Mn, Fe, Co, Ni, Cu, Zn, As, Se, Cd, Hg, and Pb). In addition, these materials have been certified for methylmercury. LUTS-1 and TORT-2 are both lobster (*Homarus americanus*) hepatopancreas (tomalley) from Canadian east coast edible grade lobster tomalley. TORT-2 was homogenized, spray dried, acetone extracted (to partially remove fat), and vacuum dried to a powder before bending and bottling. TORT-2 is certified for 23 trace and minor elements (Na, Mg, P, S, Cl, K, Ca, V, Cr, Mn, Fe, Co, Ni, Cu, Zn, As, Se, Sr, Mo, Cd, Sn, Hg, and Pb) plus methylmercury. LUTS-1 was processed in a different manner from TORT-2, and in its final form is a slurry with none of the lipid materials removed.[17] The raw material was homogenized by comminution, and a small amount of ethoxyquin was added as an antioxidant. The water content was increased slightly, and homogenization was repeated using high pressure homogenization. LUTS-1 is known as a "second generation" biological reference material, much more closely related to the "real" samples that monitoring laboratories routinely analyze in comparison to dry powder CRMs. LUTS-1 is certified for 17 trace metals (Mg, K, Ca, Cr, Mn, Fe, Co, Ni, Cu, Zn, As, Se, Sr, Ag, Cd, Hg, and Pb) as well as methylmercury; values for this material are reported in two ways, as a dry weight and as packaged. The NRC biological CRMs are all certified for methylmercury.

Food and Agriculture Products

IRMM produces five materials of food origin that are environmentally related. Wholemeal flour (CRM 189) and brown bread (CRM 191) are certified for Cd, Cu, Fe, Mn, Pb, and Zn; nine noncertified values are given. A natural skim milk powder (CRM 063) is certified for 12 elements, Ca, Cd, Cl, Cu, Fe, Hg, K, Mg, N, Na, P, and Pb; four noncertified values are reported. Two spiked skim milk powders are also available, CRM 150 and CRM 151. Six certified values (Cd, Cu, Fe, Hg, I, Pb) and four noncertified values (Mn, Ni, Se, Zn) are reported. The trace metal levels for CRM 150 are approximately five to ten times higher than in the natural skim milk powder (for the certified values only), and CRM 151 is spiked approximately five times higher than CRM 150.

IAEA produces two milk powder CRMs, A-11 and IAEA-152. A-11 is certified for 20 elements (As, Ca, Cd, Cl, Co, Cr, Cu, Fe, Hg, I, K, Mg, Mn, Mo, Na, P, Pb, Rb, Se, Zn) and IAEA-153 is certified for nine (Br, Ca, Fe, K, Mg, Na, P, Rb, Zn). A rye flour (V-8) is certified for 11 elements (Br, Ca, Cl, Cu, Fe, K, Mg, Mn, P, Rb, Zn), and a whey powder (IAEA-155) is certified for 17 elements (Br, Cd, Cl, Co, Cr, Cs, Hg, Mg, Mn, Na, Ni, P, Pb, Rb, Sc, Se, Zn).

Four food and agriculture SRMs from NIST are available. SRM 1568a was prepared from rice flour and is certified for 17 trace elements (Na, Mg, Al, P, S, K, Ca, Mn, Fe, Cu, Zn, As, Se, Rb, Mo, Cd, and Hg), and reference values are given for 11 other elements.[7,8] A wheat flour, 1567a, is certified for Al, Cd, Ca, Cu, Fe, Mg, Mn, Mo, P, K, Ru, Se, Na, S, and Zn. This material also has ten noncertified concentrations listed; both these materials are packaged in 80-g quantities. A total diet reference material (SRM 1548), representative of the daily adult intake of the U.S. population, was produced from a mixture of 200 foods.[18] The material was blended, freeze dried, radiation sterilized, and bottled in 6-g quantities. One unit consists of two bottles. Certified values are reported for seven major elements (N, Cl, Na, K, P, S, Ca) and seven trace elements (Mg, Fe, Zn, Mn, Cu, Se, Cd), as well as some nutritional parameters (dietary fiber, cholesterol, and caloric content). A nonfat milk powder (SRM 1549) is also available from NIST; certified concentrations are reported for Cd, Ca, Cl, Cr, Cu I, Fe, Pb, Mn, Hg, P, K, Se, Na, S, and Zn.

Waters

IRMM produces two materials (CRM 398 and CRM 399) for the quality control of major elements determinations in fresh water.[19] These two reference materials were prepared from silica-free deionized water (0.1 mol/L nitric acid), to which the compounds of interest were added. CRM 398 is of low element content and CRM 399 of high element content. Certified concentrations are reported for Al, Ca, Cl, Fe, Mg, Mn, K, and Na in both materials, plus S and P in CRM 398 and CRM 399, respectively. CRM 403 is a reference material for trace elements in seawater. The sample was filtered, acidified, and bottled in 2-L quantities. Certified concentrations for Ni, Cu, Zn, Mo, Cd, and Pb and three noncertified concentrations (Al, V, As) are reported. IRMM has two simulated rainwaters, CRM 408 and CRM 409, certified for five major elements and NH_4, NO_3 and SO_3. Two freshwater CRMs (CRM 479 and 480) are certified for nitrate content.

NIST has two CRMs in preparation: SRM 1643d is a spiked solution in water and SRM 1640 is a natural water CRM. Two simulated rainwater SRMs, 2694-I and 2694-II, were developed to aid in the analysis of acidic rainwater. SRM 2694-I at pH 4.30 and SRM 2694-II at pH 3.60 are certified for pH, electrolytic conductivity, acidity (meq/L) fluoride, nitrate, sulfate, Na, K, Ca, and Mg.

The NRC has a suite of four natural water reference materials for trace metals. NASS-4 is an open ocean sample with a salinity of 31.3 parts per thousand (‰), certified for As, Cd, Cr, Co, Cu, Fe, Pb, Mn, Mo, Ni, Se(IV), U, and Zn. CASS-3, a coastal seawater collected 400 to 800 meters off the shore of Nova Scotia, Canada has a salinity of 29.2‰ and is certified for As, Cd, Cr, Co, Cu, Fe, Pb, Mn, Mo, Ni, and Zn. SLEW-1, an estuarine water, salinity 11.6‰, was collected in the St. Lawrence estuary and is certified for As, Cd, Cr, Co, Cu, Fe, Pb, Mn, Ni, and Zn. These materials were filtered through 0.45-µm filters, acidified, homogenized, and bottled in 2-L polyethylene bottles. SLRS-3 is a riverine water, collected in the Ottawa River and processed in the same manner as the seawater samples, with an additional final filtering through 0.2-µm filters and radiation sterilization to inhibit bacterial action. Certified values are reported for 21 elements, Al, Sb, As, Ba, Cd, Cr, Co, Cu, Fe, Pb, Mn, Mo, Ni, Sr, U, V, Zn, Ca, Mg, K, and Na.

Sediments

Two sediment reference materials for trace metals are available from BCR. CRM 277 and CRM 320 are estuarine and river sediment, respectively. Certified concentrations are reported for As, Cd, Cr, Cu, Hg, Mn, Ni, Pb, Se, and Zn, with 13 noncertified values reported for each material. BCR CRM 462 is a coastal sediment from Arcachon Bay (France) certified for tributyltin and dibutyltin.[24]

NIES No. 2 is a sediment collected from a pond within the grounds of the University of Tokyo; certified values are reported for 13 elements (Na, Al, K, Ca, Fe, Cr, Co, Ni, Cu, Zn, As, Cd, Pb), and 12 noncertified values are reported.

IAEA produces four sediments, SD-M-2/TM, SL-1, SL-3, and IAEA-356. SD-M-2/TM is a marine sediment certified for Al, As, Ba, Br, Ca, Cd, Ce, Co, Cr, Cs, Cu, Eu, Fe, Hf, Hg, K, La, Lu, Mn, Na, Nd, Ni, Pb, Rb, Sb, Sc, Sm, Sr, Tb, Th, U,

V, Yb, Zn, and Zr. SL-1 is a lake sediment collected from the Sardis Reservoir in Mississippi, certified for As, Ba, Br, Cd, Ce, Co, Cr, Cs, Cu, Dy, Eu, Fe, Hf, La, Lu, Mn, Na, Nd, Ni, Pb, Rb, Sb, Sc, Sm, Th, Ti, U, V, Yb, and Zn.

An estuarine sediment from Chesapeake Bay (SRM 164a) is available from NIST. The material was freeze dried, radiation sterilized, sieved, ball milled, blended, and bottled in 75-g aliquots. Certified concentrations are reported for Mg, Al, P, Ca, V, Cr, Mn, Fe, Co, Ni, Cu, Zn, As, Cd, Hg, and Pb. Noncertified concentrations are reported for Li, Be, Na, Si, K, S, Ti, Sc, Ge, Se, Rb, Mo, Sb, Te, Cs, Ce, Eu, Tl, and Th. Buffalo River Sediment (SRM 2704) was collected at Buffalo, NY, and bottled in 50-g units. Certified values are reported for Al, Ca, C, Fe, Mg, P, K, Si, Na, Ti, Sb, As, Ba, Cd, Cr, Co, Cu, Pb, Mn, Hg, Ni, Tl, U, V, and Zn. Noncertified concentrations are reported for 22 elements.

Three sediments are available from the NRC, MESS-2, BCSS-1, and PACS-1. MESS-2 is an estuarine sediment from the Beaufort Sea; it is certified for Sb, As, Cd, Cr, Co, Cu, Pb, Li, Mn, Hg, Ni, Se, Sn, V, and Zn. BCSS-1, from the Baie des Chaleurs in the Gulf of St. Lawrence, is certified for Sb, As, Be, Cd, Cr, Co, Cu, Pb, Mn, Ni, Se, Sn, V, and Zn. A sediment with elevated trace metal levels is PACS-1, from Esquimalt Harbour, British Columbia. Certified concentrations are reported for Sb, As, Cd, Cr, Co, Cu, Pb, Mn, Hg, Mo, Ni, Se, Sn, V, and Zn. In addition, PACS-1 is certified for three organo-tin species, monobutyltin, dibutyltin, and tributyltin. All three NRC sediments are certified for the following major constituents: Al_2O_3, C, CaO, Cl, Fe_2O_3, K_2O, MgO, Na_2O, P_2O_5, S, SiO_2, TiO_2.

Soils

BCR produces a soil reference material for trace metals (CRM 141R). This material is a powder with particles smaller than 90 μm. It is certified for Cu, Zn, Cd, Hg, and Pb, with noncertified values reported for 14 other elements. CRM 142R is a light sandy soil certified for Cd, Cu, Hg, Ni, Pb, and Ce, with three noncertified values. CRM 143R is a sewage sludge amended soil. Values for the aqua regia soluble fraction are reported for all three materials.

SOIL-7, distributed by the IAEA, is certified for 27 elements: Sc, V, Cr, Mn, Co, Cu, Zn, As, Rb, Sr, Y, Zr, Sb, Cs, La, Ce, Nd, Sm, Eu, Tb, Dy, Yb, Hf, Ta, Pb, Th, and U. Three soil SRMs have currently been made available by NIST; SRM 2709 is an agricultural soil from the San Joaquin Valley, CA, with elevated levels of Se and Hg (with respect to the average crustal abundances). SRM 2710 is a highly contaminated Northern Great Plains soil collected in a flood plain near Butte, MT. SRM 2711 is a moderately contaminated soil from Montana. Certified or noncertified values are given for Na, Mg, Al, Si, P, S, K, Ca, Ti, V, Cr, Mn, Fe, Co, Ni, Cu, Zn, As, Se, Sr, Ag, Cd, Sb, Ba, Au, Hg, Tl, Pb, and Th in all three materials.

Sewage Sludge

Three sewage sludge CRMs are produced by BCR; CRM 144R is a sewage sludge of domestic origin, CRM 146R is a sewage sludge of industrial origin, and CRM 145R is a sewage sludge of mixed origin. Each material has certified values for nine elements and three noncertified values. The aqua regia soluble fraction is also reported for these samples.

Miscellaneous Materials

BCR CRM 038 is a coal ash from pulverized coal and is certified for Na, Mn, Fe, Co, Cu, Zn, As, Cd, Hg, and Pb. CRM 176 is an ash from city waste incineration, and certified values are reported for Cr, Fe, Co, Ni, Cu, Zn, Se, Cd, Sb, Hg, Tl, and Pb. NIES 8 is a vehicle exhaust particulate, collected from the ventilators of a highway tunnel by electrostatic precipitators. This material contains approximately 80% carbon and is certified for Na, Mg, Al, K, Ca, V, Cr, Co, Ni, Cu, Zn, As, Sr, Cd, Sb, and Pb.

SRM 1633a is a fly ash obtained from a coal-fired power plant and is distributed by NIST. Certified values are reported for Na, Mg, Al, Si, Ca, V, Cr, Mn, Fe, Ni, Cu, Zn, As, Se, Rb, Sr, Cd, Sb, Hg, Tl, Pb, Th, and U. SRM 1648 is an urban particulate material collected at St. Louis, MO. Certified concentrations for Na, Al, K, V, Cr, Fe, Ni, Cu, Zn, As, Se, Cd, Pb, and U are reported. Three coal ashes from different coal fired power plants are also available from NIST. SRM 2689 from a plant using bituminous coal from western Kentucky, SRM 2690 from a Colorado plant using sub-bituminous coal, and SRM 2691 from a Wyoming plant also using sub-bituminous coal are all certified for Na, Mg, Al, Si, P, S, K, Ca, Ti, and Fe.

Radionuclides

Four reference materials are available from IAEA that have been affected by the Chernobyl nuclear power plant accident. IAEA-307, a sea plant (*Posidonia oceanica*), and IAEA-308 (Mediterranean seaweed) both were collected along the shore in the vicinity of the Principality of Monaco in October 1986.[20] Nine certified values are given for IAEA-307; ^{40}K,^{106}Ru,^{110m}Ag,^{134}Cs,^{226}Ra,$^{239+240}$Pu, and ^{241}Am and ten for IAEA-308, ^{40}K,^{106}Ru,^{110m}Ag,^{134}Cs,^{137}Cs,^{210}Pb,^{228}Th,^{238}Pu,$^{239+240}$Pu, and ^{241}Am. IAEA-352 is a tuna homogenate described in the section entitled "Biological Tissue Materials" in this chapter; it is certified for four radionuclides (^{40}K,^{137}Cs,^{210}Pb,^{210}Po). A fish (*Belona belona*) from the Baltic Sea (MA-B-3/RN described in the same) has been certified for two radionuclides (^{40}K,^{137}Cs).

SL-2 and SD-N-2 are a lake sediment and a marine sediment respectively, certified for ^{40}K and ^{137}Cs in the lake sediment and ^{40}K, ^{137}Cs, ^{232}Th, and $^{239/240}$Pu in the marine sediment. Two stream sediments, IAEA-313 and IAEA-314, are both certified for ^{226}Ra at 342 and 732 Bq/kg, respectively. IAEA-312, a soil sample, is also certified for ^{226}Ra at 269 Bq/kg. Soil-6 is certified for the five radionuclides ^{226}Ra, ^{90}Sr, ^{137}Cs, ^{226}Ra, and ^{239}Pu. A Pacific Ocean surface water (IAEA-298) collected outside the Mururoa atoll is certified for ^{90}Sr, ^{137}Cs, and $^{239+240}$Pu.

Sediment and soil samples with approximate activities at the time of calibration are also supplied by NIST. Fresh water lake sediment, SRM 4354, contains approximately 25 g of freeze-dried, pulverized sample. NIST intends this CRM to be used in tests of measurements of environmental radioactivity in a calibrated real sample matrix. Columbia River sediment, SRM 4350B, was collected downstream from a nuclear reactor facility and is certified for $^{239/240}Pu$ and ^{241}Am. The fission and activation products are elevated over typical world-wide levels. Two soil samples are also available from NIST. Peruvian soil, SRM 4355, has nonmeasurable activity and is intended as a blank. In contrast, concentrations of $^{239/240}Pu$ and ^{241}Am in Rocky Flats Soil Number 1, SRM 4353, which was collected near Rocky Flats, CO, are at elevated levels. Two CRMs related to human health are also available from NIST, SRM 4351 human lung and SRM 4352 human liver; both materials have been freeze dried, cryogenically ground, homogenized, and packed in a glass bottle under vacuum. Radioactivity concentrations for both samples are on the order of 10^{-4} Bq/g^{-1}.

CRMs CERTIFIED FOR ORGANIC CONTAMINANTS

The number of reference materials certified for organic constituents is steadily increasing; however, the principle of independent methodologies necessary for certification becomes extremely difficult when only a few developed methodologies are available. For this reason many more noncertified or method-dependent values are reported for organic constituents than for trace metals. In addition to the natural matrix CRMs, most suppliers also distribute calibration solutions. These are useful to the analyst for calibration of the instrumentation for retention times and detector response factors, spiking samples, recovery studies, and method response factors.[20] Many CRMs are certified for more than one type of compound and are listed under these five subsections: polychlorinated biphenyls (PCBs), polycyclic aromatic hydrocarbons (PAHs), chlorinated pesticides, shellfish toxins, and miscellaneous organics.

Polychlorinated Biphenyls

There are sixteen available CRMs for polychlorinated biphenyls that are relative to environmental analysis. A solution of polychlorinated biphenyls, iso-octane CRM 365, is distributed by IRMM. Certified concentrations are reported for the following PCBs: IUPAC No. 8, 20, 28, 35, 52, 101, 118, 138, 153, and 180. These same congeners are individually available in powder form under separate CRM labels. BCR CRM 349 is a cod liver oil and BCR CRM 350 is a mackerel oil. These materials are provided in sealed glass ampules in units of approximately 2 g and are each certified for seven PCB congeners (28, 52, 101, 118, 135, 153, and 180); four noncertified values are reported (138+163, 44, 128, and 194). A sewage sludge, CRM 392, also is certified for six PCBs (24, 52, 101, 118, 153, 180). A mixture of waste lubricating oils (CRM 420 and CRM 449) have been certified for nine PCBs; the oil was filtered and spiked with a commercial PCB mixture. CRM 450 is a natural milk powder certified for six PCBs. CRM 481 is an industrial soil certified for eight PCBs.

IAEA MA-B-3/OC is a garpike fish (*Belona belona*), the same material used for the trace metal and radionuclide CRMs described in the previous sections. It is certified for two congeners (101 and 153). IAEA MA-3/OC is a shrimp homogenate certified for arochlor 1254 and 1260 at 33 and 5 ng/g, respectively. IAEA MA-A-1/OC, a dried copepoda described in the section entitled, "Biological Tissue Materials," is certified at 120 ng/g and 140 ng/g for arochlor 1242 and arochlor 1254, respectively. IAEA-351 is a tuna homogenate from the Mediterranean Sea (similar to IAEA-350 in the same section). Certified values are given for arochlor 1254 and 1260 and PCB congeners 101, 118, 138, 153, and 180. A sediment sample (IAEA-357) collected from the Lagoon of Venice, Italy, is available from the IAEA. It is labeled as a "hot spot" sediment, suggesting the levels are higher than average. Certified values are reported for congeners 52, 101, 105, 118, 138, 153, 156, 170, 180, and 194.

Chapter 12 describes in detail the organic CRMs available from NIST. SRM 1493 and SRM 2262 are calibration solutions of PCBs in 2,2,4-trimethylpentane contain analytes at two concentration levels (approximately 0.2 and 2 μg/ml). The certified congeners are 8, 18, 28, 44, 52, 66, 77, 101, 105, 118, 126, 128, 138, 153, 170, 180, 187, 195, 206, and 209 in both SRMs, with the additional congeners 1, 29, 50, 87, 104, 154, 188, and 201 in SRM 2262. A CRM in cod liver oil is SRM 1588.[20] This material was analyzed using gas chromatography with electron capture detection and gas chromatography with mass spectrometric detection and is certified for five congeners (101, 138, 153, 170, and 180). Recently NIST has provided noncertified values for 20 additional PCB congeners for this material. A chromatogram for information purposes is supplied with each certificate, along with a summary of analytical results by the different methods. SRM 1939, Polychlorinated Biphenyls in a River Sediment, collected from the Hudson River, NY, is certified for three congeners (26, 28, 44). Noncertified concentrations of 12 additional PCB congeners, determined by a method employing Soxlet extraction and single column gas chromatography, are also reported. Certified concentrations for 21 PCBs ranging between 1 and 25 ng/g are reported in SRM 1941a,[21] a marine sediment collected from the Baltimore Harbor, MD. A mussel tissue (*Mytilus edulis*), SRM 1974a, also reports noncertifed concentrations for 13 PCB congeners. A whale blubber SRM is available from NIST that is certified for 27 PCB congeners.[25]

NRC has produced a set of four mixtures of pure, synthetic chlorinated biphenyls (CLB-1) in iso-octane covering 51 individual chlorinated biphenyl congeners of established identity. The congeners are IUPAC 15, 18, 31, 40, 44, 49, 52, 54, 60, 77, 86, 87, 101, 103, 105, 114, 118, 121, 128, 129, 137, 138, 141, 143, 151, 153, 154, 156, 159, 170, 171, 173, 180, 182, 183, 185, 187, 189, 191, 194, 195, 196, 200, 201, 203, 205, 207, 208, and 209. The concentration of the congeners range from 1.2 μg/mL (IUPAC 159) to 139.5 μg/mL (IUPAC 15). Each 1-mL ampule mixture in the set contains 14 or 15 congeners. CLB-2 is a solution of ^{13}C-labeled 2,2′4,4′,5,5′-hexachlorobiphenyl (IUPAC 153) in iso-octane. This material is intended to aid the analyst in quantitative analysis of polychlorinated biphenyls by combined gas chromatography/mass spectrometry.

Three sediment CRMs for polychlorinated biphenyls are also available from NRC. CS-1 is a coastal sediment, and HS-1 and HS-2 are Nova Scotia harbor sediments. The suite of samples provides a range of PCB levels from below 2 ppb in CS-1 to 112 ppb in HS-2. The matrix varies from a clean carbonaceous clay to an organic and sulfur-rich material from a commercially active harbor. These materials have been freeze dried, homogenized, and subsampled into solvent-rinsed pint-sized steel cans; each sample contains approximately 200 g of sediment material. Certified values are reported for congeners 101, 138, 151, 153, 170, 180, 194, 196, 199, and 209.

Polycyclic Aromatic Hydrocarbons (PAH)

BCR distributes individual units of 48 different polycyclic aromatic compounds to be used as reference substances for mutagenicity and carcinogenicity tests. Supplied in solid form, they are packaged in 20 of 100-mg quantities. In addition, eight nitro- and seven oxygenated-polycyclic aromatic hydrocarbons are available as individual compounds. CRM 088 is a dried sewage sludge certified for eight PAHs, and CRM 524 is a contaminated industrial soil certified for nine analytes.

A sediment, IAEA-357, from the lagoon of Venice, is certified for anthracene, benz[*a*]anthracene, chrysene, benzo[*ghi*]perylene, phenanthrene, phytane, pyrene. benzo[*a*]pyrene, benzo[*e*]pyrene, and indeno[*1,2,3-cd*]pyrene.

NIST produces three calibration solutions to be used to determine retention times and detector response for PAH determinations by LC and GC. CRMs NIST SRM 1647c is a an acetonitrile solution of the 16 PAHs (approximately 1 to 20 μg/mL) on the U.S. Environmental Protection Agency's (EPA) priority pollutant list (naphthalene, acenaphthylene, acenapthene, fluorene, phenanthrene, anthracene, fluoranthene, benz[*a*]anthracene, chrysene, benzo[*b*]fluoranthene, benzo[*k*]fluoranthene, benzo[*a*]pyrene, perylene, indeno[*1,2,3-cd*]pyrene, dibenz[*a,h*]anthracene, and benzo[*ghi*]perylene. SRM 1491 and SRM 2260 contain these same PAHs plus eight additional PAHs (2-methylnaphthalene, 1-methylnaphthalene, biphenyl, 2,6-dimethylnaphthalene, 1,6,7-trimethylnaphthalene, 1-methylphenanthrene, pyrene, and benzo[*e*]pyrene), in hexane and toluene, respectively. These solvents are more compatible with GC and normal-phase LC analyses and for spiking into nonaqueous solutions. SRM 1587, nitrated polycyclic aromatic hydrocarbons in methanol, is certified for 2-nitrofluorene, 9-nitroanthracene, 3-nitrofluoranthene, 1-nitropyrene, 7-nitrobenz[*a*]anthracene, 6-nitrochrysene, and 6-nitrobenzo[*a*]pyrene.

NIST SRM 1974a, a blue mussel tissue (*Mytilus edulis*), is certified for the nine selected PAHs (phenanthrene, anthracene, fluoranthene, pyrene, perylene, benzo[*b*]fluoranthene, benzo[*a*]pyrene, benzo[*ghi*]perylene, and ideno[*1,2,3-cd*]pyrene, with additional noncertified concentrations for another 19 PAHs. The material was collected from Dorchester Bay within Boston harbor. The two methods used for certification were GC-MS and reversed-phase liquid chromatography with fluorescence detection. Samples are provided as a frozen tissue homogenate and concentrations are provided on both a wet weight and dry weight basis for user convenience. SRM 1941a, the Baltimore Harbor sediment,[21] is certified for phenanthrene, anthracene, fluoranthene, pyrene, benz[*a*]anthracene, benzo[*b*]fluoranthene, benzo[*k*]fluoranthene, benzo[*a*]pyrene, perylene, indeno[*1,2,3-cd*]pyrene, and benzo[*ghi*]perylene. Coal tar SRM 1597 has certified or noncertified values for the same PAHs as SRM 1941 plus naphthalene and triphenylene. In addition, PAHs are certified in shale oil (SRM 1580), petroleum crude oil (SRM 1582), air particulate matter (SRM 1648), urban dust (SRM 1649), and diesel particulate (SRM 1650). PAH values are also reported in river sediment SRM 1939 and SRM 1941a.

HS-3, HS-4, HS-5, and HS-6 are four harbor marine sediment research materials distributed by NRC for determination of PAHs. This suite of samples covers a range of concentrations for 16 PAHs designated as priority pollutants by the U.S. EPA. These PAHs are naphthalene, acenaphthylene, acenaphthene, fluorene, phenanthrene, anthracene, fluoranthene, pyrene, benz[*a*]anthracene, chrysene, benzo[*a*]pyrene, benzo[*b*]fluoranthene, benzo[*k*]fluoranthene, benzo[*ghi*]perylene, dibenz[*a,h*]anthracene, and indeno[*1,2,3-cd*]pyrene. SES-1 is a spiked natural estuarine sediment material with information only values for 15 of the 16 priority pollutant PAHs. These materials have been freeze dried, passed through a No. 120 sieve, homogenized, and subsampled in 200-g aliquots into stainless steel cans. NRC also distributes deuterated polycyclic aromatic compounds (PACs) in toluene. DPAC-1 contains 21 compounds and DPAC-2 contains six PACs.

Chlorinated Pesticides

A total of 11 CRMs are available for chlorinated pesticide analysis. Spiked and unspiked milk powder CRMs are available from BCR. CRM 187, a natural milk powder, is certified for four chlorinated pesticides (HCB, α-HCH, γ-HCH, p,p′-DDE), and noncertified values are reported for two compounds (β-HEPO, dieldrin). A spiked milk powder, CRM 188, is certified for the six certified and noncertified compounds in CRM 187, plus β-HCH, endrin, and p,p′-DDT. CRM 430 is a pork fat CRM, originating from pigs fed a diet with enriched PCB concentrations. Certified concentrations for all the above PCBs are reported, ranging between 20 and 3400 ng/g for endrin and p,p′-DDT, respectively.

IAEA MA-A-3/OC is a shrimp homogenate certified for eight pesticides (aldrin, HCB, heptachlor, α-HCH, lindane, pp′DDD, pp′DDE, and pp′DDT). IAEA MA-B-3/OC, a garpike fish tissue (described earlier in this chapter in the section Biological Tissue Materials) is also certified for eight pesticides (aldrin, dieldrin, HCB, α-HCH, lindane, pp′DDD, pp′DDE, and pp′DDT). IAEA MA-A-1/OC, a dried copepoda also described earlier in that section, is certified for six pesticides (aldrin, α-HCH, lindane, pp′DDD, pp′DDE, and pp′DDT). The IAEA tuna homogenate (IAEA-351) is certified for six chlorinated pesticides (aldrin, HCB, lindane, pp′DDD, pp′DDE, pp′DDT). The IAEA "hot spot" sediment, IAEA-357, is certified for pp′DDD, pp′DDE, and pp′DDT.

Two calibrating solutions for chlorinated pesticides are available from NIST. SRM 1492 and SRM 2261 each contains 15 compounds in hexane (hexachlorobenzene, γ-HCH, heptachlor, aldrin, heptachlor epoxide, *cis*-chlordane, 4-*trans*-nonachlor, dieldrin, mirex, 2,4′-DDE, 4,4′-DDE, 2,4′-DDD, 4,4′-DDD, 2,4′-DDT, and 4,4′-DDT). The concentrations of these compounds are at approximate concentration levels of 0.2 and 2.0 μg/mL for SRM 1492 and SRM 2261, respectively. These SRMs replace SRM 1583, five chlorinated pesticides in 2,2,4-trimethylpentane. NIST cod liver oil (SRM 1588) is certified for ten pesticides (dieldrin, α-HCH, 4,4′-DDD, 4,4′DDE, 2,4′-DDT, 4,4′-DDT, hexachlorobenzene, *trans*-chlordane, *trans*-nonachlor, and *cis*-chlordane). These compounds are all at naturally occurring concentrations. Noncertified concentrations are given for 12 chlorinated pesticides (heptachlor, γ-BHC, heptachlor epoxide, *cis*-chlordane, 4 *trans*-nonachlor, dieldrin, 2,4′-DDE, 4,4′-DDE, 2,4′-DDD, 4,4′-DDD, 2,4′-DDT, and 4,4′-DDT) in a mussel tissue (SRM 1974a). River sediment SRM 1939 contains high levels of chlorinated pesticides (60 to 550 ng/g); however, only noncertified values are reported for five compounds (heptachlor epoxide, *cis*-chlordane, *trans*-nonachlor, 4,4′-DDE, and 4,4′-DDD). In contrast, the concentrations of these chlorinated pesticides is lower in marine sediment SRM 1941a.[21] Fifteen chlorinated pesticides are reported in whale blubber SRM 1945.[25]

Shellfish Toxins

Two certified reference materials for domoic acid were prepared by NRC following an incident in December 1987 in which consumed mussels containing domoic acid (the toxin responsible for Amnesic Shellfish Poisoning (ASP)) resulted in two deaths and a number of illnesses in Canada.[22] MUS-1[23] is a reference material prepared from precooked mussel tissue. Batches of contaminated and clean mussels were blended to give a domoic acid concentration of 126 mg/kg. An antioxidant (0.02% ethoxyquin) was added to preserve the sample. The moisture content was adjusted to 87%, and the homogenate was bottled in 15-mL polypropylene vials before thermal sterilization. The level of domoic acid is suitable for determination by use of HPLC or GC-MS after sample extraction and work-up. DACS-1B[23] is a domoic acid calibration solution an acetonitrile/water mixture suitable for determination by means of LC-UVD and LC-MS. MUS-2 is a homogenized blend of mussel (*Mytilus edulis* L.) digestive gland and the microalga *Prorocentrum lima* containing okadaic acid and DTX-1, toxins responsible for Diarrhetic Shellfish Poisoning (DSP). MUS-2 is distributed in a set of four polypropylene bottles, each containing 4 g of material. The concentrations of the two compounds are suitable for determination by LC-FLD or by LC-MS after sample extraction and work-up. A calibration solution containing okadaic acid in methanol, OACS-1, is also available from NRC. This material is intended for use as an instrument calibration solution to aid in the determination of DSP toxins in plankton and shellfish. NRC also distributes a set of three calibration solutions for the analysis of toxins associated with Paralytic Shellfish Poisoning (PSP), PSP-1B. The individual solutions containing saxitoxin (STX-1), neosaxitoxin (NEO-1), and a mixture of gonyautoxin 2 and 3 (GTX2β-1) are suitable for analysts working in the area of seafood testing.

Dioxins and Furans

IRMM supplies several CRMs certified for dioxins and furans. CRM 429 is a fly ash extract containing 1.2 mL of toluene extract in iso-octane and CRM 490 consists of approximately 30 g of fly ash. A set of spike and unspiked milk powders are also available. BCR RM 532 is unspiked and RM 533 and RM 534 are spiked to low and higher levels, respectively. NIST SRM 1588, Organics in Cod Liver Oil, is certified for α-tocopherol at 112 μg/g using normal-phase and reverse-phase HPLC with UV detection. This cod liver oil CRM has also been fortified with six chlorinated dibenzo-*p*-dioxins and a dibenzofuran. Although these compounds are not certified in this material, information values are given. SRM 1614 is two solutions of labeled and unlabeled 2,3,7,8-tetrachlorodibenzo-*p*-dioxin in iso-octane. A fish homogenate (CARP-1) certified for dioxins, furans, and PCBs is distributed by NRC and is described in detail in Chapter 8.

REFERENCES

1. Terms and Definitions Used in Connection with Reference Materials, ISO Guide 30-1981, International Standards Organization, Geneva, Switzerland (1981).
2. **J. K. Taylor,** Handbook for SRM Users, NBS Special Publication 260-100 (1985).
3. **A. Y. Cantillo,** Standard and Reference Materials for Marine Science, NOAA Technical Memorandum NOS ORCA 68, 3rd ed. (1993).
4. Institute for Reference Materials and Measurements, European Commission-JRC, B-2440, Geel, Belgium.
5. **J. J. Belliardo and P. J. Wagstaffe,** BCR reference materials for food and agriculture analysis: an overview, *Fresenius Z. Anal. Chem.*, 332 (1988) 533–538.
6. **K. Okamoto,** Biological reference materials from the National Institute for Environmental Studies, *Fresenius Z. Anal. Chem.*, 32 (1988) 524–527.
7. Standard Reference Materials Catalog 1995–1996, National Institute of Standards and Technology, NIST Special Publication 260.
8. **S. D. Rasberry,** NBS activities in biological reference materials, *Fresenius Z. Anal. Chem.*, 332 (1988) 528–532.
9. **P. Quevauviller, K. Vercoutere, H. Muntau, and B. Griepink,** Certified Reference Material (CRM414) for the quality control of trace element analysis in plankton, *Fresenius Z. Anal. Chem.*, 345 (1993) 12–17.
10. IAEA Analytical Quality Control Services Intercomparison Runs and Reference Materials 1992.
11. **R. M. Parr, R. Schelenz, and S. Ballestra,** IAEA Biological Reference Materials, *Fresenius Z. Anal. Chem.*, 332 (1988) 518–523.
12. IAEA, MA-A-1/TM, Information sheet.
13. IAEA-350, Information sheet.
14. **K. Okamoto and K. Fuwa,** Mussel tissue powder, a certified reference material, *Analyst*, 110 (1985) 785.
15. **K. Okamoto,** Isotope dilution/inductively coupled plasma mass spectrometric determination of total tin in NIES fish tissue reference material, *Spectrochim. Acta.*, 46B (1991) 1615–1622.
16. National Research Council of Canada, Institute for National Measurement Standards, Ottawa, Ontario, Canada.
17. **S. S. Berman and R. E. Sturgeon,** A new approach to the preparation of biological reference materials for trace metals, *Fresenius Z. Anal. Chem.*, 332 (1988) 546–548.
18. **W. Wolf, G. V. Iyengar, and J. T. Tanner,** Mixed diet reference materials for nutrient analysis of foods: preparation of SRM 1548 total diet, *Fresenius Z. Anal. Chem.*, 338 (1990) 473–475.
19. **P. Quevauviller, K. Vercoutere, and B. Griepink,** Certified reference materials (CRM 398 and CRM 399) for the quality control of major element determination in freshwater, *Mikrochim Acta*, 108 (1992) 195–204.
20. **S. A. Wise, M. M. Schantz, R. M. Parris, R. E. Rebbert, B. A. Benner, Jr., and T. E. Gills,** Standard reference materials for trace organic contaminants in the marine environment, *Analysis*, 20 (1992) M57–M61.

21. **M. M. Schantz, B. A. Benner, Jr., M. J. Hays, W. R. Kelly, R. D. Vocke, Jr., R. Demiralp, R. R. Greenburg, S. B. Schiller, G. G. Lauenstein, S. A. Wise,** Certification of Standard Reference Material (SRM) 1941a, organics in marine sediment, *Fresenius Z. Anal. Chem.*, 352 (1995) 166–173.
22. **J. L. C. Wright, R. K. Boyd, A. S. W. deFreitas, M. Falk, R. A. Foxall, W. D. Jamieson, M. V. Laycock, A. W. McCulloch, A. G. McInnes, P. G. Odense, V. Pathak, M. A. Quilliam, M. Ragan, P. G. Sim, P. Thibault, J. A. Walter, M. Gilgan, D. J. A. Richard, and D. Dewar,** Identification of domoic acid, a neuroexcitatory amino acid, in toxic mussels for eastern P.E.I., *Can. J. Chem.*, 67 (1989) 481–490.
23. **W. R. Hardstaff, W. D. Jamieson, J. E. Milley, M. A. Quilliam, and P. G. Sim,** Reference materials for domoic acid, a marine neurotoxin, *Fresenius Z. Anal. Chem.*, 338 (1990) 520–525.
24. **P. Quevauviller, M. Astruc, L. Ebdon, H. Muntau, W. Cofino, R. Morabito, and B. Griepink,** A programme to improve the quality of butyltin determinations in environmental matrices, *Mikrochim. Acta.*, 123 (1996) 163–173.
25. **M. M. Schantz, B. J. Koster, L. M. Oakley, S. B. Schiller, S. A. Wise,** Certification of polychlorinate pesticides in a whale blubber Standard Reference Material, *Anal. Chem.*, 67 (1995) 901–910.

Index

A

D

E

F

I

J

K

L

M

N

O

P

Q

R

S